# Celestial Calculations

# Celestial Calculations

## A Gentle Introduction to Computational Astronomy

J. L. Lawrence

The MIT Press
Cambridge, Massachusetts
London, England

© 2018 Massachusetts Institute of Technology

All rights reserved. No part of this book may be reproduced in any form by any electronic or mechanical means (including photocopying, recording, or information storage and retrieval) without permission in writing from the publisher.

This book was set in Times by Westchester Publishing Services. Printed and bound in the United States of America.

Source code and executables for the programs described in this book can be downloaded from https://CelestialCalculations.github.io.

Library of Congress Cataloging-in-Publication Data

Names: Lawrence, J. L. (Jackie L.), author.
Title: Celestial calculations : a gentle introduction to computational
   astronomy / J. L. Lawrence.
Description: Cambridge, MA : The MIT Press, [2019] | Includes bibliographical
   references and index.
Identifiers: LCCN 2018026935 | ISBN 9780262536639 (pbk. : alk. paper)
Subjects: LCSH: Astronomy—Amateurs' manuals. | Astronomy—Data processing.
Classification: LCC QB64 .L39 2019 | DDC 523—dc23 LC record available at
https://lccn.loc.gov/2018026935

10  9  8  7  6  5  4  3  2

When I consider thy heavens, the work of thy fingers, the moon and the stars, which thou hast ordained; What is man, that thou art mindful of him? and the son of man, that thou visitest him?
—Psalm 8:3-4, KJV

Although space travel on a limited scale is an accomplished fact, we are presently an Earthbound race whose minds yearn to explore the cosmos. Perhaps well before the end of this century some lucky explorers will have visited Mars or some other planet within our Solar System. Unfortunately, most of us will have to be content with imaginary journeys. This book is dedicated to all of us "armchair explorers" who look up into the night sky and are filled with curiosity and a burning desire to understand a little better how the universe in which we live appears to operate. Accordingly, this book was not written by a professional or even by an amateur astronomer, but rather it was written from the perspective of an amateur "amateur astronomer."

This book is most especially dedicated to Mom and Dad for all your years of encouragement and sacrifice. Because you rescued my brothers and me, I know the great debt I owe and can only offer only gratitude in return. Mary Ann, as always, your steadfast support has been the single most important factor contributing to the completion of this project. David, what would you have thought about the amazing universe in which we live?

*J. L. Lawrence*

# Contents

| | Preface | xiii |
|---|---|---|
| **1** | **Introduction** | 1 |
| | 1.1 Accuracy | 4 |
| | 1.2 Other Notes | 6 |
| | 1.3 Layout of the Book | 8 |
| | 1.4 Program Notes | 9 |
| **2** | **Unit Conversions** | 11 |
| | 2.1 Some Preliminaries | 11 |
| | 2.2 Measuring Large Distances | 14 |
| | 2.3 Decimal Format Conversions | 15 |
| | 2.4 Program Notes | 19 |
| | 2.5 Exercises | 19 |
| **3** | **Time Conversions** | 21 |
| | 3.1 Defining a Day | 22 |
| | 3.2 Defining a Month | 25 |
| | 3.3 Defining a Year | 26 |
| | 3.4 Defining Time of Day | 27 |
| | 3.5 Calendar Systems | 38 |
| | 3.6 Julian Day Numbers | 40 |
| | 3.7 Some Calculations with Dates | 44 |
| | 3.8 LCT to UT | 46 |
| | 3.9 UT to LCT | 46 |
| | 3.10 UT to GST | 47 |
| | 3.11 GST to UT | 48 |
| | 3.12 GST to LST | 49 |
| | 3.13 LST to GST | 50 |
| | 3.14 Program Notes | 50 |
| | 3.15 Exercises | 51 |

| | | | |
|---|---|---|---|
| **4** | **Orbits and Coordinate Systems** | | **53** |
| | 4.1 | Trigonometric Functions | 54 |
| | 4.2 | Locating Objects on a Sphere | 56 |
| | 4.3 | The Celestial Sphere | 62 |
| | 4.4 | Ellipses | 65 |
| | 4.5 | Orbital Elements | 68 |
| | 4.6 | Equatorial Coordinate System | 84 |
| | 4.7 | Horizon Coordinate System | 88 |
| | 4.8 | Ecliptic Coordinate System | 91 |
| | 4.9 | Galactic Coordinate System | 96 |
| | 4.10 | Precession and Other Corrections | 101 |
| | 4.11 | Program Notes | 105 |
| | 4.12 | Exercises | 106 |
| **5** | **Stars in the Nighttime Sky** | | **109** |
| | 5.1 | Locating a Star | 111 |
| | 5.2 | Star Rising and Setting Times | 115 |
| | 5.3 | Creating Star Charts | 119 |
| | 5.4 | Program Notes | 123 |
| | 5.5 | Exercises | 124 |
| **6** | **The Sun** | | **125** |
| | 6.1 | Some Notes about the Sun | 125 |
| | 6.2 | Locating the Sun | 131 |
| | 6.3 | Sunrise and Sunset | 138 |
| | 6.4 | Equinoxes and Solstices | 140 |
| | 6.5 | Solar Distance and Angular Diameter | 144 |
| | 6.6 | Equation of Time | 147 |
| | 6.7 | Program Notes | 149 |
| | 6.8 | Exercises | 149 |
| **7** | **The Moon** | | **151** |
| | 7.1 | Some Notes about the Moon | 151 |
| | 7.2 | Lunar Exploration | 158 |
| | 7.3 | Locating the Moon | 161 |
| | 7.4 | Moonrise and Moonset | 169 |
| | 7.5 | Lunar Distance and Angular Diameter | 172 |
| | 7.6 | Phases of the Moon | 173 |
| | 7.7 | Eclipses | 181 |
| | 7.8 | Program Notes | 184 |
| | 7.9 | Exercises | 185 |

# Contents

**8  Our Solar System** — 187
- 8.1  The Search for Planets — 189
- 8.2  The Inner Planets — 193
- 8.3  The Outer Planets — 203
- 8.4  The Dwarf Planets — 219
- 8.5  Belts, Discs, and Clouds — 225
- 8.6  Locating the Planets — 231
- 8.7  Planet Rise and Set Times — 244
- 8.8  Planetary Distance and Angular Diameter — 245
- 8.9  Perihelion and Aphelion — 247
- 8.10  Planet Phases — 250
- 8.11  Planetary Magnitude — 251
- 8.12  Miscellaneous Calculations — 253
- 8.13  Program Notes — 263
- 8.14  Exercises — 263

**9  Satellites** — 265
- 9.1  Vectors — 269
- 9.2  Ellipses Revisited — 271
- 9.3  Geocentric and Topocentric Coordinates — 276
- 9.4  Satellite Orbital Elements — 284
- 9.5  Categorizing Satellite Orbits — 302
- 9.6  Locating a Satellite — 307
- 9.7  Satellite Rise and Set Times — 311
- 9.8  Satellite Distance — 313
- 9.9  Other Flight Dynamics — 315
- 9.10  Program Notes — 326
- 9.11  Exercises — 327

**10  Astronomical Aids** — 331
- 10.1  Recommended Authors — 332
- 10.2  Star Charts — 333
- 10.3  Star Catalogs — 336
- 10.4  Ephemerides and Almanacs — 340
- 10.5  Astronomical Calendars — 344
- 10.6  Online Resources — 345
- 10.7  High-Accuracy Resources — 348

Glossary — 351

Index — 365

# Preface

Introductory astronomy books can generally be placed into 1 of 2 categories. Books in the first category are almost devoid of any mathematics. Such books are descriptive in nature and are typically filled with photographs and artistic impressions of the awe-inspiring, otherworldly vistas that can be found scattered throughout the cosmos. By contrast, books in the second category approach astronomy from a mathematical standpoint and are filled with complex equations for describing celestial motion. Such books can be quite challenging to read because of the level of mathematics and physics required to understand the concepts being discussed.

This book is an attempt to bridge both categories. While mathematical calculations are central, this book concerns itself with applying concepts instead of deriving formulas. As a noncalculus introduction to computational astronomy, it requires relatively little mathematical skill. Rest assured, high school algebra and a little trigonometry are more than sufficient for following the various methods presented in this book! Taking a simplified mathematical approach comes at a cost, however. By avoiding more advanced mathematics, the algorithms and equations presented herein will not produce results that are accurate enough for professional astronomers. Even so, the methods presented are generally accurate to within a few minutes of time or a few arcminutes, which should be sufficient for most amateur astronomers.

Books about celestial mechanics often assume that a reader understands why a calculation is necessary and needs only to be shown how to derive and apply the proper equations. By contrast, this book emphasizes understanding *what* calculations are required, *why* they are needed, and *how* all the pieces fit together. As this book will also demonstrate, the very same principles that describe the motion of the planets and stars can be readily applied to track the man-made satellites and other objects, such as the International Space Station, that orbit Earth. The space around our planet has become increasingly crowded since Sputnik, the world's first satellite, was hurled into orbit in the fall of 1957. Today more than 500,000 objects over 10 cm in size orbit Earth. Many of those objects are useful satellites while others are debris and remnants of the rockets used to carry payloads into space. Locating and tracking those objects can be as entertaining a hobby as amateur astronomy itself.

The chapters ahead do more than merely present the mathematics necessary to explain a particular concept or predict an astronomical event. A computer is also used to demonstrate the topics and guide the reader through the incredible maze of technical details necessary to locate a star or a planet, or predict when sunrise will occur and what the phase of the Moon will be. By employing the computational power of modern personal computers, the tedium of the lengthy calculations required for virtually every task has been eliminated. When the principles discussed in this book are combined with the use of a computer, the result is a powerful and stimulating environment for enjoying the wonders of astronomy. Using a computer allows readers to concentrate on major concepts rather than getting lost in myriad technical details, and it allows a chapter review as often as necessary while a computer serves as a patient guide.

Source code is provided for all of the book's example programs, although no claim is made that their implementation is the best or most suitable for the problem being solved. In particular, the reader is forewarned that relatively little error checking is done, especially during data entry, so that by virtue of brevity the programs will be clearer and easier to follow. In addition, implementation decisions were sometimes made to simplify porting the programs from one programming language to another; these are decisions that might not have been made if software portability was not also an objective.

**Acknowledgments**

A special note of thanks is due to the staff of the MIT Press. Without their support and professionalism, this project would have been impossible. I also extend my thanks to the reviewers whose insightful comments did much to improve this work. Despite everyone's collective best efforts, any mistakes that remain must be attributed solely to the author.

*J. L. Lawrence*
*July 2018*

# 1 Introduction

For as long as we humans have been staring up into the starry night sky, we have pondered the mysteries of the universe. The majesty of a still, dark night instills a sense of wonder and awe at the vastness of the universe in which we live. The quiet beauty of a moonlit night, the magical appearance of a shooting star, or the wispy strands of the Milky Way may well cause us to ponder how we humans fit into the grand scheme of things. Perhaps it is the ethereal beauty of the nighttime sky and its propensity for making us wax philosophical that first entice us and generate our interest in astronomy.

It is impossible to say when astronomy really began. Certainly it began before recorded history, making it one of the oldest sciences. Archaeological evidence suggests that our ancestors placed great emphasis upon celestial events with examples abounding throughout the world. Some believe that Stonehenge is the world's oldest astronomical observatory and may once have been used to predict eclipses. We can look back in time to the ancient Babylonians and see that they had carefully recorded the position of the planet Jupiter and had derived a calendar based upon astronomical events. Farther to the west, the Mayas were intrigued by Venus and left religious monuments that reveal a great deal about their understanding of that planet. Even in modern times, some of our holidays are based upon celestial events like the phase of the Moon. Easter, for example, is the first Sunday following the Full Moon that occurs on or after the vernal equinox (around March 21). In turn, Whitsun Sunday and Trinity Sunday are movable dates because they are tied to Easter.

With the help of only a moderately sized telescope or a good pair of binoculars, looking up into the nighttime sky reveals a breathtaking panorama of galaxies, twinkling stars, colorful nebulae, and mysterious planets. Even when viewed with the naked eye alone, the universe is an enchanting wonderland, but we "armchair explorers" can do a lot more than merely appreciate the splendors of the nighttime sky. For example, learning the constellations is a rewarding

**Figure 1.1** The Milky Way Galaxy
Astronomy is a visual science that requires little more than a willingness to be observant. Even without the aid of a telescope, the heavens on a dark night are a stunning sight to behold, as attested by this picture of the Milky Way Galaxy. Measuring some 100,000–120,000 light years across, the Milky Way is home to 200 billion stars, in addition to our Sun, the Earth, and all of the objects in our Solar System. (Image courtesy of Dylan O'Donnell)

experience that requires nothing more than some memorization and practice at stargazing. The ancient Greeks divided the sky into 48 constellations whereas modern astronomers have divided the sky into 88 constellations. The Greeks cataloged the visible stars within each constellation, and even today, stars are still referred to by the constellation in which they are located as a quick and easy way to approximate their position.

The science of astronomy, especially in modern times, is changing at a rapid rate. New discoveries and advances in related sciences such as physics and chemistry require that we periodically reevaluate our theories about the universe, and even our most fundamental understanding of time and space itself. A striking example is the controversy over Pluto. In 1992 astronomers discovered a region of space beyond Neptune that is filled with trillions of icy objects, many of which are too large to be considered as asteroids but not large enough to be considered as planets. This region of space is the Kuiper Belt, which is estimated to contain thousands of objects that are more than 100 kilometers (62 miles) in diameter. Astronomers at the Palomar Observatory photographed this region of space, and in 2005 they discovered

# Introduction

an object approximately the size of Pluto. The discovery of this object, named Eris after the Greek goddess of strife and discord, forced astronomers to revisit the very notion of what it means to be a planet. If Pluto is called a planet, then should Eris take its place as the tenth planet in our Solar System? If Pluto and Eris are designated as planets, then should Ceres, the largest asteroid in the region of space called the Asteroid Belt, also be reclassified as a planet?

To resolve the controversy, in 2006 the International Astronomical Union (IAU) met to formally define a planet. The IAU defines a planet as a celestial body that (a) orbits the Sun, (b) has sufficient mass and gravity to be nearly round in shape, and (c) has cleared the neighborhood around its orbit. The phrase "cleared the neighborhood" means that the celestial body is gravitationally strong enough to have collected all other nearby matter into its own mass or into moons that orbit the celestial body. When a celestial body has cleared the neighborhood, no other planet-forming debris remains in the vicinity of the celestial body's orbit.

Alas! Under this internationally accepted definition, Pluto is no longer designated as a planet! Pluto, Eris, and Ceres all fail to meet the "cleared the neighborhood" criteria. The celestial body formerly known as the *planet* Pluto is now relegated to the newly created category of *dwarf planets*. Besides Pluto, 4 other objects within our Solar System are presently classified as dwarf planets: (1) Eris, which has 1 known moon named Dysnomia, (2) Haumea, a football-shaped object that rotates every 4 hours and has 2 known moons, (3) Makemake, which is about two-thirds the size of Pluto, and (4) Ceres, the largest known asteroid and the smallest of our Solar System's 5 known dwarf planets. The number of objects classified as dwarf planets may well increase in the near future as space probes continue to explore the outer reaches of our Solar System.

It is an exciting time in astronomy with new discoveries being made at a rapid pace as spaceborne instruments scan the far reaches of the cosmos and as space probes and robots arrive at the remotest regions of our Solar System. Most exciting of all, in this author's opinion, is that we may be on the verge of returning to space with manned missions that will far exceed the historic manned trips to the Moon. The purpose of this book is to create a foundation that will allow us aspiring amateur astronomers to join in the fun. The chapters ahead will do so by augmenting our natural ability to observe with the ability to calculate and predict, even if our contributions must remain limited to explorations that can be carried out from our armchairs.

## 1.1 Accuracy

Before we continue, let us briefly digress to discuss accuracy. Accuracy is a statement of how well a measurement agrees with the actual or accepted value of the entity being measured. For example, suppose a ruler with divisions every inch is used to measure the length of a book, and the result is 8 inches. This number is really only an approximation of the book's true length. In reality, is the book closer to 8.1 inches in length, 7.9 inches, or 8.02 inches?

A better measurement could be obtained by using a ruler that has divisions every tenth of an inch. With such a ruler, the book's length could be measured to the closest tenth of an inch rather than to the closest inch. Similarly, a ruler could be constructed with even finer divisions, say every hundredth of an inch, to give an even better measure of the length. However, it is easily seen that physical considerations limit how finely a ruler may be subdivided.

Any measurement of a physical entity (such as length, time, or temperature) is by necessity an approximation. Astronomy requires making many different kinds of measurements (e.g., a star's location, the instant a planet passes through some point in space, the distance to the Sun) that must be understood as approximations. This is not to say that all numbers encountered are approximations. The length of a mile is exactly 5,280 feet by definition. There are exactly 1,000 meters in a kilometer. These examples are not approximations because they are exact by definition rather than being the result of some measurement. However, measurements that *use* these exact definitions are still only approximations.

Approximations arise when measurements are made for several reasons. First, as we have already demonstrated, the accuracy of the instrument used places a limit on the accuracy of the resulting measurement. Second, a human must typically judge how closely a measurement falls within the limits of an instrument. (When measuring with a ruler, what should we do when the length falls somewhere between 2 divisions?) Third, approximation formulas may be used to derive other measurements. For example, the chapter on locating planets uses an approximation to solve Kepler's equation. That approximation is in turn used in another formula to estimate the position of a particular planet. Hence, the final solution, which is based on approximate measurements and an approximation formula, will not give the *exact* location of a planet but only an *approximate* location.

Why is a discussion of accuracy important? Because the answers obtained from approximations and subsequently displayed as a result of a computer's calculation may imply greater accuracy than is really the case. To illustrate,

# Introduction

suppose a ruler subdivided every tenth of an inch is used to find the area enclosed by a rectangle. Assume that the length of the rectangle is measured to be 3.1 inches, and its width is measured to be 0.5 inches. The area enclosed by the rectangle is thus

$Area = (3.1 \text{ inches})(0.5 \text{ inches}) = 1.55 \text{ square inches}.$

This result implies that the area is known to the nearest hundredth of an inch when our original measurements were known only to the nearest tenth of an inch! Clearly, we cannot know the area of the rectangle to the nearest hundredth of an inch when our measuring instrument (a ruler) was only accurate to the nearest tenth of an inch.

In the context of the algorithms presented in this book, a calculation might imply that an event, such as sunrise, can be computed to the nearest second when in reality the result may be accurate only to the nearest 5 minutes. The situation is compounded because a computer can easily perform calculations with many more digits of accuracy than the original measurements warrant.

The convention normally used in scientific measurements to specify the accuracy of a measurement is to give all the accurate digits plus a single digit of uncertainty. For instance, if a ruler is graduated in millimeter (mm) increments and a stated measurement is 21.3 mm, there are 3 digits of accuracy. By convention the digit 3 is a digit of uncertainty because an estimate had to be made of where the actual length lies between 21.0 and 22.0 mm.

When performing arithmetic operations on numbers, the result obtained cannot be more accurate than the least accurate measurement. Suppose the measurements 21.5 mm and 0.003 mm are multiplied together. Since the least accurate measurement is 21.5 (its accuracy is known only to the first decimal place whereas 0.003 is known to 3 decimal places), the resulting product should be rounded to have only 1 digit to the right of the decimal point. Thus, $(21.5)(0.003) = 0.0645$, which should be rounded to the nearest tenth, giving an approximate answer of 0.1 mm.

In writing this book, it was difficult to strike a good balance between giving the correct number of digits of accuracy and providing enough digits to allow readers to compare their calculations with the examples. The approach generally taken in both the text and the computer programs is to show intermediate calculations to 6 decimal places while final results may be shown only to 2 decimal places. Using extended precision for intermediate calculations *does not* imply an accuracy of 6 digits. In fact, the algorithms used in this book are often accurate only to a few minutes of arc or a few minutes of time.

## 1.2 Other Notes

If we compare results obtained from this book with published results in star charts or astronomy journals, there will likely be discrepancies for several reasons. Differences may occur because of round-off errors and because the formulas used in this book are less accurate (and less complex!) than the methods used by other sources. However, the results in this book should be sufficient for general use. After all, a result accurate to a few minutes of arc is probably more than adequate for most amateur astronomers' purposes.

Besides round-off errors and using less accurate approximations, errors can be introduced by an inaccurate observer's location. That is, if an observer has only an approximate latitude and longitude for their location, the results produced by the programs may differ significantly from what is observed because they are based on an inaccurate location. Furthermore, the programs and techniques contained in this book are not intended to work for all possible dates. The algorithms should be reasonably accurate from about 1800 AD to 2100 AD, but not all algorithms will work well for years outside that range.

Some readers may wish to use a hand calculator to follow along with the text, modify the programs for another application, or even convert the programs to another programming language. Several points of caution are in order.

1. As we already stated, 6 decimal places are generally shown to guide you in comparing your computations with those of a computer, not to indicate a high degree of accuracy.

2. Carefully note whether angles are expressed in radians or degrees. In this book, angles will always be expressed in degrees unless otherwise noted. We will also assume the trigonometric functions accept degrees rather than radians and that the inverse trigonometric functions return degrees rather than radians. When an equation with an inverse trigonometric function requires radians, the factor $\frac{\pi}{180}$ will be included to convert inverse trigonometric function results to radians.

3. Beware of inverse trigonometric functions. Computers and calculators usually return a result between plus or minus $90°$, but it is often necessary to adjust the result to be sure the answer is in the correct quadrant. This will be explained more fully in chapter 5.

4. Beware of the difference between FIX and INT. This problem surfaces only when negative numbers are considered. Throughout this book, FIX will be understood to be a function that returns the integer part of an argument

# Introduction

while `INT` will be a function that returns an integer that is less than or equal to the argument. Some examples will illustrate the difference.

$INT(1.5) = 1$,
$INT(1.4) = 1$,
$INT(-1.5) = -2$,
$INT(-1.4) = -2$,
$FIX(1.5) = 1$,
$FIX(1.4) = 1$,
$FIX(-1.5) = -1$,
$FIX(-1.4) = -1$.

In addition to `INT` and `FIX`, a few other functions will be useful in the chapters ahead.

- The `ABS` function (absolute value) returns the nonnegative value of a number without regard to its sign. Mathematicians denote this function with 2 vertical bars, as in $|x|$. For example,

$ABS(-5.0) = 5$,
$ABS(5.4) = 5.4$,
$ABS(0) = 0$.

- The `FRAC` function returns the fractional part of a number as a positive decimal value. It is obtained by ignoring the integer part of the number and whether the number is positive or negative. Thus,

$FRAC(1.5) = 0.5$,
$FRAC(-1.5) = 0.5$.

Some programming languages may not have a `FRAC` function that works as described here. This deficiency can be overcome by implementing the function

$FNFRAC(x) = ABS(x - FIX(x))$.

- The modulo function gives the remainder after 1 number is divided by another. For example, 9 modulo 5 is 4 because 9 divided by 5 is 1 with a remainder of 4. 9 modulo 3 is 0 because 9 divided by 3 is 3 with a remainder of 0. We will most often use the modulo function to adjust angles to be in the range $[0°, 360°]$, where the angle to be adjusted could be greater than $360°$ or even negative. Some programming languages provide a modulo function, but check to see how they handle negative numbers. This book requires that the result is always positive when the divisor is positive. For example,

$-100 \text{ MOD } 8 = 4,$

$-400 \text{ MOD } 360 = 320,$

$270° \text{ MOD } 180° = 90°,$

$-270.8° \text{ MOD } 180° = 89.2°,$

$390° \text{ MOD } 360° = 30°,$

$390.5° \text{ MOD } 360° = 30.5°,$

$-400° \text{ MOD } 360° = 320°.$

- The ROUND function returns an integer by rounding the given number to the nearest integer. For example,

$\text{ROUND}(1.4) = 1,$

$\text{ROUND}(1.8) = 2,$

$\text{ROUND}(-1.4) = -1,$

$\text{ROUND}(-1.8) = -2.$

## 1.3 Layout of the Book

This book begins with a discussion of some basic principles required for computing the location of celestial objects and builds toward a climax at about chapter 5. Chapter 2 shows how to perform a few unit conversions that are required in the remaining chapters. Chapter 3 discusses the important topic of time conversions, which is probably the most difficult topic to understand of all those presented in this book. The time conversion techniques are simple enough to perform, but the reason for doing them is not immediately obvious. In chapter 4, coordinate system conversions are discussed. Coordinate system conversions are necessary to account for the location of an observer with respect to the Earth and to account for changes in perspective, such as viewing the Solar System as geocentric (Earth-centered) rather than heliocentric (Sun-centered).

The real fun begins with chapter 5, which combines the concepts from preceding chapters into a computer program that will calculate the location of a star for an observer at a given date, time, and location. Furthermore, the program presented in chapter 5 will produce a star chart for the date, time, and location in question. Chapters 6, 7, and 8 apply the same principles to predict the location of the Sun, Moon, and planets, respectively. In addition, techniques are described for calculating interesting items such as the phase of the Moon, distance to the planets, time of sunrise and sunset, and an object's weight on different planets. Chapter 9 shows how to apply the concepts from

# Introduction

preceding chapters to locate the multitude of man-made satellites that now encircle Earth. Finally, chapter 10 completes the book by discussing how to use various astronomical aids such as an ephemeris.

## 1.4 Program Notes

Example programs are provided with this book to illustrate the algorithms and techniques presented in the text. Source code for the programs, the compiled executables, and supporting data files can be downloaded from the publisher's website, which is given on this book's copyright page. Once the programs have been downloaded from the publisher's website, no special steps are required to install the example programs; they may be copied to any convenient location on your computer's disk drive. The programs do not modify any operating system files, create any files in user or system directories, or create any Microsoft Windows registry entries. The programs and data can be completely removed from your computer by merely deleting them from the location you copied them to on your computer's disk drive.

This book's example programs follow the naming convention **RunChapX**, where **X** is the chapter to which the program applies. Thus, **RunChap1** is the program for this chapter. Refer to the README.TXT file included with the programs for details about the source code and data files.

All of this book's programs are menu driven and operate in the same manner. Select the desired operation from a menu of possible options, set check boxes for items such as whether you wish to see intermediate calculations, enter any data required, and see the results in a scrollable area within the program's application window. When entering data, avoid using commas. That is, the value five thousand should be entered into a program as 5000 rather than as 5,000. Entering a comma may work in some cases but generate an error in others, so it is best to simply avoid using commas.

This chapter's program (**RunChap1**) allows viewing and manipulating already-built data files containing the names and locations of various celestial objects. Among other features, this program will list stars and their locations from a star catalog, list the constellations and the brightest star in each constellation, and determine in what constellation a given location lies. For the star catalogs provided with this book, a celestial object's location is given in equatorial coordinates referenced to the standard epoch J2000, although constellation boundaries are given relative to the 1875 epoch. The equatorial coordinate system and the concept of a standard epoch will be explained later in chapter 4. For the moment, it suffices to know that equatorial coordinates tell astronomers where to point their telescopes to see an object whereas an epoch tells them when that object's coordinates were measured.

The star catalogs provided with this book are in an "XML-like" format and can be viewed with any ordinary text editor, such as Microsoft's Notepad program, to understand their format in case you want to build your own catalog of favorite celestial objects. See the README.TXT file, or open and view any of the star catalog data files with a text editor for more details about the required data format.

Chapter 5 uses the data files supplied with this book to produce star charts. Besides stars extracted from *Sky Catalog 2000.0*, several other data files are provided with data about celestial objects as taken from publicly available NASA data sources. The Messier catalog, the Henry Draper catalog, and the SAO J2000 catalog, among others, are provided.

# 2 Unit Conversions

Converting from one system of measurements to another is often necessary in science and mathematics. The metric system (meters, grams, liters, Celsius) is used more frequently in science than the English system (feet, pounds, gallons, Fahrenheit), so it is important to be able to convert between them, for example, to convert between miles and kilometers when describing the distance to a celestial object.

Besides converting *between* systems of measurements, conversions are frequently done *within* the same system as a matter of convenience. For example, we often convert between inches and feet or feet and miles to express a measurement in more convenient units. That is, it is far more convenient to state that point $A$ is 30.5 miles away from point $B$ than it is to say that the distance between them is 1,932,480 inches!

After reviewing a few preliminaries pertinent to unit conversions and handling large numbers, this chapter will define some units that astronomers use when measuring the vast distances between Earth and the stars. The chapter will conclude with some time- and angle-related conversions, and some practice exercises.

## 2.1 Some Preliminaries

It is assumed that the reader already knows how to convert between measurement units. To be sure, several practice exercises are provided at the end of this chapter. Moreover, this chapter's program uses the cross-multiplication technique and the relationships shown in table 2.1 to perform various unit conversions. (The letter E in the table indicates a number written in scientific notation, which is described later in this section.) The program for this chapter also shows how to convert between degrees Celsius and degrees

**Table 2.1** Conversion Factors
The unit on the left is equal to the unit on the right. Cross-multiplication can be used with these relationships to easily convert between measurement units.

| Unit Conversion Relationships | |
|---|---|
| 25.4 mm | 1 inch |
| 0.3048 m | 1 foot |
| 1.609344 km | 1 mile |
| 1 light year | 5.87E12 miles |
| 1 light year | 0.3068 parsecs |
| 1 AU | 9.29E7 miles |
| 180° | 3.14159 radians |
| 360° | $24^h$ |

Fahrenheit.[1] Converting between Celsius and Fahrenheit cannot be done by cross-multiplication, but instead requires the equations

$$°C = \frac{5}{9}(°F - 32) \tag{2.1.1}$$

and

$$°F = 32 + \frac{9}{5}°C. \tag{2.1.2}$$

These equations are provided only for completeness' sake. Temperature conversions will not be needed for the remainder of this book.

Besides converting between units and systems of measurements, the reader should be comfortable with using scientific notation to express very large and very small numbers. For example, astronomers measure the wavelength of light reaching Earth from distant stars to determine the materials that make up those stars. A typical wavelength is on the order of 1 one-hundred-millionth of a centimeter (0.00000001 cm). At the other extreme, astronomers measure vast distances that often exceed trillions of miles (1 trillion miles is 1,000,000,000,000 miles). Writing down so many zeros to express numbers such as these is inconvenient and error prone. Inadvertently dropping three zeros changes a trillion miles to a mere billion miles, which is a significant difference. Expressing numbers in scientific notation is one way to avoid making such order-of-magnitude mistakes.

---

1. Another temperature scale, the Kelvin scale, is frequently used in science to describe very cold temperatures. 0 °K is −459.67 °F, or −273.15 °C.

**Unit Conversions** 13

A number expressed in scientific notation is written as the product of 2 numbers. The first number is between 1 and 10; that is, a number with a single nonzero digit to the left of the decimal point. The second number is a scaling factor, written as a power of 10, which indicates where to place the decimal point. For example, $9.3 \times 10^7$ is the proper way to express the approximate average distance from Earth to the Sun (93 million miles) in scientific notation. The $\times$ symbol means to multiply; it does *not* mean a variable named $x$. In the scaling factor $10^7$, the number 10 is called the base while the number 7 is called the exponent. More generally, we say that the number $a^b$ has a base $a$ and an exponent $b$, which is shorthand for saying that the number $a$ is to be multiplied by itself $b$ times. Thus, $2^3$ means to multiply 2 by itself 3 times, giving the value $2^3 = 2 \times 2 \times 2 = 8$. Applying this to our example, we have

$$10^7 = 10 \times 10 \times 10 \times 10 \times 10 \times 10 \times 10 = 10,000,000,$$

and so

$$9.3 \times 10^7 = 9.3 \times 10,000,000 = 93,000,000.$$

Consider the number $1.5 \times 10^{-4}$. What does a negative exponent mean? The number $a^{-b}$ is shorthand for expressing the fraction $\frac{1}{a^b}$. So $2^{-3} = \frac{1}{2^3} = \frac{1}{8}$. Applying the meaning of negative exponents to our example, we have

$$10^{-4} = \frac{1}{10^4} = \frac{1}{10 \times 10 \times 10 \times 10} = 0.0001,$$

which then means that

$$1.5 \times 10^{-4} = 1.5 \times 0.0001 = 0.00015.$$

The easiest way to deal with scientific notation is to remember that the power of 10 exponent indicates how many digits to the left (for negative exponents) or to the right (for positive exponents) to place a decimal point. So in our example of expressing the average distance to the Sun in scientific notation, the exponent 7 tells us that there are 7 digits to the right of the decimal point. This knowledge allows us to quickly write down the number 93 followed by 6 zeros (not 7 because the "3" digit counts as one of the numbers to the right of the decimal point). Similarly, we easily see that

$$4.239 \times 10^4 = 42,390$$

because the exponent 4 tells us that there are to be 4 digits to the right of the decimal point, with 239 being the first 3 of those 4 digits.

Learning to express a number in scientific notation is easy. Consider the number 830,600. Place a decimal point after the number 8 and note that there are then 5 digits remaining after the decimal point that we just inserted. This means that 5 will be the exponent of our scaling factor. Drop the two extraneous trailing zeros (but *not* the zero between 3 and 6!) in our number to give

$$830{,}600 = 8.306 \times 10^5.$$

Handling numbers less than 1 in scientific notation is also easy. To convert a number from scientific notation, such as $4.203 \times 10^{-3}$, first write down the number before the scaling factor without any decimal point (i.e., 4203 for this example). Then add zeros to the *left* of the number we just wrote down equal to the number of zeros indicated by the exponent as if the exponent were a positive number. Since our exponent in this case is $-3$, we write down a total of 3 zeros, giving us 0004203. Lastly, put a decimal point after the first 0, which in this case gives us

$$4.203 \times 10^{-3} = 0.004203.$$

Converting a number less than 1 to scientific notation is even easier. First, move the decimal point to the right to just past the first nonzero digit in the number that we wish to express in scientific notation. Count how many places the decimal point was moved, and that value, expressed as a negative number, becomes our exponent. If we use 0.00003089 as an example, the digit 3 is the first nonzero digit to the right of the decimal point. Moving the decimal point to the right just after the digit 3 requires moving the decimal point 5 places. Hence we have

$$0.00003089 = 3.089 \times 10^{-5}.$$

It is common practice in computer programming to use the letter E, which stands for exponent, to indicate that a number is expressed in scientific notation (e.g., 9.29E7 represents the number $9.29 \times 10^7$). This technique will be used frequently throughout this book to conform to common practice in programming languages.

## 2.2 Measuring Large Distances

Even when large distances are expressed in scientific notation, they are unwieldy to manipulate. Therefore astronomers have defined other measurement units for dealing with vast distances. For distances within the Solar System, the astronomical unit (AU) is often used. One AU is defined to be the distance from Earth to the Sun, but that distance varies as Earth goes around the Sun. To avoid having a measurement unit that varies as Earth orbits the

Sun, 1 AU is formally defined to be exactly 149,597,870,700 meters, which is approximately 92,900,000 miles, and it is the value shown in table 2.1.

When objects are as far away as the stars, even the AU measurement unit is cumbersome to use. So astronomers defined another unit of measurement, the light year, for measuring such vast distances. A light year is what the term implies—the distance that light travels in 1 year. Using the conversion factors that relate light years and miles (1 light year is $5.87 \times 10^{12}$ miles) and miles and AUs (1 AU is $9.29 \times 10^7$ miles), it is easy to show that 1 light year is approximately 63,186 AU. (Hint: first convert 1 light year to miles, and then convert the resulting miles to AUs.) Clearly, light years are more convenient measurement units than AUs for expressing stellar distances!

Another unit, the parsec, is sometimes used to measure distances that are of the same magnitude as light years. A parsec is approximately 3.26 light years. We will not have occasion to use parsecs in this book, but you may encounter parsecs when dealing with stars and other objects in the far reaches of space.

## 2.3 Decimal Format Conversions

It is common practice to express time in terms of hours, minutes, and seconds. One might, for example, say that the time is 4 hours, 32 minutes, and 29 seconds Central Standard Time. This format, which is called the HMS format, is written as $4^h 32^m 29^s$.[2] To avoid difficulties with knowing whether the time is a.m. or p.m., a 24-hour clock will be used throughout this book. Thus, 1:30 p.m. is expressed in HMS format as $13^h 30^m 00^s$.

Angles are often expressed in the DMS format, which is similar to the HMS format in that it uses superscripts to represent degrees, arcminutes, and arcseconds. For example, an angle that is 24 degrees, 13 minutes, and 18 seconds of arc would be written in DMS format[3] as $24°13'18''$. Note that degrees are subdivided into minutes and seconds, which are also units for measuring time. When confusion may arise as to whether minutes and seconds refer to time or angles, the terms *arcminutes* and *arcseconds* are used to distinguish between time and angles.

With respect to astronomy, time and angles can be thought of as being related. Earth rotates once in 24 hours through an angle of 360°, which gives

---

2. This example would normally be written as 4:32:29, but this book will use the superscript notation to conform to how time is normally expressed in astronomy publications. However, this book's computer programs will use the : character to indicate HMS time (e.g., 4 :32:29) because superscripts are cumbersome to produce in computer programs.

3. The computer programs for this book use "d," "m," and "s" to indicate DMS format (e.g., 24d 13m 18s).

**Table 2.2** Converting Time and Angles
Converting between time and angles can be done by noting that the Earth rotates 360° in 24 hours.

| Unit of Time | Equivalent Angle | Angle | Equivalent Time |
|---|---|---|---|
| $24^h$ (1 day) | 360° | 1 radian | $3.82^h$ |
| $1^h$ | 15° | 1° | $4^m$ |
| $1^m$ | 15' | 1' | $4^s$ |
| $1^s$ | 15'' | 1'' | $0.067^s$ |

us a simple relationship between time and degrees. That is, $24^h = 360°$. We will frequently use this relationship in later chapters to convert between time and angles. Table 2.2 expands on this relationship to show some additional relationships between time and angles. The relationships on the left side of the table are exact whereas some of those on the right have been rounded. For practice, use table 2.2 to show that $5^h = 75°$.

The HMS and DMS formats are not very convenient for computational purposes. Instead, time (and angles) are usually converted to a decimal format in which the minutes and seconds are expressed as a fractional part of hours (or degrees). Once expressed in decimal format, arithmetic operations such as addition and subtraction are much easier to perform since there is no need to separately manipulate the hours, minutes, and seconds units in the HMS format (or degrees, minutes, and seconds units for the DMS format). For instance, the decimal format for $4^h30^m0^s$ is $4.5^h$ since 30 minutes is 0.5 hours. Adding $1.5^h$ ($1^h30^m00^s$) to $4.5^h$ gives $6.0^h$ by simply adding the two numbers without the need to consider minutes and seconds separately.

It is important to note that $4^h30^m00^s$ is *not* the same as $4.3000^h$, nor is the time 12:30 the same as $12.30^h$ because three-tenths of an hour is 18 minutes, not 30 minutes. Also note that valid ranges for the decimal format depend on whether a number represents an angle or time. When expressed in decimal format, time is in the range $[0^h, 24^h]$ while angles are in the range $[-360°, 360°]$.

The procedure for converting time expressed in HMS format to decimal format is the same as that for converting angles expressed in DMS format to decimal format. Likewise, converting time expressed in decimal format to HMS format is identical to converting angles expressed in decimal format to DMS. However, dealing with angles is slightly more complicated than dealing with time because an angle can be negative, and care must be taken to account for the sign of the resulting number. The conversion procedures below deal only with angles. Everywhere degrees are mentioned, substitute hours (as well as minutes for arcminutes and seconds for arcseconds) and the procedure will work for time conversions as well.

# Unit Conversions

Converting DMS to decimal format requires 7 steps. Assume that *Degrees*, *Minutes*, and *Seconds* are variables that contain the DMS format of an angle. For example, if we wish to convert 24°13'18" to decimal format, then

*Degrees* = 24°,
*Minutes* = 13',
*Seconds* = 18".

The steps required to convert to decimal format are then:

1. If the DMS value entered is negative (i.e., *Degrees* < 0), let *SIGN* = −1 else let *SIGN* = 1. (Be sure to notice that *SIGN* is 1 if *Degrees* is 0. Many programming languages and calculators have a function that returns the sign of a number, but they normally return a result of zero when the number is zero.)

2. Let *Degrees* = ABS(*Degrees*).

3. Convert arcseconds to decimal arcminutes (*dm*) by applying the equation

$$dm = \frac{Seconds}{60}.$$

4. Add the results of step 3 to *Minutes* to obtain the total number of arcminutes. Thus,

*Total Minutes* = *dm* + *Minutes*.

5. Convert the total arcminutes to decimal degrees by dividing by 60.

$$Decimal\ Degs = \frac{Total\ Minutes}{60}.$$

6. Add the results of step 5 to the results of step 2.

*Decimal Degs* = (*Step* 2) + (*Step* 5).

7. Account for the possibility that the angle was negative, which can be done by simply multiplying the results of step 6 by *SIGN*. Thus,

*Decimal Degs* = *SIGN* ∗ (*Step* 6).

The ∗ symbol in step 7 is the conventional way that computer languages represent the multiplication operation; this symbol will be used frequently throughout this book.

Converting an angle from decimal format to DMS format requires 6 steps. Assume that *Dec* is an angle in decimal format that we wish to convert to DMS format. The steps required are:

1. If *Dec* is negative, let *SIGN* = −1 else let *SIGN* = 1.
2. *Dec* = ABS(*Dec*).
3. *Degrees* = INT(*Dec*).
4. *Minutes* = INT[60 ∗ FRAC(*Dec*)]. The computation FRAC(*Dec*) gives the total number of arcminutes expressed as a fractional part of the degrees, so multiplying by 60 converts this value to arcminutes.
5. *Seconds* = 60 ∗ FRAC[60 ∗ FRAC(*Dec*)].
6. To account for a negative angle being converted, use the *SIGN* from step 1. Multiply step 3 by *SIGN* if step 3 is positive, and merely append "−" to step 3 if the result of step 3 is 0.

The procedures for converting between DMS and decimal format can be tricky to implement because of the need to consider negative angles. Implementation problems arise when an angle is negative, but its integer part is 0 (e.g., −0.586°, −0°35′09.6″). Consider using the procedure just presented to convert the angle −0.586° to DMS format. The value obtained for *Degrees* in step 3 will be 0, so step 6 cannot just multiply step 3 by *SIGN* because doing so will lose the fact that the resulting angle should be negative. So step 6 appends a "−" to the answer to obtain the correct result.

Similarly, care must be taken when implementing a procedure to convert an angle in DMS format to decimal format. In this case, the implementation difficulty encountered is that one must allow a user to enter a "negative 0" value for the integer degrees in the DMS format. In both procedures presented, the approach taken to ensure that negative angles are properly handled is to capture whether the angle is positive or negative in the first step, and then ignore whether the angle is positive or negative until the very last step.

Another implementation issue that arises when doing HMS/DMS conversions is round-off error. Consider steps 4 and 5 of the procedure for converting an angle in decimal format to DMS format. Round-off errors may cause the *Minutes* or *Seconds* (or both) calculated in those steps to be greater than 60. For example, suppose that step 5 results in a value of 59.99995 for *Seconds*. Round this value to two decimal places and *Seconds* becomes 60.00, which is not a valid value for seconds of time or seconds of arc. In such a situation, *Seconds* must be set to 0, and 1 minute must be added to *Minutes*. Having done so, one must then ensure that the value for *Minutes* is still valid (i.e., in the range [0, 59]) and handle accordingly.

The program for this chapter properly handles negative angles. The code also handles the round-off error problem just described so that the results displayed do not cause *Minutes* or *Seconds* to exceed 60 for angles or time.

# Unit Conversions

## 2.4 Program Notes

The program **RunChap2** uses cross-multiplication to perform the conversions shown in table 2.1. It also implements Celsius/Fahrenheit temperature conversions and conversions between HMS/DMS and decimal formats. You may select the program's "Show Intermediate Calculations" check box to see intermediate calculations, or see only the final result by unchecking that check box.

## 2.5 Exercises

In the following practice problems, remember that the letter E is used to express a number in scientific notation. The answers given here may differ slightly from what you will obtain when using a calculator, but if the results are not close, check your arithmetic to see if an error has been made.

1. Convert 5 mm to inches.
   (Ans: 0.196850 inches.)
2. Convert 10 inches to mm.
   (Ans: 254 mm.)
3. Convert 30 meters to feet.
   (Ans: 98.425197 feet.)
4. Convert 25 feet to meters.
   (Ans: 7.62 meters.)
5. Convert 100 miles to kilometers.
   (Ans: 160.9344 km.)
6. Convert 88 km to miles.
   (Ans: 54.680665 miles.)
7. Convert 12 light years to miles.
   (Ans: 7.044E13 miles.)
8. Convert 9.3E7 miles to light years.
   (Ans: 1.5843E-5 light years.)
9. Convert 5 light years to parsecs.
   (Ans: 1.534 parsecs.)
10. Convert 3 parsecs to light years.
    (Ans: 9.7784 light years.)
11. Convert 2 AU to miles.
    (Ans: 1.8580E8 miles.)

12. Convert 10,000 miles to AU.
    (Ans: 1.076426E-4 AU.)
13. Convert 180° to radians.
    (Ans: 3.141593 radians.)
14. Convert 2.5 radians to degrees.
    (Ans: 143.239449°.)
15. Convert $2^h$ to degrees.
    (Ans: 30.0°.)
16. Convert 156.3 to hours.
    (Ans: $10.42^h$.)
17. Convert $10^h 25^m 11^s$ to decimal hours. (Hint: enter the time as 10:25:11.)
    (Ans: $10.419722^h$.)
18. Convert $20.352^h$ to HMS format.
    (Ans: $20^h 21^m 07.2^s$.)
19. Convert 13°04′10″ to decimal degrees. (Hint: enter the angle as 13d 04m 10s.)
    (Ans: 13.069444°.)
20. Convert −0.508333° to DMS format.
    (Ans: −0°30′30.00″.)
21. Convert 300°20′00″ to decimal degrees.
    (Ans: 300.333333°.)
22. Convert 10.2958° to DMS format.
    (Ans: 10°17′44.88″.)
23. Convert 100 °C to °F.
    (Ans: 212.00 °F.)
24. Convert 32 °F to °C.
    (Ans: 0.00 °C.)

# 3 Time Conversions

Everyone has at least a rudimentary idea of what time is. Yet explaining what this abstract concept means is like trying to explain the color red to a blind person who has never seen *any* colors. Referring to a dictionary will only compound the problem! Time, whether or not we can adequately describe it, plays an important role in our daily lives. Time is also an important quantity in astronomy and science in general.

This chapter describes how to perform some important time-related conversions that are required to locate stars, celestial objects such as nebulae, and Solar System objects. This will become clearer later in the book, but the essential idea is this: if we know (a) where a celestial object was at some instant in time, (b) how much time has elapsed since we last knew the object's location, and (c) the characteristics of the object's orbit, then Kepler's laws and the power of mathematics can be applied to calculate the object's current position. Among other things, determining an object's position requires converting calendar dates into a more convenient form and converting between "solar" time and "star" time.

The subject of time can be very confusing. This chapter will briefly mention 8 ways to define a year, 4 ways to define a month, and over a dozen ways to refer to the time of day. Additionally, just to make things interesting, there are 2 different calendar systems to worry about. Don't panic! We'll actually trim this morass down to only a few basic definitions that we will use for the remainder of this book.[1] For the moment, concentrate more on *how* to perform the various time system conversions rather than *why* they are needed. The "why" will become apparent as you progress through this chapter and especially as you work through the chapters ahead.

---

1. We'll mostly be concerned with mean solar and sidereal days, civil months, civil years, and the Gregorian calendar system. Unfortunately, we must deal with 4 time of day definitions: local civil time, Universal Time, local "star" time, and Greenwich "star" time.

## 3.1 Defining a Day

Among the smaller units of time that can be gauged by astronomical events is the day. In modern times (no pun intended!), we consider a day as beginning and ending at midnight. Choosing midnight as the starting point for a day is really an arbitrary decision. Another choice might be to count a day as beginning and ending at sunset, which is how the Jews in biblical times counted days. Even today, Orthodox Jews reckon the Sabbath as beginning at sunset on Friday and ending at sunset on Saturday. In any case, a day is normally defined in terms of the Sun's position; that is, as the time it takes for the Sun to return to the same location in the sky that it was in the day before.

Choosing what position of the Sun to use as the "same location"—highest point in the sky (noon), rising above the horizon (sunrise), dipping below the horizon (sunset), and so forth—for defining a day is somewhat arbitrary. We will shortly see that choosing starting and ending points to define a month and year is arbitrary, too, but do we also have multiple choices for selecting the celestial object itself that we will use to measure intervals of time? The answer is yes! In this section we will use the motion of the Sun, Moon, and stars as the basis for describing intervals of time. In fact, we are free to choose whatever starting and ending points we wish as well as whatever celestial object we deem most convenient as references for defining intervals of time in terms of astronomical events. This flexibility is a major reason why defining time is such a complicated undertaking.

A solar day is defined as the time it takes for the Sun to return to the same apparent location in the sky as it was the day before. Time measured by the solar day is called apparent or solar time, which means that we are gauging time by where the Sun appears to be in the sky. A sundial is an example of an instrument that measures time in terms of the apparent position of the Sun.

Unfortunately, a solar day is not uniform in length because the Sun's apparent motion across the sky is not uniform. For some days during the year, it takes the Sun more than 24 hours to return to the same apparent location in the sky that it was the day before, while on other days it takes less than 24 hours. Solar days are too imprecise for almost any astronomical calculations.

Why are there irregularities in the Sun's apparent motion? For one reason, the Earth-Sun orbit is an elliptical orbit. Thus, as explained by Kepler's laws, Earth moves faster as it get closer to the Sun and slower as it gets farther away from the Sun. This causes the rate of the Sun's apparent motion along its path in the sky to vary throughout the year. Moreover, Earth moves around the Sun at a rate of about 1° per day. This causes the Sun to appear to have moved by about 1° per day against the background of stars.

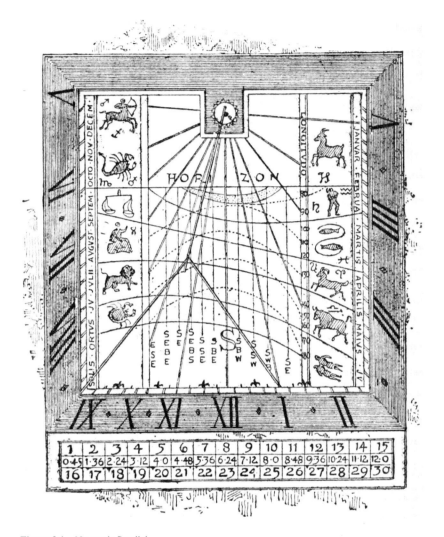

**Figure 3.1** Newton's Sundial
This engraving from Robert Ball's 1895 *Great Astronomers* is the sundial that a young Isaac Newton carved on a stone wall of his family's home in Woolsthorpe, England. The signs of the zodiac are along the sides of the sundial, and the days of the month are at the bottom. Newton did indeed carve a sundial on the family home, but this depiction is probably more ornate than the one he actually created.

To avoid difficulties caused by irregularities in the Sun's apparent motion, astronomers define a fictitious Sun, called the mean Sun,[2] which moves in a perfectly circular orbit at a uniform rate along Earth's equator. A day defined by the motion of this fictitious Sun is called a mean solar day. A mean solar day is exactly 24 hours in length whereas, as we have already indicated, a solar day can vary by a few seconds from 1 day to the next and up to about 30 seconds over the course of a year. Our wristwatches and the clocks in our homes measure time relative to a mean solar day. The difference between time as measured by the apparent motion of the Sun and time as measured by the motion of the mean Sun is called the equation of time. This concept, which we will discuss in chapter 6, allows us to convert between time measured by a sundial (apparent solar time) and time measured by our wristwatches (mean time).

Formally speaking, a solar day is the time interval between 2 successive transits[3] of the true Sun across an observer's meridian.[4] A mean solar day is the time interval between 2 successive transits of the mean Sun across an observer's meridian. Carefully note that the only difference between these 2 definitions is whether we're talking about the Sun's actual elliptical orbit or a fictitious Sun's circular orbit.

One advantage of the mean solar day is that the mean Sun returns to the same point in the sky in exactly 24 hours. However, a star does not return to the same position in the sky in 24 hours as measured by a mean solar day clock. This is true because at the end of a mean solar day, Earth has advanced in its orbit around the Sun by about 1°. The effect noticed by an Earthbound observer is that the stars appear to move with respect to the Sun. This apparent motion of the stars with respect to the Sun can be avoided if a day is measured with respect to the stars instead of the Sun, which is in fact what astronomers have done. Astronomers define a sidereal day as the time interval between 2 successive transits of a fixed star across an observer's meridian. Time measured according to a sidereal day is called sidereal time, or is sometimes more loosely

---

2. We will encounter the adjective "mean" throughout this book in reference to various orbital characteristics of some celestial object, such as the Moon's mean position or a planet's mean anomaly. The adjective mean indicates that we're talking about how an object would behave if it moved in a fictitious circular orbit. The adjective "true" is used, as in true position and true anomaly, when we wish to talk about how an object behaves in its actual elliptical orbit. More will be said in chapter 4 about why we bother with a fictitious circular orbit.

3. In the context of our present discussion, a transit means that the Sun has crossed overhead an observer located at some stated position on Earth.

4. The concept of a meridian will be explained more fully in chapter 4. For now think of an observer's meridian as a semicircle that passes through 3 points: Earth's North and South Poles, and the point directly overhead the observer. An observer's meridian depends on where the observer is located on the Earth's surface.

called "star" time. At the end of exactly 24 sidereal hours (1 sidereal day), a star returns to the same position in the sky that it was 1 sidereal day earlier.

Clocks can be built to measure time according to a solar day, mean solar day, or sidereal day. Sundials, which measure time based on an apparent solar day, require adjustments[5] throughout the year to make them correspond to mean time, which is what our wristwatches and the clocks in our homes measure. Although we do not adjust our wristwatches and clocks during the year (except for the special case of daylight saving time) to account for irregularities in the Sun's motion, it is readily apparent from our clocks that events such as sunrise and sunset occur at different times throughout the year.

Sidereal clocks are specially designed for astronomical purposes and are regulated by the motion of the stars. A sidereal day is slightly shorter than a mean solar day with approximately $23^h56^m$ of mean solar time being equal to $24^h$ of mean sidereal time. A sidereal day is shorter than a mean solar day because Earth's rotation moves an observer's meridian, which is what we're using to define the beginning and ending of a sidereal day, at a rate of about $1°$ per day. Henceforth, unless otherwise noted, a 24-hour day will be understood to refer to a mean solar day to correspond with how we keep time with our wristwatches.

## 3.2 Defining a Month

Just as the apparent motion of the Sun can be used to define a day and subsequently subdivided into smaller time intervals to measure hours, minutes, and seconds, the Moon's orbit can be used as the basis for defining a month. Not surprisingly, there are several ways to define a month. At a simple level, one can say that a day is the time it takes for the Sun to complete 1 trip around Earth.[6] At a similarly simplistic level, one can say a month is the time it takes for the Moon to complete 1 orbit around Earth.

To measure a month, it is necessary for the Moon to complete 1 orbit around Earth with respect to some reference point. If the reference point is a star, the Moon completes 1 orbit in 27.3217 days. A month defined in this way is a sidereal month. When the Sun is used as a reference point, the Moon takes

---

5. The amount of the adjustment that has to be made is the equation of time.
6. Actually, a day is the time it takes for Earth to rotate once on its axis, but this is conceptually equivalent to the Sun revolving about Earth. We will often find it convenient to assume that the Sun orbits around a stationary Earth even though in reality Earth orbits the Sun. (Well, sort of! See chapter 6.) Also, the Sun itself is not stationary. It rotates on its own axis as it travels in an elliptical orbit around the Milky Way Galaxy.

29.5306 days to complete an orbit. Using the Sun as a reference defines the synodic month, which is the "phase" month: New Moon to New Moon, Full Moon to Full Moon, and so on. The phrase "lunar month" refers to a synodic month with some specific phase of the Moon (e.g., Full Moon) as the agreed-upon start of the month. Lunar months are rarely used today except as the basis for some religious calendars, such as the Jewish and Muslim calendars. In calendar systems based on lunar months, a given religious festival will always occur during the same lunar phase.

The definition of a month with which we are most likely familiar is the civil calendar month, which is not based on an orbital reference point at all. Instead, a civil year (defined in the next section) is divided into 12 months with the number of days in a month varying from 28 to 31 days. Calculations based on civil calendar months are complicated because the number of days in a month varies according to which month it is. For the remainder of this book, when we refer to a month we will mean a civil calendar month.

## 3.3  Defining a Year

To define a year in terms of astronomical events, we again turn our attention to the Sun's motion only to find that there are multiple ways to define a year, just as there are multiple ways to define a day or a month. We briefly mention some of the different ways to define a year to give a greater appreciation for how difficult the concept of time really is.

The first definition to consider is the tropical year, for which it is convenient to assume that the mean Sun orbits Earth. Visualizing the Earth-Sun relationship in this way, the mean Sun crosses the plane of Earth's equator twice a year. These 2 times are called the equinoxes, and they occur around March 21 (vernal equinox) and September 22 (autumnal equinox). On these 2 dates the lengths of daylight hours and nighttime hours are approximately equal. The equinoxes provide convenient reference points for defining a tropical year as the time interval between 2 successive vernal equinoxes. A tropical year is equal to about 365.2422 mean solar days.

A tropical year suffers from the disadvantage of having a fractional number of days. To circumvent this difficulty, a civil year is defined to be exactly either 365 or 366 days in length, depending upon whether it is a leap year. We are perhaps most familiar with civil years, as they are the basis for the calendars that we use each day. The price paid for eliminating the fractional number of days that occur in a tropical year is that civil years vary in length because of the necessity to include the concept of leap years.

# Time Conversions

The Besselian year is sometimes used in astronomical calculations, and it is identical to a tropical year except that it begins when the mean Sun reaches an ecliptic longitude of 280°, which occurs on approximately January 1.[7] A Besselian year is the same length as a tropical year because both are based upon the time interval between leaving from and returning to a given celestial reference point (vernal equinox for a tropical year, ecliptic longitude 280° for a Besselian year). Because the vernal equinox is the reference point for a tropical year, a tropical year technically begins and ends around March 21 rather than January 1. By contrast, the reference point of ecliptic longitude 280° was deliberately chosen so that a Besselian year will begin and end at approximately January 1.

It is possible to define a year in many other ways by simply choosing a different reference point. When a star is used as a reference point, the sidereal year is defined, and it is approximately 365.2564 days. Using the point at which Earth is closest to the Sun as a reference point, the anomalistic year is defined, and it is approximately 365.2596 days in length. The draconic year combines the motion of the Sun and Moon to create a reference point for defining a year that is useful in predicting eclipses; it is approximately 346.6201 days in length. The Julian year is exactly 365.25 days in length, and it is the average length of a year in the Julian calendar system, which we will discuss in section 3.5. The Milky Way Galaxy, in which Earth and the Solar System reside, can also be used as a reference point and thereby defines a Galactic year. A Galactic year is the time it takes for the Solar System to orbit once around the center of the Milky Way Galaxy. A Galactic year is a very long time, approximately 225–250 million years, no matter which of the preceding Earth-based definitions we use for a year!

Fortunately, we need not concern ourselves with all these ways to define a year. In the sections and chapters ahead, we will generally need to worry only about civil years.

## 3.4 Defining Time of Day

At this point, the astute reader may have noted that the various ways for defining a day (solar, mean solar, and sidereal), a month (sidereal, synodic, lunar, and calendar), and a year (tropical, civil, Besselian, sidereal, anomalistic, draconic, Julian, and Galactic) create units of time that are independent of where

---

[7]. The ecliptic coordinate system is presented in the next chapter. For now, think of ecliptic longitude as similar to measuring longitude on Earth, except that we're measuring longitude with respect to a specific reference point in the sky.

an observer is actually located on Earth's surface. However, an observer's location *is* a crucial factor for calculating where in the sky a celestial object will appear. This is obvious when considering the location of an object such as the Sun. When the Sun is directly overhead (i.e., noon) for an observer in Europe, it most certainly is not overhead at that same instant in time for someone on the West Coast of the United States. Thus, a major concept to understand is how to adjust time based on an observer's location, which we can do by more precisely defining what we mean by time of day.

With a basic understanding of the various ways to define a day, month, and year, we must now contemplate what it means to answer the question, *What time is it?* (We'll consider the question in the next section.) This deceptively simple question requires a rather lengthy answer that is even more complex than the preceding discussion about defining a day, month, and year!

Defining the time of day is a complex undertaking for at least 3 reasons. First, the time of day depends on how a day is defined; that is, what reference point (Sun, mean Sun, or stars) is being used. Second, because Earth rotates on its axis, the time of day with respect to where the Sun is in the sky for 2 different observers depends on where each observer is located. In essence, this means that when some astronomical event occurs, such as sunrise, it will be observable at *precisely the same instant in time* for 2 observers *only if* they are both located at exactly the same longitude. Third, with the advent of atomic clocks and the ability to use radio signals originating from space as time references, there are several precise, but very different, ways to measure the passage of time. Increasingly precise methods for measuring the passage of time is a necessity in our modern world of globally distributed computer networks and navigation based on Global Positioning System (GPS) satellites. Because of such inventions, there is a need to synchronize clocks more precisely than was feasible with earlier methods for measuring the passage of time. However, we will limit our discussion in this section by worrying only about defining time of day for use in the science of astronomy.

Because the instant in time at which an astronomical event can be observed depends upon an observer's location, how can 2 observers at different longitudes coordinate time, without directly communicating with each other, so that they both will know when the event can be seen from their respective locations? More specifically, how can 2 observers at different longitudes (a) agree upon the time of day and (b) know when an astronomical event will be observable by each of them given that they know when it will be observable by one of them? To answer this two part question, consider first how a day should be defined for our 2 observers.

# Time Conversions

As we pointed out earlier, the length of a solar day is affected by seasonal variations in the apparent motion of the Sun as Earth progresses in its orbit around the Sun. Moreover, using a device such as a sundial to measure time in a solar day means that observers at different longitudes will always obtain different time of day readings. For example, if 1 of our 2 observers is only 50 miles due west of the other, the 2 observers will disagree on the time of day, as measured by a sundial, by approximately 3 minutes. Such a relatively small difference in time is probably unimportant for scheduling a meeting between our 2 observers,[8] but the difference *is* important in the context of astronomical events. Apparent time (sundial time) is simply too imprecise for most uses, including astronomy, and therefore it is not appropriate for our 2 observers to base time of day on a solar day.

We can avoid the daily and seasonal variations inherent in a solar day by using a mean solar day instead. This is a significant improvement because it makes time regular from 1 day to the next and makes each day exactly 24 hours in length regardless of where Earth is in its orbit around the Sun. This advantage is important enough that our 2 observers should agree to use a mean solar day as their basis for measuring time of day. A mean solar day is in fact what we implicitly assume in the modern world when we refer to the time of day.

Unfortunately, if we do nothing more than agree to use a mean solar day, time of day is still relative to each observer's location. For example, we may think of noon as when the mean Sun is "directly overhead." But when the mean Sun is "directly overhead" for someone in Boston, it certainly is not for someone in Los Angeles or even for someone only a few hundred miles east or west of Boston. Simply agreeing to define noon to be with respect to the position of the mean Sun does nothing to alter this physical reality.

We must find 1 more missing piece of the puzzle before our 2 observers can agree on the time of day: the need for the observers to synchronize their clocks. Time zones, which we will now endeavor to explain, are a key element for achieving this synchronization. By combining time zones and a mean solar day, not only do we ensure that every day is exactly 24 hours in length, but the observers' locations can be accounted for by making a simple adjustment that takes into account their geographic locations (i.e., what time zone they fall within). Let's see how this is possible.

---

8. Although we have not yet described time zones, 2 observers at opposite ends of the same time zone will disagree on the apparent time of day by as much as an hour! An hour is clearly a significant difference, whether we are scheduling a meeting or predicting when an astronomical event will occur.

**Figure 3.2** Greenwich Observatory
Notice the ball on the left cupola of this circa 1850 depiction (from Robert Ball's *Great Astronomers*) of the Flamsteed House at the Greenwich Royal Observatory. This "time ball" was installed in 1833 and could be seen by ships in the nearby harbor. The ball was drawn up the pole and then lowered at precisely the same time each day. This gave ships an accurate method for setting their clocks, which were needed for navigation. This and other methods were used historically to synchronize clocks. Weather permitting, the Royal Observatory still raises and lowers the time ball each day at precisely 13:00 Greenwich Mean Time.

Since Earth rotates 360° in a mean solar day, we can subdivide Earth longitudinally into 24 equal geographic areas, called time zones, that are 15° wide in longitude. Geographical longitude is determined relative to Greenwich, England, so time zones are also defined relative to Greenwich (longitude 0°). Earth rotates 15° in an hour, so in terms of time each time zone is 1 hour in width.

We can directly tie time zones to the motion of the mean Sun by synchronizing all clocks within a time zone so that noon is the precise moment at which the mean Sun transits that time zone's central meridian. The central meridian is the meridian whose longitudinal location is the geographic center of a time zone, which means there are 7.5° of longitude within a time zone on either side of the central meridian. The central meridian at Greenwich is 0° longitude, the central meridian for the first time zone due west of Greenwich is 15° W longitude, the central meridian of the next westward time zone is 30° W longitude, and so on. Time zones east of Greenwich are handled in the same way. The central meridian for the first time zone east of Greenwich is 15° E longitude, the second central meridian is at 30° E longitude, and so on.

When we synchronize time within a time zone in this fashion, the time at the central meridian is called that time zone's Standard Time. Standard Time means that all clocks are synchronized to the same standard for everyone in a given geographic region regardless of their location. Every observer in that time zone always has the same time of day as everyone else within their time zone. This, then, is a method whereby our 2 observers can agree on the precise time of day regardless of their respective locations.

Let's take the clock synchronization idea a little further by agreeing to make Standard Time for each time zone relative to Standard Time at Greenwich. This simplifies matters because rather than each time zone having to determine when the mean Sun transits a central meridian, it can be done in 1 place (Greenwich), and then all time zones can synchronize their local time with Greenwich. When we define time zones relative to Greenwich and synchronize each time zone's Standard Time with Standard Time at Greenwich, the resulting system for synchronizing time around the world is referred to as civil time. Time within a given time zone is called that time zone's local civil time (LCT), or simply local time.

By agreeing to base Standard Time for all time zones relative to Greenwich and adjusting Standard Time by 1 hour for each time zone away from Greenwich, we have just achieved 2 very important results. First, Standard Time in adjacent time zones differs from each other by exactly 1 hour of mean solar time. Second, determining the Standard Time for an observer in any time zone is a simple matter of adding (or subtracting) an hour for each time zone that separates the observer from the Greenwich time zone. Because Earth rotates from west to the east, Standard Time in time zones west of Greenwich is earlier than Standard Time at Greenwich while Standard Time in time zones east of Greenwich is later.

For example, suppose Standard Time in the Greenwich, England, time zone is precisely $12^h00^m00^s$, and an observer is 2 time zones west of Greenwich. Then the Standard Time for that westward observer is exactly $10^h00^m00^s$ (subtract hours when going from east to west). Suppose another observer is 4 time zones east of Greenwich. Then the Standard Time for that eastward observer's time zone is $16^h00^m00^s$ (add hours when going from west to east).

Let's return to our 2 observers: if they are in the same time zone and using Standard Time, they can both agree on a precise instant in time even though the Sun will be at a different position in the sky for both of them. They can agree on a precise instant in time by merely adding or subtracting an hour for each time zone that separates them as explained in the previous paragraph.

There's a bit more information to consider in the important story of time zones. In practice, time zone boundaries are irregular because allowances are

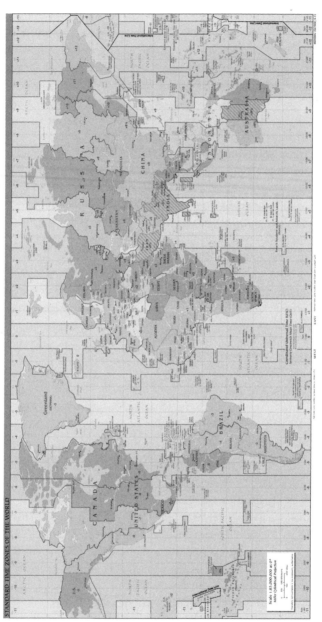

**Figure 3.3** US Time Zones
This time zone map of the United States and Canada shows that political boundaries often influence how we keep time; otherwise the north–south dividing lines would be straight!

# Time Conversions 33

made to accommodate state, country, and other man-made boundaries. For instance, the continental United States is divided into 4 time zones with very irregular boundaries, as can be seen from any map that shows US time zones. Going from east to west, the 4 US time zones are the Eastern Standard Time (EST), Central Standard Time (CST), Mountain Standard Time (MST), and Pacific Standard Time (PST) zones.

The fact that time zone boundaries are irregular in practice does not change how they work. Clocks are still synchronized for all observers within a time zone, adjacent time zones are 1 hour apart, and observers in different time zones can still agree on a precise time by adding or subtracting an hour for each time zone that separates them. Note that in some regions of the world, this is not strictly true because adjacent time zones may in fact differ by some amount other than an hour. For example, Canada's Newfoundland time zone differs from the immediately adjacent Atlantic time zone by 30 minutes rather than an hour. Consult a map of time zones to be sure that you apply the correct time zone adjustment when you are in an area that adjusts time between time zones by some amount other than an hour.

It is customary in many countries to add or subtract an hour to Standard Time depending on the season. This is called daylight saving time (DST). Most areas of the United States have adopted DST and add an hour to the clock during the spring while subtracting an hour during the fall. (This can be remembered by the adage "spring forward, fall back," which describes whether to set the clock ahead or back.) Remember to account for DST during time of day conversions!

The concept of time zones is relatively recent in human history, having been introduced in only the late 19th century. Prior to establishing time zones and Standard Time, time at a particular locality was established relative to a locally chosen meridian and some well-known time standard, such as the Big Ben clock for the city of London. Time was measured by the transit of the mean Sun across that locally chosen meridian, and the locally chosen time standard was synchronized with that transit. This meant that the time of day would often differ from city to city because meridians were selected by local authorities rather than by some central authority for a region. Time defined locally in this fashion was historically called Local Mean Time (LMT). It should be clear that Standard Time is simply an improved way to establish LMT over a geographic region larger than a city—namely, a time zone.

From this point forward, we will use LCT rather than Standard Time or LMT [9] to refer to time of day within a local time zone. This conforms with

---

[9]. Standard Time, LMT, and LCT are technically not the same because of their precise definitions and how they arose historically. For the purposes of this book, however, we do not need to distinguish between Standard Time, LMT, and LCT.

current usage, although the acronym LMT does help to emphasize that local time of day is defined with respect to the motion of the mean Sun.

Astronomical calculations are usually based on time relative to the time zone in which Greenwich, England, is situated. LMT for Greenwich has historically been called Greenwich Mean Time (GMT), but this terminology has been superseded by Universal Time (UT). We can easily convert between UT and a particular time zone's LCT by adding or subtracting an hour for each time zone that separates that time zone from Greenwich. Technically, UT and GMT are not the same thing because they differ in when a day begins. Astronomers originally chose to define a GMT day as beginning at noon because that is when the mean Sun transits the central meridian. This can be confusing because we normally think of a day as starting at midnight rather than at noon. UT, which was introduced to avoid this confusion, defines a day as beginning at midnight.

Although GMT and UT define the start of a day differently, they both refer to the same instant in time. That is, $8^h00^m00^s$ is the same instant in time, whether we are using GMT or UT to establish when a day begins. One can think of GMT and UT as being the same even though technically they are not. Some authors continue to use GMT rather than UT to emphasize that time of day is being measured with respect to the motion of the mean Sun at the Greenwich central meridian. We will conform with common usage and use UT instead of GMT to refer to time in the Greenwich time zone. Also note that UT and GMT are often referred to as Zulu Time.

Modern timekeepers no longer establish a time standard at Greenwich by making astronomical observations of the Sun to determine its precise location. Quasars, which are distant galaxies that emit radio signals, can be monitored through a worldwide network of radio telescopes and used to provide a very precise time standard. Timekeepers can also use atomic clocks to establish an extremely precise time standard.

Differences in the various timekeeping methods have given rise to a plethora of ways to define and measure time of day, such as Coordinated Universal Time (abbreviated as UTC as a compromise between English-speaking and French-speaking peoples), UT1, and UT2. For technical reasons, the time of day reported by each of these methods differs, but the differences are sufficiently small that they matter only when making precise (tenths of a second) time measurements. We will ignore these differences and assume that UTC, UT, UT1, UT2, GMT, and so on, all refer to the same time of day at the Greenwich central meridian.

Now that our 2 observers have a reliable mechanism for agreeing on the time of day, how can they know when an astronomical event will be observable for their location, assuming that they know the LCT at which the event will occur?

**Time Conversions**                                                                 35

The answer is surprisingly simple. We merely note that because Earth rotates 15° per hour, we only have to add (or subtract) an amount of time that is proportional to the distance an observer is from his or her time zone's central meridian. Let's look at an example.

Suppose the Sun will rise at precisely $6^h30^m28^s$ with respect to the central meridian for the Eastern Standard Time (EST) zone. Assume observer #1 is 2° in longitude east of the EST central meridian, and observer #2 is 5° in longitude west of the EST central meridian. Since observer #1 is east of the central meridian, the Sun will rise *earlier* for him than it will at the central meridian while for observer #2 the Sun will appear to rise *later* than it will at the central meridian. Earth rotates 15° per hour, which is equivalent to 1° every 4 minutes, so we merely have to adjust the stated LCT for sunrise by 4 minutes for every 1° in longitude that an observer is from the time zone's central meridian. Hence, for observer #1, sunrise will occur $2*4 = 8$ minutes earlier ($6^h22^m28^s$ EST) while for observer #2 sunrise will occur $5*4 = 20$ minutes later ($6^h50^m28^s$ EST).

Publications that list the times for various astronomical events are often based on UT as the reference point for time. If we know the UT for a particular event, it can be converted to the proper LCT for any observer by adding/subtracting 4 minutes for every 1° of longitude that the observer is from longitude 0°. Alternatively, we can convert UT to the LCT for the time zone in which an observer is situated and then adjust by 4 minutes for every 1° that the observer is from his or her time zone's central meridian.

Earlier we pointed out that astronomers use mean solar time to avoid the irregularities caused by Earth's orbit around the Sun. However, mean solar time is also affected by Earth's rotation about its own axis. Earth's rotation is not uniform and is impacted by the gravitational influence of the planets in the Solar System. Consequently, in truth a mean solar day is not *really* as regular in length as we have so far assumed.

The IAU has defined the Terrestrial Time (TT) standard to account for irregularities in Earth's rotation.[10] TT is based on an atomic clock and is independent of any irregularities in Earth's rotation. In chapter 7 we will make adjustments for the difference between UT and TT in order to make better predictions about lunar events. For most of this book we will blissfully ignore the impact on time of day caused by irregularities in Earth's rotation. The error

---

10. The IAU defined Terrestrial Time (TT) in 1991 to replace Terrestrial Dynamic Time (TDT), which was defined in 1976 as a successor to Ephemeris Time (ET). TT, TDT, and ET all have the objective of accounting for irregularities in Earth's rotation. The technical details of the differences between these and other systems of timekeeping are outside the scope of this book. We will discuss only TT and will assume that all current timekeeping systems are essentially the same.

incurred in doing so is unlikely to amount to more than a few minutes for this century, which will generally be sufficient for the level of accuracy aspired to in this book. In 1980 the error was slightly more than 50 seconds, although that error increased to about 66 seconds in 2010.

Before we complete our discussion of time of day, we should mention that there is 1 other important way that time of day can be defined. Just as the definition of a day can be tied to the stars (sidereal day), time of day can also be measured by the stars. Time of day defined with respect to the position of a fixed star is called sidereal or "star" time. Sidereal time at Greenwich, England, is called Greenwich Sidereal Time (GST) while sidereal time for a specific observer is called that observer's Local Sidereal Time (LST). The sidereal time for 2 observers will differ depending on their respective locations. Unlike LCT, where all observers in the same time zone agree on the mean time of day, 2 observers will not agree on the LST, even if they are in the same time zone, unless they are also at the same longitude.

Because of wobble in Earth's rotation, sidereal time is not uniform, just as mean solar time is not uniform. Astronomers have defined a system of measuring sidereal time, called mean sidereal time, to account for wobble in Earth's axis. What has been described so far is apparent sidereal time. The difference between apparent sidereal time and mean sidereal time is only a few seconds and will be ignored henceforth. We will assume that apparent sidereal time and mean sidereal time are equal.

Whew! Clearly, understanding time of day is far more complex than it first appears. After a brief discussion of calendar systems and Julian day numbers, we will describe how various time conversions are done. For the moment, concentrate more on *how* to do the various conversions rather than on *why* they are necessary. Generally speaking, most of the routines in later chapters will ask for the LCT for a given location. Given an observer's LCT, it is usually desirable to convert it to the observer's LST. The process involved is to convert LCT to UT, UT to GST, and then finally to convert GST to LST. Converting LST to LCT works by just reversing this process.

In the time conversion procedures presented in later sections, it will often be necessary to compute a time zone adjustment to account for an observer's location. Table 3.1 gives the time zone adjustments (relative to UT) in hours for each of the 4 time zones (each 1 hour apart) in the continental United States. If an observer is not located in 1 of these time zones, a time zone adjustment, expressed in hours, can be calculated as

$$Adjustment = \text{ROUND}\left(\frac{\psi}{15°}\right), \tag{3.4.1}$$

# Time Conversions

**Table 3.1** Time Zone Adjustments
The continental United States is divided into four time zones, each 1 hour apart. The adjustment shown in this table is the time adjustment relative to UT.

| Time Zone | Adjustment |
|---|---|
| EST | −5 hours |
| CST | −6 hours |
| MST | −7 hours |
| PST | −8 hours |

where $\psi$ is the observer's longitude. This equation should be obvious because it merely uses the fact that time zones are 15° wide in longitude. An observer's geographic longitude in this equation is expressed in decimal degrees and is positive for locations east of Greenwich while negative for west longitudes. For example, the time zone adjustment for 30° W longitude is $-2^h$ while the time zone adjustment for 45° E longitude is $+3^h$.

Note that table 3.1 and equation 3.4.1 account only for what *time zone* an observer is in relative to Greenwich and provide an adjustment to the LCT for all observers within that time zone. The table and equation do *not* account for *where* an observer is within a time zone (e.g., how far away the observer is from the time zone's central meridian). When it is necessary to account for an observer's longitude within a time zone, the observer's LCT will be adjusted by 4 minutes per 1° of longitude away from the prime meridian at Greenwich. This is accomplished by the equation

$$Adjustment = \frac{\psi}{15°}. \tag{3.4.2}$$

This equation is necessary so that an observer at a particular location within a time zone can know the actual local time at which a predicted astronomical event will occur.

To summarize the salient points, we will use LCT to refer to the local civil time within an observer's time zone. LCT is what an observer's wristwatch measures, and it is synchronized with UT, which is the local civil time at the Greenwich, England, prime meridian (longitude 0°). LST and GST are based on sidereal time rather than mean solar time and are analogous to LCT and UT. Additionally, the world is divided into 24 longitudinal time zones starting from the prime meridian at Greenwich. Time zone boundaries are typically irregular to account for man-made boundaries, such as national borders. Despite irregular time zone boundaries, adjacent time zones are usually 1 hour apart

(15° longitude). By using time zones and synchronizing clocks with Greenwich, all observers within the same time zone will agree on the same LCT, regardless of the position of the mean Sun.

## 3.5 Calendar Systems

Because there are so many ways to define a day, month, year, and time of day, it should be no surprise that there are also several different ways to create a calendar. Early calendars were based on the apparent motion of the Sun and Moon, but they generally failed to account for the fact that Earth orbits the Sun in a fractional number of days. Thus, if a calendar defined a year to be 365 solar days in length, a calendar year was actually short by 0.2422 days. After 10 years, such a calendar would be wrong by a little more than 2 days. This was hardly noticeable to early man, and it would probably not be noticeable to the average person today. However, after 100 years, such a calendar is wrong by 24 days and wrong by 242 days after 1,000 years. With this kind of cumulative error, it would soon become apparent that the seasons were occurring in the wrong months (winter in July, summer in December, etc.) when compared to recorded history. Such a calendar system is very inconvenient indeed.

The Roman emperor Julius Caesar (100 BC–44 BC) introduced an improved calendar in about 46 BC. His calendar, called the Julian calendar, assumes a year is exactly 365.25 days in length so that an additional day must be added to the calendar every fourth year. Therefore any year that is evenly divisible by 4 is called a leap year and given an extra day. The Julian calendar is off by about 0.0078 days (approximately $11^m 14^s$) a year and by a little over a week (7.8 days) every 1,000 years. Close, but not close enough.

In 1582, Pope Gregory XIII was informed that the Julian calendar, which was in common usage, was already incorrect by nearly 2 weeks. To remedy the situation, a new calendar system was proposed that retained the concept of a leap year but modified how a leap year is defined. In the calendar system that Pope Gregory officially established, called the Gregorian calendar in his honor, a century year (i.e., a year ending in 2 zeros) is a leap year only if it is also evenly divisible by 400. The length of a year in a Gregorian calendar is 365.2425 days, with an error of about 0.0003 days (25.92 seconds) per year. The Gregorian calendar is the most commonly used civil calendar in the world today. Because of the improvements made by Pope Gregory, our modern calendar is only off by 3 days every 10,000 years, making it unlikely that another calendar reform will occur for quite some time.

**Time Conversions**  39

It is worth noting an interesting historical sidelight: the Gregorian calendar was first put into effect on October 15, 1582. At that time, the 10 days between October 5 and October 14 were simply abolished. (In one sense, there never was an October 6, 1582!) However, the Gregorian calendar was not widely adopted outside the Holy Roman Empire until the 18th century. In fact, England and the American colonies did not accept the new calendar system until 1752. In that year, it is said, rioting protesters in England demanded that they be given back their "missing" days, although stories of actual riots are likely untrue. However, even as late as England and America were in adopting the Gregorian calendar, Russia did not follow suit until 1918!

Several other calendar systems have been developed in the past, some of which are still in use today. The Chinese calendar began in 2397 BC and is based on 60-year cycles. The ecclesiastical calendar is used by some Roman Catholic and Protestant countries to reckon years as beginning with Advent Sunday (the Sunday closest to the last Sunday in November, which means Advent Sunday is the fourth Sunday before Christmas). Calendars used by Hindus, Hebrews, and Muslims also base years on important religious events. Hindus divide the year into 12 months with the months based on the signs of the zodiac. The Hebrew calendar begins in the year 3761 BC with a cycle of 19 years in which the 3rd, 6th, 8th, 11th, 14th, 17th, and 19th years are leap years. The year 3761 BC was chosen as the beginning of the Hebrew calendar because Jewish scholars in the Middle Ages calculated October 7, 3761 BC as the date when God created Adam and Eve. Muslim countries often use a calendar that dates from July 16, 622 AD, which is the year that the prophet Mohammed fled from Mecca to Medina. Only the Julian and Gregorian calendars will be of further interest in this book.

Determining whether a year is a leap year in the Gregorian calendar is a simple matter. To be a leap year, the year must satisfy 2 conditions.

1. The year must be evenly divisible by 4.
2. If the year is a century year, it must be evenly divisible by 400.

For example, the year 1906 is not a leap year because it is not evenly divisible by 4. The year 1908 is evenly divisible by 4 but is not a century year, so it is a leap year. The year 1800 is evenly divisible by 4 and is a century year, but because it is not evenly divisible by 400, it is not a leap year. The year 1600 is a leap year because it is evenly divisible by 4, and it is a century year that is also evenly divisible by 400.

## 3.6 Julian Day Numbers

It is necessary in many astronomical calculations to know the number of days that have elapsed between 2 events. Julian day numbers, typically abbreviated as JD, are used to facilitate calculating elapsed days. The Julian day number for a given date is the number of days, including fractional days, that have elapsed since noon at Greenwich, England, on January 1, 4713 BC. It is important to realize that a Julian day number begins at $12^h$ UT (exactly noon) and *not* at midnight ($0^h$ UT), as we normally reckon the beginning of a new day. This may seem unusual when first working with Julian day numbers, but the reason for starting a Julian day number at noon is that astronomers historically used the transit of the mean Sun across an observer's meridian to define the beginning of a day, which of course corresponds to noon at that meridian.

When a calendar date is converted to a Julian day number, the result is sometimes called the Julian date. This can be confusing because the phrase "Julian date" suggests some connection with the calendar system named after Julius Caesar. Julian day numbers do *not* mean that they are given with respect to the Julian calendar because Julian day numbers cover all years from 4713 BC forward, regardless of whether a date is from the Julian or Gregorian calendar system. To avoid confusion, in this book we will use "Julian date" to exclusively mean a date in the Julian calendar system and "Julian day number" to mean the number of elapsed days since noon UT on January 1, 4713 BC.[11]

Before we describe how to convert a calendar date to a Julian day number, 1 more detail must be explained. The Julian day number also accounts for the time of day at Greenwich (UT). To accomplish this, the time of day is expressed as a fractional part of the day and added to the day of the month. A couple of examples will illustrate this point.

Express 6:00:00 UT on February 14 as a fractional day.

1. Use the techniques from chapter 2 to express the time in decimal format. In this case, 6:00:00 becomes $6.0^h$ in decimal format.

2. Divide the decimal hours by 24 to get the time as a fractional part of the day. For this example,

$$\frac{6.0}{24} = 0.25 \text{ days.}$$

---

[11]. Many astronomy books and websites use "Julian Date" when a Julian day number is meant. You must determine from context whether a date in the Julian calendar system is meant, or (more likely) a Julian day number.

# Time Conversions

3. Add the fractional part of the day to the day. Doing so for this example gives February 14.25 as the day we must use for converting a date to its Julian day number.

As another example, express $14^h 33^m 36^s$ on March 21 as a fractional day.

1. In decimal format, $14^h 33^m 36^s = 14.56^h$.
2. Dividing by 24, we have $\frac{14.56}{24} = 0.606667$ days.
3. Adding this to the day gives March 21.606667 as the day we must use for converting a date to its Julian day number.

Converting a calendar date to a Julian day number requires 5 steps. Assume that the calendar date is *Month/Day/Year* where *Month* is an integer ranging from 1 to 12 with 1 = January, 2 = February, and so on. *Day* is the day of the month, including the time of day as a fractional part of the day as explained in the previous 2 examples. *Year* is the calendar year, and it is positive to indicate AD dates and negative to indicate BC dates.

Let's convert January 1, 2010, at $0.0^h$ UT (i.e., midnight) to its corresponding Julian day number. The necessary steps are listed here with the result for this sample problem given in parentheses at the end of each step.

1. If *Month* > 2, set $y = Year$, and $m = Month$. Otherwise, $y = Year - 1$, and $m = Month + 12$.
   (Ans: $y = 2009$, $m = 13$.)
2. If *Year* < 0, set $T = 0.75$ else $T = 0$.
   (Ans: $T = 0$.)
3. Determine if the date is a Gregorian date. Dates before October 15, 1582, are not, while all other dates are Gregorian.
   (Ans: 1/1/2010 is Gregorian.)
4. If the date is Gregorian, compute $A = \mathtt{FIX}\left(\frac{y}{100}\right)$ and $B = 2 - A + \mathtt{FIX}\left(\frac{A}{4}\right)$. If the date is not Gregorian, set $A = 0$ and $B = 0$.
   (Ans: $A = 20$, $B = -13$.)
5. Compute

$$JD = B + \mathtt{FIX}(365.25y - T) + \mathtt{FIX}[30.6001(m+1)] + Day + 1{,}720{,}994.5$$

to complete the calculation.
   (Ans: $JD = 2{,}455{,}197.50$.)

As another example, convert March 21, 2015, at $12^h 00^m 00^s$ UT (noon) to its corresponding Julian day number. (Fractional days must be included, so we must convert 3/21.50/2015 because noon is half of a day.) The resulting Julian day number is 2,457,103.0.

Carefully note from these 2 examples that whenever the fractional part of a Julian day number is 0.0, the time of day was noon (UT) for the calendar date converted whereas whenever the fractional part is 0.5, the time of day was midnight (UT). More generally, the time of day (UT) can be retrieved from a Julian day number by multiplying the fractional part of the Julian day number by 24 and adding $12^h$ to the result. Note that $12^h$ must be added because Julian day numbers start at noon (UT), not midnight. If adding $12^h$ produces a result that is greater than $24^h$, then $24^h$ must be subtracted to ensure the resulting time of day is in the range $[0^h, 24^h]$.

For example, assume the fractional part of a Julian day number is 0.27. Then the corresponding time of day (UT) is $0.27 * 24 + 12 = 18.48^h = 18^h 28^m 48^s$. As another example, assume the fractional part of a Julian day number is 0.78. Then the corresponding time of day (UT) is $0.78 * 24 + 12 = 30.72^h$. Since this result is greater than $24^h$, 24 must be subtracted to put the time of day into the proper range. Thus, the time of day is $30.72^h - 24^h = 6.72^h = 6^h 43^m 12^s$.

Converting a Julian day number back to its corresponding calendar date requires 10 steps. Assume *JD* is the Julian day number. We will use the Julian day number 2,400,000.5 to illustrate the process.

1. Add 0.5 to the Julian day number. That is, let $JD1 = JD + 0.5$.
   (Ans: $JD1 = 2,400,001.0$.)
2. Compute $I = \text{FIX}(JD1)$ and $F = \text{FRAC}(JD1)$.
   (Ans: $I = 2,400,001$, $F = 0.0$.)
3. If $I > 2,299,160$, set $A = \text{FIX}[(I - 1,867,216.25)/36,524.25]$ and $B = I + 1 + A - \text{FIX}(A/4)$. Otherwise set $B = I$.
   (Ans: $A = 14$, $B = 2,400,013$.)
4. Set $C = B + 1524$.
   (Ans: $C = 2,401,537$.)
5. Compute
$$D = \text{FIX}\left(\frac{C - 122.1}{365.25}\right).$$
   (Ans: $D = 6574$.)
6. Compute $E = \text{FIX}(365.25D)$.
   (Ans: $E = 2,401,153$.)

7. Compute

$$G = \text{FIX}\left(\frac{C-E}{30.6001}\right).$$

(Ans: $G = 12$.)

8. The day is given by $Day = C - E + F - \text{FIX}(30.6001G)$.

(Ans: $Day = 17$, $0^h$ UT since $Day$ is an integer value.)

9. The month is given by $Month = G - 1$ if $G < 13.5$, and $Month = G - 13$ if $G > 13.5$. (Note that $G$ cannot equal 13.5 because the result obtained in step 7 will always be an integer.)

(Ans: $Month = 11$.)

10. The year is given by $Year = D - 4716$ if $Month > 2.5$, and $Year = D - 4715$ if $Month < 2.5$. (Note that $Month$ cannot equal 2.5 because the result obtained in step 7, and consequently in step 9, is always an integer.) If the $Year$ obtained is negative, then the resulting date is BC. Otherwise, the date is AD.

(Ans: $Year = 1858$ AD.)

The calendar date corresponding to the Julian day number 2,400,000.5 is thus November 17, 1858, at $0^h$ UT. We immediately know that the time of day is midnight (UT) because the fractional part of the Julian day number is 0.5.

There are a couple of peculiarities to note about these algorithms for performing Julian day number conversions. First, if you convert the Julian day number 0.0 to a calendar date, the resulting year is –4712, not –4713 as might be expected. The reason this happens is that astronomers number the year immediately preceding 1 AD as 0 whereas we normally consider the preceding year to be 1 BC.

The second peculiarity is that the Julian day number computed by the above algorithm for the pair of dates October 5.0, 1582, and October 15.0, 1582, is exactly the same ($JD = 2,299,160.50$), as is the pair of dates October 6.0, 1582, and October 16.0, 1582 ($JD = 2,299,161.50$). A moment's reflection reveals why this occurs. The Gregorian calendar was instituted on October 15, 1582. The "lost days" between October 5 and 14 are treated as Julian dates in the algorithm for converting a calendar date to a Julian day number, but as Gregorian dates in the reverse process.

We need not concern ourselves with either of these peculiarities because in this book we won't be performing astronomical calculations for the 16th century or earlier! Some websites, such as for the US Naval Observatory, account for both of these peculiarities by limiting their algorithms to consider only Gregorian dates.

The modified Julian day number (MJD) is sometimes used to avoid the large numbers produced from the previous algorithms for handling Julian day numbers. MJD is defined to be the number of days that have elapsed since $0^h$ (UT) on November 17, 1858, and is given by the equation

$$MJD = JD - 2{,}400{,}000.5, \tag{3.6.3}$$

where $JD$ is the Julian day number for the date being converted. The reason for choosing $0^h$ (UT) November 17, 1858, should be obvious because its Julian day number is 2,400,000.5, thus giving a modified Julian day number of 0.0 for November 17, 1858. For our earlier example (March 21, 2015, at $12^h00^m00^s$ UT), the MJD is 57,102.5.

Note that when the fractional part of a modified Julian day number is 0.0, the time of day for the calendar date being converted is midnight (UT), whereas when the fractional part is 0.5, the time of day was noon (UT). This is exactly the opposite of what the fractional part of a Julian day number means. Also, note that the time of day (UT) can be extracted from a modified Julian day number by simply multiplying the fractional part of the modified Julian day number by 24. There is no need to add $12^h$ to the result, which must be done in the case of a Julian day number.

## 3.7 Some Calculations with Dates

Julian day numbers are convenient for several calculations involving dates. For example, to calculate the number of elapsed days between 2 dates, simply subtract their corresponding Julian day numbers. One reason for doing so is to calculate how many days have elapsed since the beginning of the year. Given a calendar date, some authors reference the number of elapsed days since the beginning of the year as that date's day number. To avoid confusion with the Julian day number, we will refer to this as the "days into the year," which is a more descriptive phrase anyway.

It is possible to compute the number of elapsed days into a year without resorting to Julian day numbers. Only 2 steps are required.

1. If the year is a leap year, set $T = 1$ else set $T = 2$.
2. The number of days into the year is given by the equation

$$N = \mathrm{FIX}\left(\frac{275 * Month}{9}\right) - T * \mathrm{FIX}\left(\frac{Month + 9}{12}\right) + Day - 30.$$

For example, March 9, 2005, was 68 days into the year 2005 while March 9, 2000, was 69 days into 2000 because 2000 was a leap year.

Converting the number of days into a year to a specific calendar date is only slightly more complex. Assume that $N$ is the number of days into the year. To illustrate the process, convert $N = 68$ for the year 2005 back to its corresponding calendar date. The required steps are:

1. If *Year* is a leap year, set $A = 1523$, otherwise set $A = 1889$.
   (Ans: $A = 1889$.)
2. Let

$$B = \text{FIX}\left(\frac{N + A - 122.1}{365.25}\right).$$

   (Ans: $B = 5$.)
3. Let $C = N + A - \text{FIX}(365.25B)$.
   (Ans: $C = 131$.)
4. Let $E = \text{FIX}(C/30.6001)$.
   (Ans: $E = 4$.)
5. If $E < 13.5$, then *Month* $= E - 1$, otherwise *Month* $= E - 13$.
   (Ans: *Month* $= 3$.)
6. *Day* $= C - \text{FIX}(30.6001E)$.
   (Ans: *Day* $= 9$.)

Sometimes it is interesting to know what day of the week a certain date falls on. For instance, it might be amusing to determine the day of the week on which someone was born. Julian day numbers provide a simple way to calculate the day of the week on which a given date falls. Using February 7, 1985, as an example, the required steps are:

1. Convert the date to its Julian day number, *JD*, at $0^h$ UT. (Be sure to *exclude* the time of day as a fractional part of the day in the date. Thus, for this example, use *Day* $= 7$ and *not* 7.xxxx.)
   (Ans: $JD = 2{,}446{,}103.5$.)
2. Calculate $A = (JD + 1.5)/7$.
   (Ans: $A = 349{,}443.57143$.)
3. Let $B = 7 * \text{FRAC}(A)$.
   (Ans: $B = 4.000000$.)
4. Let $N = \text{ROUND}(B)$.
   (Ans: $N = 4$.)

The number $N$ resulting from the this calculation gives the corresponding day of the week where $N=0$ is Sunday, $N=1$ is Monday, and so on. So, February 7, 1985, fell on a Thursday ($N=4$).

## 3.8  LCT to UT

We now turn our attention to performing conversions between local and Greenwich time zones, and between solar time and sidereal time. We will begin by showing how to convert LCT to UT. Converting between LCT and UT is independent of the date because it is merely a matter of making a time zone adjustment.

For example, convert $18^h00^m00^s$ LCT to UT for an observer in the Eastern Standard Time zone. Assume that this is not daylight saving time.

1. Convert *LCT* to decimal format.
   (Ans: $LCT = 18.0^h$.)

2. If necessary, adjust for daylight saving time. If the *LCT* given is on DST, subtract $1^h$, otherwise do nothing in this step.
   (Ans: no adjustment needed, $T = 18.0^h$.)

3. Using equation 3.4.1 or table 3.1 as appropriate, calculate a time zone adjustment.
   (Ans: $Adjustment = -5^h$.)

4. Subtract the time zone adjustment in step 3 from the result of step 2.
   (Ans: $UT = 23.0^h$.)

5. If the result of step 4 is negative, add $24^h$. If the result of step 4 is greater than 24, subtract $24^h$. (Note that if $24^h$ must be added to step 4, the resulting time is on the previous date whereas if $24^h$ must be subtracted, the resulting time is for the next day.)
   (Ans: no adjustment, $UT = 23.0^h$.)

6. Convert the result of step 5 to HMS format if desired.
   (Ans: $UT = 23^h00^m00^s$.)

Assuming the observer is at 45° E longitude rather than in the Eastern Standard Time zone, the result is $UT = 15^h00^m00^s$ because the observer is 3 time zones east of Greenwich (the adjustment from equation 3.4.1 is $+3^h$).

## 3.9  UT to LCT

Convert $23^h30^m00^s$ UT to LCT for an observer within the Eastern Standard Time zone, and assume daylight saving time.

1. Convert *UT* to decimal format.
   (Ans: $UT = 23.5^h$.)

2. Using equation 3.4.1 or table 3.1 as appropriate, calculate a time zone adjustment.
   (Ans: $Adjustment = -5^h$.)

3. Add the time zone adjustment from step 2 to the result of step 1.
   (Ans: $LCT = 18.5^h$.)

4. If the result of step 3 is negative, add $24^h$. If the result of step 3 is greater than 24, subtract $24^h$. (Note that adding $24^h$ means that the resulting LCT is for the next day whereas subtracting $24^h$ means that the resulting LCT is for the previous day.)
   (Ans: no adjustment, $LCT = 18.5^h$.)

5. If necessary, adjust for daylight saving time. If the individual is on DST, add $1^h$, otherwise do nothing in this step.
   (Ans: $LCT = 19.5^h$.)

6. Convert the result of step 5 to HMS format if desired.
   (Ans: $LCT = 19^h30^m00^s$.)

If the observer was at 45° E longitude and on daylight saving time, the result would be $LCT = 3^h30^m00^s$ on the next day.

## 3.10  UT to GST

To convert UT to GST, the date must be known. Convert $23^h30^m00^s$ UT to GST for February 7, 2010.

1. Convert the given date to its Julian day number at $0^h$ UT (i.e., do not express the time of day as a fractional part of the day for this step).
   (Ans: $JD = 2,455,234.5$.)

2. Calculate the Julian day number for January 0.0 of the given year. Let this Julian day number be $JD_0$.
   (Ans: $JD_0 = 2,455,196.5$.)

3. Subtract step 2 from step 1 to get the number of elapsed days into the year. Let *Days* be this number.
   (Ans: $Days = 38$.)

4. Let $T = \frac{JD_0 - 2,415,020.0}{36,525.0}$.
   (Ans: $T = 1.099973$.)

5. Let $R = 6.6460656 + 2400.051262T + 0.00002581T^2$.
   (Ans: $R = 2646.636775$.)

6. Let $B = 24 - R + 24(Year - 1900)$.
   (Ans: $B = 17.363225$.)
7. Let $T_0 = 0.0657098 Days - B$.
   (Ans: $T_0 = -14.866252$.)
8. Convert the UT given into decimal format.
   (Ans: $UT = 23.5^h$.)
9. $GST = T_0 + 1.002738 UT$.
   (Ans: $GST = 8.698091^h$.)
10. If the GST from the previous step is negative, add $24^h$. If the GST from the previous step is greater than 24, subtract $24^h$.
    (Ans: $GST = 8.698091^h$.)
11. Convert the result of step 10 to HMS format if desired.
    (Ans: $GST = 8^h 41^m 53^s$.)

## 3.11 GST to UT

Converting GST to UT also requires that the date be known. Calculate the UT for $8^h 41^m 53^s$ GST on February 7, 2010.

1. Convert the given date (at $0^h$) to its Julian day number.
   (Ans: $JD = 2,455,234.5$.)
2. Calculate the Julian day number for January 0.0 of the given year. Call this Julian day number $JD_0$.
   (Ans: $JD_0 = 2,455,196.5$.)
3. Subtract step 2 from step 1 to get the number of days into the year. Call this number $Days$.
   (Ans: $Days = 38$.)
4. Let $T = [JD_0 - 2,415,020.0]/36,525.0$.
   (Ans: $T = 1.099973$.)
5. Let $R = 6.6460656 + 2400.051262T + 0.00002581T^2$.
   (Ans: $R = 2646.636775$.)
6. Let $B = 24 - R + 24(Year - 1900)$.
   (Ans: $B = 17.363225$.)
7. Let $T_0 = 0.0657098 Days - B$.
   (Ans: $T_0 = -14.866252$.)

# Time Conversions

8. If the result of step 7 is negative, add $24^h$. If the result of step 7 is greater than 24, subtract $24^h$.
   (Ans: $T_0 = 9.133748$.)
9. Convert the GST given to decimal format.
   (Ans: $GST = 8.698056^h$.)
10. Let $A = GST - T_0$.
    (Ans: $A = -0.435692$.)
11. If $A$ is negative, add $24^h$. Otherwise make no adjustment.
    (Ans: $A = 23.564308$.)
12. $UT = 0.997270 A$.
    (Ans: $UT = 23.499977^h$.)
13. Convert the result of step 12 to HMS format if desired.
    (Ans: $UT = 23^h 30^m 00^s$.)

Note that steps 1–7 for converting GST to UT are identical to steps 1–7 for converting UT to GST.

## 3.12  GST to LST

Converting GST to LST requires knowing an observer's longitude, but it is independent of the date. Assume that the GST is $2^h 03^m 41^s$ for an observer at 40° W longitude. Calculate the corresponding LST.

1. Convert the *GST* to decimal format.
   (Ans: $GST = 2.061389^h$.)
2. Calculate a time zone adjustment using equation 3.4.2. Remember that east longitudes are positive while west longitudes are negative. Also note that this adjustment is *almost* the same as the time zone adjustment in equation 3.4.1, but it includes the fractional part of the time zone adjustment whereas equation 3.4.1 does not.
   (Ans: $Adjust = -2.666667^h$.)
3. $LST = GST + Adjust$.
   (Ans: $LST = -0.605278$.)
4. If *LST* is negative, add $24^h$. If *LST* is greater than 24, subtract $24^h$. Otherwise make no adjustments.
   (Ans: $LST = 23.394722^h$.)

5. Convert *LST* to HMS format if desired.
   (Ans: $LST = 23^h23^m41^s$.)

Notice that the LST is simply the GST with an adjustment that takes into account an observer's longitude. This adjustment is the difference, expressed in hours, between the observer's longitude and that of Greenwich, England.

## 3.13 LST to GST

Converting LST to GST is very similar to converting GST to LST. The difference arises in step 3 in which an adjustment is *subtracted* from the LST to calculate the GST whereas an adjustment is *added* to GST to calculate the LST. Assume that an observer at 50° E longitude calculates the LST to be $23^h23^m41^s$. Convert this LST to GST.

1. Convert *LST* to decimal format.
   (Ans: $LST = 23.394722^h$.)
2. Calculate a time zone adjustment using equation 3.4.2. Remember that east longitudes are positive while west longitudes are negative.
   (Ans: $Adjust = +3.333333^h$.)
3. $GST = LST - Adjust$.
   (Ans: $GST = 20.061389^h$.)
4. If *GST* is negative, add $24^h$. If *GST* is greater than 24, subtract $24^h$. Otherwise make no adjustment.
   (Ans: no adjustment.)
5. Convert *GST* to HMS format if desired.
   (Ans: $GST = 20^h30^m41^s$.)

## 3.14 Program Notes

The program **RunChap3** does all the time and date conversions described in this chapter. When doing mean time conversions for time zones inside the continental United States, the program uses table 3.1 to make time zone adjustments. For other time zones, the program allows a longitude to be entered and then uses equation 3.4.1 to compute a time zone adjustment. This applies only to mean time conversions. Sidereal time conversions require including fractional parts of an hour and use equation 3.4.2 to compute time zone adjustments whether in a US time zone or not.

# Time Conversions 51

It is frequently necessary to enter an observer's latitude and/or longitude. Although this book's programs do not allow entering a numeric sign to indicate latitude/longitude direction, for convenience they do allow omitting the "E/W" (longitude) and "N/S" (latitude) designators. Thus, one may enter a longitude as 48.5, 48.5E, or 48.5W (or their equivalent HMS forms). However, one *may not* enter −48.5, +48.5, −48.5E, or +48.5W because numeric signs are not permitted for latitude/longitude. When the direction designator is omitted, it is assumed that the latitude or longitude entered is positive (i.e., N latitude, E longitude). To avoid potential confusion, you should get in the habit of always specifying the direction, particularly for longitudes. The difference *is* significant, as you can easily see by converting a value such as $5^h$ LCT to UT for 75° W longitude and comparing the results to the same conversion for 75° E longitude!

## 3.15 Exercises

1. Was 1984 a leap year?
   (Ans: yes.)
2. Was 1974 a leap year?
   (Ans: no.)
3. Was 2000 a leap year?
   (Ans: yes.)
4. Was 1900 a leap year?
   (Ans: no.)
5. Convert midnight UT on November 1, 2010, to its Julian day number.
   (Ans: 2,455,501.5.)
6. Convert $6^h$ UT on May 10, 2015, to its Julian day number.
   (Ans: 2,457,152.75.)
7. Convert $18^h$ *UT* on May 10, 2015, to its Julian day number.
   (Ans: 2,457,153.25.)
8. Convert 2,369,915.5 to its corresponding calendar date.
   (Ans: 7/4/1776 at midnight UT.)
9. Convert 2,455,323.0 to its corresponding calendar date.
   (Ans: 5/6/2010 at noon UT.)
10. Convert 2,456,019.37 to its corresponding calendar date.
    (Ans: 4/1/2012 at $20^h 52^m 48^s$ UT.)

11. On what day of the week did 7/4/1776 fall?
    (Ans: Thursday.)
12. On what day of the week did 9/11/2011 fall?
    (Ans: Sunday.)
13. How many days into the year was 10/30/2009?
    (Ans: 303 days.)
14. If the date was 250 days into 1900, what was the date?
    (Ans: 9/7/1900.)
15. Assume that the date is 12/12/2014, and an observer in the Eastern Standard Time zone is at 77° W longitude. Assume that it is not daylight saving time. If LCT is $20^h00^m00^s$, what are the corresponding UT, GST, and LST times?
    (Ans: $UT = 1^h00^m00^s$ (next day!), $GST = 6^h26^m34^s$ (12/13/2014), and $LST = 1^h18^m34^s$ (12/13/2014).)
16. Assume that the date is 7/5/2000 for an observer at 60° E longitude and that it is daylight saving time. If LST for the observer is $5^h54^m20^s$, what are the corresponding GST, UT, and LCT times?
    (Ans: $GST = 1^h54^m20^s$, $UT = 7^h00^m00^s$, and $LCT = 12^h00^m00^s$.)

# 4 Orbits and Coordinate Systems

Chapter 3 discussed the time element of positional astronomy. The methods presented there accounted for differences in various methods for measuring time and, at least as an initial start, for an observer's location on Earth. With the techniques from chapter 3 as background, only 2 major items are missing before we can predict the position of a celestial object. Those 2 missing items are the ability (a) to locate an object on a sphere and (b) to describe where an object is in its orbit.

To supply these missing items, we need to understand some of the properties of spheres and ellipses. Spherical geometry can be applied to uniquely and unambiguously describe the location of an object on a sphere, which provides us with the first missing item. Johannes Kepler discovered that celestial objects move in elliptical orbits and so, as should be expected, the geometric properties of ellipses will be used to describe where an object is in its orbit. An understanding of ellipses will therefore provide the second missing item.

Besides spheres and ellipses, this chapter will also discuss orbital elements and some coordinate systems that astronomers use to locate objects on a sphere. An understanding of orbital elements is not required to understand spherical coordinate systems, but orbital elements follow quite naturally from a discussion of ellipses, and orbital elements *are* required for later chapters. This chapter will also include techniques for converting between different coordinate systems, most of which will be repeatedly applied in succeeding chapters where we will apply an understanding of orbital elements and coordinate systems to actually compute the location of various celestial objects.

This chapter will describe 5 different coordinate systems and conversion techniques, all based on spherical geometry, that are frequently used in astronomy. We will begin with the terrestrial latitude-longitude system, which is actually used to locate an object on Earth's surface rather than in the sky. The terrestrial latitude-longitude coordinate system is important because in order to properly describe where a celestial object will appear, it is necessary to know an observer's location since where an object appears in the sky at some instant in time differs from 1 place on Earth to the next.

A second coordinate system, the horizon coordinate system, is an easy-to-use coordinate system for locating objects in the sky. Unfortunately, coordinates expressed in the horizon coordinate system constantly change as Earth rotates. An object's horizon coordinates are different for observers at different locations on Earth as well as for different times during the day.

To avoid coordinates that constantly change with respect to the motion of the Earth, time of day, and an observer's location, a third coordinate system, the equatorial coordinate system, is defined in which an object's position is the same regardless of an observer's location or time of day. Because celestial objects such as stars, galaxies, and nebulae are so very, very far away, their equatorial coordinates change very slowly over relatively long periods of time. For most purposes, the unimaginably vast distances involved mean that such celestial objects can be considered as stationary with respect to Earth and therefore have fixed equatorial coordinates. Adjustments can be made to a distant object's equatorial coordinates to account for its motion, but such adjustments are usually made only when high accuracy is required.

On the other hand, objects within the Solar System move much more rapidly along their orbits with respect to Earth than do distant celestial objects. Consequently, equatorial coordinates for objects so relatively close to Earth change daily. For convenience, the positions of Solar System objects are calculated in the ecliptic coordinate system, the fourth coordinate system we will consider, for some instant in time and then converted to equatorial coordinates.

The final coordinate system we will describe is the galactic coordinate system. Calculations involving objects within the Milky Way Galaxy are often done within the galactic coordinate system and then converted to equatorial coordinates. The galactic coordinate system is presented for completeness but will not be used beyond this chapter.

## 4.1 Trigonometric Functions

Before turning our attention to this chapter's major topics, we must briefly digress to consider some properties of the trigonometric functions. This is necessary because dealing with spheres, ellipses, orbital elements, and spherical coordinate systems requires applying equations that are expressed in terms of angles and trigonometric functions. It is assumed that the reader already has a

# Orbits and Coordinate Systems

working knowledge of the basic trigonometric functions (sine, cosine, tangent, etc.). Based on that assumption, this section will briefly look at making adjustments so that the angles returned by the inverse trigonometric functions are in the correct quadrant.

The trigonometric functions are defined over the entire range of real numbers and, excepting the tangent and cotangent functions, always produce a real number between $-1.0$ and $+1.0$ inclusive. This fact leads to an ambiguity when inverse trigonometric functions are involved. For example, the tangent of $45°$ is 1, which is also the same as the tangent of $225°$. What, then, is the inverse tangent (also called the arctangent and denoted by $\tan^{-1}$) of 1? Should it be $45°$ or $225°$? To determine the correct answer, inverse trigonometric functions must be considered case by case in the context of the problem being solved.

By convention, the inverse cosine function (also called arccosine and denoted by $\cos^{-1}$) returns angles between $0°$ and $180°$ while the inverse sine (also called arcsine and denoted by $\sin^{-1}$) and arctangent functions return angles between $-90°$ and $+90°$. The arcsine and arccosine functions will rarely cause a problem in the algorithms presented in this book, but the arctangent function poses an added complexity.

When the arctangent is required and the resulting angle is supposed to be in the range $0°$ to $360°$, the argument to the arctangent function will be given in the form $\frac{y}{x}$. The correct angle obtained from the arctangent depends on the numeric sign of $y$ and $x$. There are 4 cases to consider as summarized in table 4.1. Remember, however, that an adjustment is needed only if the resulting angle is to be in the range $[0°, 360°]$ instead of $[-90°, +90°]$.

For example, suppose $y = 5$, $x = -2$, and the angle $\theta$ is to be in the range $0°$ to $360°$. Now $\theta = \tan^{-1}\left(\frac{5}{-2}\right) = -68.1986°$. Table 4.1 indicates that $180°$ must be added to $\theta$ to place the angle in the correct quadrant. Therefore, the correct answer is

$$\theta = -68.1986° + 180° = 111.8014°.$$

**Table 4.1** Angle Adjustments for the Arctangent
Computing the arctangent requires an adjustment to place the answer in the proper quadrant.

| y | x | Adjustment |
|---|---|---|
| + | + | 0° |
| + | − | 180° |
| − | + | 360° |
| − | − | 180° |

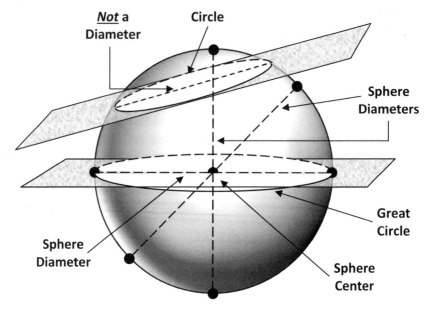

**Figure 4.1** Great Circles
Cutting a sphere with a plane results in circles of different sizes. When a circle has the same diameter as the sphere itself, it is called a great circle.

It is important to remember that when an inverse trigonometric function is needed, an angle adjustment may or may not be required. An angle adjustment may be needed for the arcsine and arctangent functions if the desired angle could fall outside the range of $[-90°, +90°]$. An adjustment may be needed for the arccosine function if the desired angle could fall outside the range of $[0°, 180°]$. Failure to consider the inverse trigonometric functions on a case-by-case basis in the context of the problem being solved can produce an incorrect result.

## 4.2 Locating Objects on a Sphere

The terrestrial latitude-longitude system is probably the most familiar technique for locating objects on Earth's surface.[1] Several concepts are needed to properly describe this coordinate system.

Visualize a sphere (see figure 4.1). The length of any line segment that passes through the center of the sphere and whose endpoints terminate at

---

1. Earth isn't really a sphere. However, we will consider it to be a perfect sphere because the error incurred in doing so is negligible unless high accuracy is required.

the sphere's boundary is equal to the length of the sphere's diameter. The converse is also true. Any line segment whose endpoints are on the sphere's boundary and whose length is equal to the length of the sphere's diameter must necessarily pass through the sphere's center. Referring to figure 4.1, the vertical, horizontal, and rotated line segments are all sphere diameters because they pass through the center of the sphere and terminate at the sphere's boundary.

Imagine passing a sheet of paper through a sphere. No matter where the sheet of paper cuts through the sphere, a circle is formed at the intersection of the paper and the sphere. Some circles formed at such intersections are larger than others. When the circle formed has the same diameter as the sphere itself, that circle is called a great circle. It should be obvious that the center of every great circle is the same as the center of the sphere on which that great circle is drawn, and that it is impossible to draw a circle on a sphere that is larger than a great circle. The circle formed when the plane shown in the middle of figure 4.1 intersects the sphere is a great circle because the circle's diameter goes through the center of the sphere. However, the circle formed at the top of figure 4.1 is not a great circle because that circle's diameter does not go through the center of the sphere.

In Earth's case, a line segment drawn from the North Pole to the South Pole (see figure 4.2) passes through the center of the Earth and is therefore the same length as Earth's diameter. Besides being a diameter, such a line segment lies on Earth's axis of rotation. Also, following from the basic properties of a sphere just discussed, any circle that goes through both the North and South Poles is a great circle. Not all great circles pass through the North and South Poles. The equator is but 1 example of a great circle that does not go through either the North or the South Pole. In fact, an infinite number of great circles can be drawn on the surface of a sphere and in any orientation.

A semicircle that passes through both the North and South Poles of Earth is called a meridian. Another way to view meridians is to say that they are half of a great circle that passes through both the North and South Poles. Great circles that pass through both poles actually form 2 meridians (semicircles), 1 on either side of the Earth.

The meridian that passes through Greenwich, England, is a special meridian called the *prime meridian*. That is, the prime meridian is the semicircle that starts at the North Pole, goes through Greenwich, and terminates at the South Pole. Choosing Greenwich as the location for the prime meridian is arbitrary and done for historical reasons. Even so, the prime meridian is of great importance because it is used as a standard reference point for locating objects on Earth's surface. In essence, as we will see, an object's position can be described in terms of its distance from the prime meridian and Earth's equator.

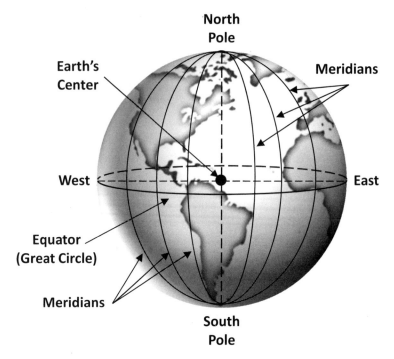

**Figure 4.2** Meridians
Semicircles that pass through both Earth's North and South Poles are called meridians. Meridians are half of a great circle that passes through Earth's North and South Poles.

Figure 4.2 shows several examples of meridians drawn on Earth's surface. Now the center of the arc that forms any meridian is the same as the center of the Earth. This is because by definition a meridian is one-half of a great circle that passes through both the North and South Poles, and we know that the center of every great circle is the same as the center of the sphere on which it is drawn. We can exploit this relationship between meridians and Earth's center to devise a simple method for locating objects on the surface of the Earth. Let's see how.

Using the equator as a convenient reference, the location at which an object falls on a meridian can be described as the angle between 2 specific line segments. One line segment is drawn from the center of the Earth to the point where the equator intersects the meridian on which the object lies (see line segment[2] $\overline{A}$ in figure 4.3). The other line segment is drawn from Earth's center

---

2. Mathematicians place a bar over a letter or group of letters to denote a line segment. Omitting the bar refers to the length of a line segment. So, $\overline{ABC}$ is a line segment that begins at endpoint $A$, goes through point $B$, and terminates at endpoint $C$ whereas $ABC$ is that line segment's length.

# Orbits and Coordinate Systems

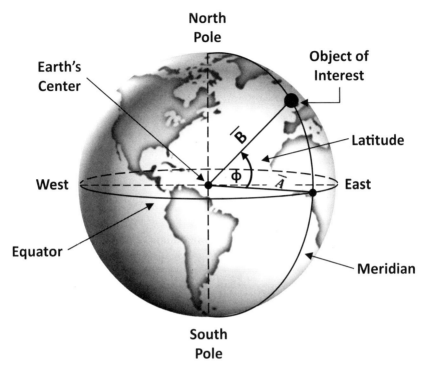

**Figure 4.3** Latitude
An object's latitude is its angular distance from the equator measured along a meridian that intersects both the equator and the object.

to the object (line segment $\overline{B}$). The angle between these 2 line segments is the object's latitude, represented by the symbol $\phi$. Latitude is thus an object's angular distance from Earth's equator. Notice that an object's latitude is independent of the meridian on which it falls.

Objects that fall precisely on the equator, regardless of which meridian they are on, are at 0° latitude. Objects located above the equator are in the range 0° to 90° N while objects below the equator are in the range 0° to 90° S. The North Pole is at precisely 90° N latitude while the South Pole is at precisely 90° S latitude. An object exactly halfway between the North Pole and the equator is at 45° N latitude while an object exactly halfway between the South Pole and the equator is at 45° S latitude.

Latitude alone is not sufficient to uniquely locate an object on Earth's surface. This is obvious because 2 objects can be precisely the same angular distance from the equator (i.e., at the same latitude) but be on opposite sides of the Earth. A second reference point and measurement is necessary to

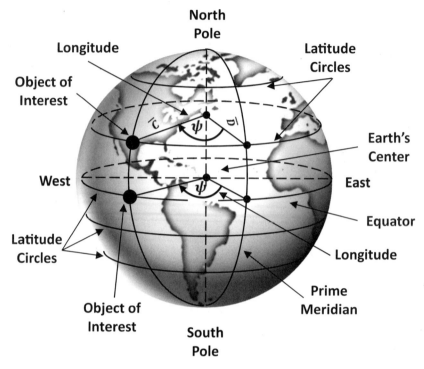

**Figure 4.4** Longitude
An object's longitude is its angular distance from the prime meridian.

distinguish between locations at the same latitude. This is where the prime meridian comes into the picture.

Again starting at the equator, imagine drawing a series of circles around the Earth in such a way that every point on the circle's boundary has the same latitude. Figure 4.4 shows several such circles, which we will call latitude circles. Latitude circles are always parallel to the equator. Note that only 1 latitude circle, the equator itself, is a great circle. All other latitude circles are smaller in size than the great circle located at the equator.

We can use latitude circles to define 2 very specific line segments. Draw 1 line segment from the object to the center of the latitude circle on which the object lies (line segment $\overline{C}$ in figure 4.4). Draw the other line segment from the center of the latitude circle on which the object lies to the point where the prime meridian intersects the latitude circle (line segment $\overline{D}$). The angle between these 2 line segments is the object's longitude, designated by $\psi$. Longitude is thus an object's angular distance from the prime meridian.

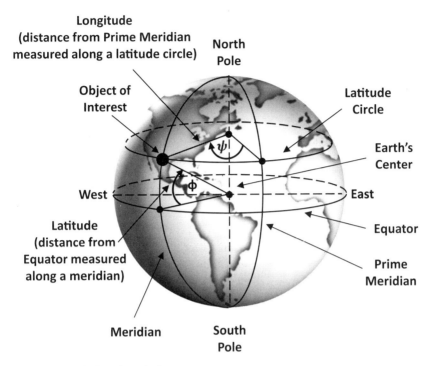

**Figure 4.5** Latitude and Longitude
Latitude combined with longitude uniquely specify an object's location on Earth.

Starting from the prime meridian, longitudes are in the range 0° to 180° W and 0° to 180° E. Objects on the prime meridian are at 0° longitude. All objects on the same meridian have the same longitude.

Although longitude is normally expressed in degrees, sometimes it is useful to express longitude in hours, minutes, and seconds. Since there are 360° in a circle and 24 hours in a day, 1 hour corresponds to 15°. So, longitude can be converted to HMS format by dividing by 15°. For example, an object located at 97°30′ W longitude is at $6^h30^m00^s$ when longitude is expressed in HMS format. That is, convert 97°30′ W to decimal format (97.5° W), divide by 15° to convert to hours ($6.5^h$), and then convert to HMS format ($6^h30^m00^s$).

Latitude and longitude are sufficient to uniquely and unambiguously describe any location on Earth's surface. Figure 4.5 shows combining the concept of meridians and latitude circles to locate any object on Earth's surface in terms of its angular distance from the equator (latitude) and the prime meridian (longitude).

## 4.3 The Celestial Sphere

A convenient way to describe the location of stars and other celestial objects is to assume they are embedded on a large sphere, called the celestial sphere, that has Earth as its center. By extending the plane of Earth's equator until it intersects the celestial sphere, a celestial equator is formed. Extending Earth's North and South Poles until they intersect the celestial sphere defines the North and South Celestial Poles. The star Polaris, often called the North Star or Pole Star, is very close to the North Celestial Pole, but the Pole Star and North Celestial Pole do *not* refer to the same location.

The celestial sphere can be subdivided into meridians in the same manner as the Earth can, but how should we define a celestial prime meridian? At first glance, it might appear that Earth's prime meridian could simply be extended outward until it intersects the celestial sphere to create a celestial prime meridian. Unfortunately, since Earth rotates on its axis, this would have the most undesirable effect of having a celestial prime meridian that changes with the time of day.

Instead of extending Earth's prime meridian until it intersects the celestial sphere, astronomers have chosen a fixed point in the sky called the First Point of Aries, denoted by the symbol $\Upsilon$, to play a role analogous to that of Greenwich, England. Using this fixed point in the sky as a reference, the celestial prime meridian is defined to be the semicircle that begins at the North Celestial Pole (again, this is *not* the same as Polaris, the Pole Star!), passes through the First Point of Aries, and terminates at the South Celestial Pole.

Where is the First Point of Aries, and why was that particular location chosen? The Greek astronomer Hipparchus defined the First Point of Aries around 130 BC. He selected that location because, at that time in history, it was the point at which the Sun first entered the constellation of Aries as the Sun traveled in its orbit around Earth. However, in the centuries that have passed since the time of Hipparchus, the location of the First Point of Aries has changed, due to the effects of precession[3], and it is now in the constellation of Pisces instead of Aries. Although the constellation in which it falls has changed, the First Point of Aries is still the point at which the Sun crosses the celestial equator from the south to the north.

Recall from chapter 3 that the equinoxes are the points at which the Sun crosses the plane of Earth's equator, which is the same thing as saying when the Sun crosses the celestial equator. For this reason, the First Point of Aries

---

3. Precession is a change in the orientation of Earth's axis similar to the change in orientation of the axis of rotation for a rapidly spinning top.

# Orbits and Coordinate Systems

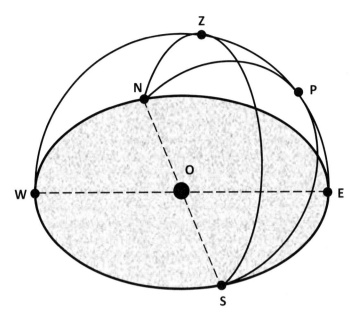

**Figure 4.6** Observer's Horizon
This figure shows an observer's horizon in which celestial meridians are defined relative to an observer. Viewed from a position outside Earth, such celestial meridians vary with the time of day and with an observer's location.

is also called the vernal equinox because it is at that point in time (around March 21) and at that location (First Point of Aries) in the Earth-Sun orbit that the Sun crosses the plane of Earth's equator.

With the First Point of Aries as a fixed reference point, a celestial prime meridian can be defined that does not change with the motion of the Earth, and celestial meridians can be defined relative to this celestial prime meridian. In fact, any semicircle that passes through both the North and South Celestial Poles and whose position is defined relative to the celestial prime meridian is a celestial meridian whose position on the celestial sphere does not vary with the Earth's rotation.

Defining celestial meridians relative to the First Point of Aries is not the only possible way, or even the only useful way, to define celestial meridians. Celestial meridians can also be defined relative to an observer. Consider figure 4.6. An observer is located at position $O$ and his horizon is extended in all directions to intersect the celestial sphere. This forms the great circle *NESW* on the celestial sphere, where these letters represent compass directions. Notice that line segment $\overline{NS}$ does *not* intersect the North and South Celestial Poles unless the observer happens to be at Earth's equator, in which case points $N$ and $S$ are the North and South Celestial Poles, respectively.

Suppose point $P$ in figure 4.6 is a star whose position we determine at some convenient time of day. The semicircle $NPS$ defines a celestial meridian relative to the observer at position $O$. Of course, relative to the stars a celestial meridian defined with respect to an observer changes as Earth rotates. However, relative to our Earthbound observer, the celestial meridian stays fixed and it is the stars themselves, including $P$, that appear to move relative to the celestial meridian we have created.

Since the star $P$ moves relative to the central meridian $NPS$, there is no advantage in using the star $P$ as a reference point for defining a celestial meridian. Instead of choosing some arbitrary star, consider point $Z$ in figure 4.6, where $Z$ is the point directly overhead an observer located at position $O$. $Z$ is called the observer's zenith, or simply the zenith, while the point on the celestial sphere that lies directly beneath an observer is called the observer's nadir. The semicircle $NZS$ also forms a celestial meridian, which is called the observer's local celestial meridian, or more often the observer's meridian. The adjective "local" makes it explicitly clear that the meridian so defined is valid locally only because it is relative to an observer's location.

When a celestial object crosses an observer's local celestial meridian, that object is said to transit or culminate. Recall from chapter 3 that a day is defined as the interval between 2 successive transits of a mean Sun (for a mean solar day) or a fixed star (for a sidereal day) across an observer's meridian. When no specific meridian is mentioned, transit (or culminate) is understood to mean with respect to an observer's local celestial meridian.

Celestial meridians defined relative to an observer, whether using an observer's zenith or a star as a reference point, will appear to move as Earth rotates when viewed from a location outside the Earth, but they are stationary when viewed from Earth regardless of an observer's location. Why one would want to define a celestial meridian relative to an observer may not be clear until we discuss the equatorial and horizon coordinate systems in more detail, which we will do in sections 4.6 and 4.7.

An important side note before we leave this discussion: be aware that the word "meridian" is frequently used without specifying whether a terrestrial or celestial meridian is meant. In most cases it should be clear from the context which type of meridian is meant. If the context is navigation or locating objects on Earth, meridian is understood to mean a terrestrial meridian. In the context of astronomy, meridian, when unqualified, will almost always mean a celestial meridian.

# Orbits and Coordinate Systems

Similarly, "prime meridian" is sometimes used without specifying whether a terrestrial or celestial prime meridian is meant. In most cases it will be clear from context which type of prime meridian is meant. Moreover, "celestial meridian" can be ambiguous because it could refer to a locally defined celestial meridian or one defined with respect to the celestial prime meridian. When used without any other modifiers, the phrase typically refers to a celestial meridian defined relative to the celestial prime meridian (i.e., First Point of Aries). When a phrase such as "observer's celestial meridian" or "observer's meridian" is encountered, that should be understood to refer to the unique celestial meridian that passes through the North and South Celestial Poles and an observer's zenith.

## 4.4 Ellipses

Kepler discovered that the planets do not move in perfect circles inscribed on the celestial sphere. Instead, they follow an elliptical orbit as they move around the Sun. In fact, any celestial object that orbits another (including the Earth, Sun, Moon, stars, and satellites) moves through the heavens in elliptical orbits because of the gravitational forces that Newton discovered. Therefore, because ellipses play a central role in helping us understand how orbiting objects move, we now examine some of their fundamental properties.

Figure 4.7 shows the geometric figure known as an ellipse. $P$ is any arbitrary point on the ellipse while the points $F_1$ and $F_2$ are called the foci. The foci are important in creating and defining an ellipse, but they are not points on the ellipse itself. The geometric center of the ellipse, point $C$, is not on the ellipse either. $C$ will always lie on the line segment $\overline{A_1 A_2}$ and will be precisely halfway between the 2 foci.

An ellipse is generated in such a way that the length of line segment $\overline{F_1 P}$ plus the length of line segment $\overline{PF_2}$ is always the same value (some constant $K$) no matter where $P$ is located on the ellipse. Thus, if $P'$ is another point located anywhere on the ellipse, then

$$F_1 P + PF_2 = K = F_1 P' + P' F_2.$$

The line segment $\overline{A_1 A_2}$ is called the major axis. Line segments $\overline{A_1 C}$ and $\overline{CA_2}$ are exactly half the length of the major axis and are the semi-major axes. Line segment $\overline{B_1 B_2}$, the minor axis, is perpendicular to the major axis. The semi-minor axes, line segments $\overline{B_1 C}$ and $\overline{CB_2}$, are exactly half the length of the minor axis and are the semi-minor axes. An ellipse is symmetric with respect to both the major and minor axes. Points $A_1$ and $A_2$ are called apsides (the

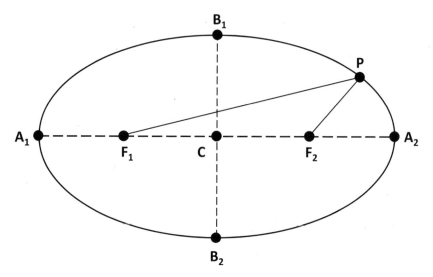

**Figure 4.7** Ellipse Defined
An ellipse is defined such that if $P$ is a point on the ellipse, and the foci are $F_1$ and $F_2$, the distance $F_1P + PF_2$ is constant no matter where $P$ is located.

singular of which is apsis), and they mark the extreme points on the ellipse with respect to the foci. Apsis $A_1$ is the point on the ellipse closest to focus $F_1$ and farthest away from focus $F_2$. Conversely, apsis $A_2$ is the point farthest away from $F_1$ but closest to $F_2$.

It should be clear from the symmetry of an ellipse that the lengths of $\overline{F_1A_1}$ and $\overline{F_2A_2}$ are the same, which means that

$$F_1A_1 + A_1F_2 = F_2A_2 + A_1F_2 = A_1A_2.$$

Now $\overline{A_1A_2}$ is the major axis while $A_1$ is just another point on the ellipse. Therefore, we see that the constant value used to generate an ellipse is the length of the major axis. That is, $K = A_1A_2$.

The eccentricity $e$ measures the "flatness" of an ellipse.[4] Mathematically, eccentricity is the ratio of the distance that a focus lies from the ellipse center to the length of a semi-major axis. That is,

$$e = \frac{F_1C}{A_1C}, \qquad (4.4.1)$$

---

4. Do not confuse the symbol for eccentricity with the constant known as Euler's number, which by convention is also denoted by the letter $e$ and whose value is $\approx 2.71828$. Euler's number is the base for the natural logarithm.

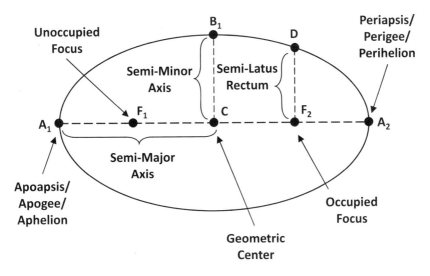

**Figure 4.8** Ellipse Attributes
This figure shows the major elements of an ellipse when applied to an orbit.

where the eccentricity will always fall within the range $0 < e < 1$. Because of symmetry, it does not matter which focus or semi-major axis is chosen to compute the eccentricity.

If both foci and the center of an ellipse coincide, the eccentricity is zero (because $F_1 C$ in equation 4.4.1 is zero), and the figure becomes a circle. With respect to figure 4.7, this means that $F_1$, $F_2$, and $C$ are all the same point, and the resulting geometric figure is indeed a circle. If $e = 1$, the geometric figure becomes a parabola. Some comets follow parabolic orbits, but we will not be discussing such orbits.

Eccentricity is an important characteristic of an object's orbit. The closer $e$ is to 0, the closer an orbit is to being a circle. The eccentricity of the Earth-Sun orbit is approximately 0.0167, which shows how close it is to a circle and why the Earth-Sun orbit can often be assumed to be a circle when low accuracy is sufficient.

Figure 4.8 shows an ellipse in the context of orbits. According to Kepler's first planetary law, the Sun is at 1 focus of the ellipse that describes a planet's orbit. If we apply this law to orbits in general, the body around which an object orbits is physically located at 1 focus of that object's elliptical orbit, which is shown as the "occupied focus" $F_2$ in figure 4.8. There is no object physically located at the other focus of an elliptical orbit, which is why $F_1$ is labeled as the "unoccupied focus."

Figure 4.8 also shows 1 of an ellipse's 2 semi-major axes (line segment $\overline{A_1C}$) and 1 of its 2 semi-minor axes (line segment $\overline{B_1C}$). $\overline{DF_2}$ is another important line segment called the semi-latus rectum. Line segment $\overline{DF_2}$ is constructed as a perpendicular line segment that extends from focus $F_2$ until it intersects the ellipse at point $D$. There is, of course, another semi-latus rectum that extends from the occupied focus $F_2$ to the bottom of the ellipse, and there are 2 more semi-latus rectums at the unoccupied focus $F_1$. The length of the semi-latus rectum is given by the equation

$$DF_2 = \frac{(B_1C)^2}{A_2C} = (A_2C)(1 - e^2), \qquad (4.4.2)$$

where $e$ is the ellipse's eccentricity.

When used in the context of orbits, apsides have multiple names. For objects orbiting the Sun, apsis $A_2$ is called "perihelion," which is the point at which an object is closest to the Sun. Apsis $A_1$ is called "aphelion" and it is the point at which the object is farthest away from the Sun. For objects orbiting Earth, the point at which an object is closest to Earth is called "perigee" (apsis $A_2$) while "apogee" (apsis $A_1$) is where the object is farthest away from Earth. In the more general case, the point at which an object is closest to the body around which it orbits is called the "periapsis" ($A_2$). "Apoapsis" ($A_1$) is the point at which an object is farthest away from the body around which it orbits.

There is a simple way to remember which apsis is closest or farthest away from the occupied focus. Think of the letter "a" in apoapsis, apogee, and aphelion as meaning "away," as in the object is "farthest away" from the body around which it orbits. Another useful memory device is that the "gee" in apogee and perigee refers to geocentric, which means that Earth is the body located at the occupied focus. Similarly, "helio" means heliocentric, so we know that the Sun is the body located at the occupied focus when we refer to aphelion and perihelion.

## 4.5 Orbital Elements

Although calculating the position of celestial objects will not be covered until later, we now have the background necessary to discuss orbital elements and how astronomers locate objects in an elliptical orbit. This is a complex undertaking! The primary reason for the complexity is Kepler's second law, a consequence of which is that the orbital speed of an object varies throughout its orbit; a planet's orbital speed slows down as it moves away from the Sun but speeds up as it approaches the Sun.

Newton's *Law of Universal Gravitation*, which relates the gravitational force between 2 objects to the distance between them, explains why an object's orbital speed changes. Consider the Sun and a planet as an example. As a planet recedes from the Sun, the distance between the Sun and the planet increases, the gravitational force between them decreases, and consequently the planet's orbital speed decreases. As a planet approaches the Sun, the distance between them decreases, the gravitational force between them increases, and the planet's orbital speed increases. What is true for the Sun and a planet is true for any object orbiting another, whether a star, a satellite, Earth, the Moon, an asteroid, or a planet. The fact that an object's orbital speed constantly changes is why a direct analysis of an elliptical orbit is so difficult.

Before getting into the details, let's begin with an overview of how to approach the problem. The idea is to first determine where an object would be if it followed a fictitious circular orbit in which the object's orbital speed is constant. Given an object's position in such an orbit, the power of mathematics can be applied to convert the object's circular orbit position to its position in its true elliptical orbit where the object's orbital speed varies.

This deceptively simple idea of mapping a circular orbit position to an elliptical orbit position requires sorting out 5 interrelated concepts: true anomaly,[5] mean anomaly, eccentric anomaly, equation of the center, and Kepler's equation. We will describe each of these in more detail shortly, and show how they are related. For the moment it suffices to know that *true anomaly* describes an object's position in its true elliptical orbit, *mean anomaly* is where the object would be if it followed a constant-speed circular orbit, and *eccentric anomaly* is where the object would be if it followed a circular orbit in which its orbital speed varies as it does in its elliptical orbit. The *equation of the center* expresses a mathematical relationship between the true and mean anomalies while *Kepler's equation* expresses a mathematical relationship between the mean and eccentric anomalies.

By applying these 5 interrelated concepts, we can determine where an object is in its elliptical orbit in either of 2 ways.

*Method 1:*

1. Compute the object's mean anomaly (subsection 4.5.2).

2. Use the equation of the center to compute the true anomaly from the mean anomaly (subsection 4.5.3).

---

5. "Anomaly" in the context of astronomy does not mean weird, unusual, or abnormal. The terminology arose historically to refer to the nonuniform (hence anomalous) apparent motions of the planets. For our purposes, think of anomaly as an angle.

*Method 2:*

1. Compute the object's mean anomaly (subsection 4.5.2).

2. Solve Kepler's equation to compute the eccentric anomaly from the mean anomaly (subsection 4.5.5).

3. Compute the true anomaly from the eccentric anomaly (subsection 4.5.4).

The astute reader may wonder why there are 2 different methods for computing the true anomaly. Why not just use the first method since it requires fewer steps? The reason, as we will see, is that the equation of the center is usually approximated, but such approximations are typically valid only for orbits with a "small" eccentricity (e.g., Sun, Moon, planets), or are valid only for a specific object. Moreover, approximating the equation of the center usually yields a less accurate true anomaly than when produced by the second method. Even so, approximating the equation of the center can still be useful because doing so is less complicated than solving Kepler's equation.

If the first method has so many shortcomings, then why not shorten the second method to start with the eccentric anomaly, and convert it directly to the true anomaly, without bothering to involve the mean anomaly at all? The answer is that directly computing the eccentric anomaly is difficult because an object's orbital speed is not constant along the circular orbit used to define the eccentric anomaly. So while a circular orbit may be easier to analyze than an elliptical one, because an object's orbital speed varies along the circular orbit used to define the eccentric anomaly, we might as well try to analyze the elliptical orbit in the first place and not bother with a fictional circular orbit. Calculating the mean anomaly is easy, so the second method starts from there and accounts for an object's varying orbital speed in step 2 (solve Kepler's equation).

Note that we can apply the second method in reverse to determine at what point in time an object will be at a given location in its true orbit. For example, if we wish to determine when a planet will be at a specific location in the sky, we apply the following basic steps:

1. Use the true anomaly to describe the planet's desired orbital position.

2. Compute the eccentric anomaly from the true anomaly (see subsection 4.5.4).

3. Use Kepler's equation to compute the mean anomaly.

Given the mean anomaly, it is a simple matter to determine when the planet will be at the desired location.

Armed with these 5 concepts and a general idea of how to proceed, let's now get into the details. To make the discussion more concrete, as Kepler did

# Orbits and Coordinate Systems

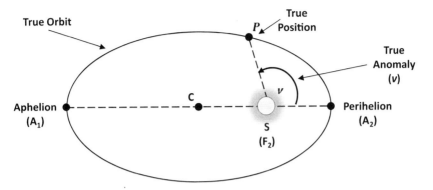

**Figure 4.9** True Anomaly
A planet's true anomaly $v$ is the angle formed by the planet's true position $P$, the Sun, and perihelion.

we'll consider the problem of locating a planet orbiting the Sun. The overall approach and equations apply to any elliptical orbit, although the terminology and exact steps may vary depending on the situation. For example, instead of talking about perihelion for objects orbiting the Sun, we would use perigee for objects orbiting Earth, or periapsis in the more general case.

### 4.5.1 True Anomaly

According to Kepler's first law, the Sun (point $S$ in figure 4.9) is at 1 focus (the occupied focus $F_2$) of a planet's elliptical orbit. Because we're dealing with elliptical orbits, as the planet $P$ moves around the Sun, the distance between the planet and the Sun varies. When the planet is closest to the Sun, it is at perihelion ($A_2$). When farthest away from the Sun, it is at aphelion ($A_1$). The true anomaly, designated by $v$, is the angle $PSA_2$. This angle[6] tells us how far the planet has progressed in its orbit from the moment of perihelion.

Besides describing where an object is in its elliptical orbit, the true anomaly can be used to calculate several useful items about the object. For example, the equation

$$r = \frac{a(1-e^2)}{1+e\cos v} \tag{4.5.1}$$

---

6. Some authors emphasize that an anomaly is not really an angle, but simply the difference between where an object is at some instant in time and a reference point. Also, because an object may have completed several orbits with respect to the reference point, the numeric value for the anomaly could exceed 360, whereas angles are typically restricted to the range [0°, 360°]. Strictly speaking, those authors are correct. Nevertheless, we will equate anomaly and angle because doing so makes anomalies easier to conceptualize and understand.

gives the distance that an orbiting object is from the occupied focus, where $e$ is the orbital eccentricity and $a$ is the length of the orbit's semi-major axis. Carefully note that this equation gives the distance from the *center* of the object at the occupied focus to the *center* of the orbiting object. Also, recall from equation 4.4.2 that $a(1-e^2)$ is the length of the semi-latus rectum because, referring to figure 4.8, $a = \overline{A_2 C}$. We will use equation 4.5.1 in the chapters ahead to compute the distance to the Sun, Moon, and planets.

For example, assume a satellite is circling Earth in an elliptical orbit whose eccentricity is 0.5 and semi-major axis is 40,000 km in length. If the true anomaly is 45°, how far away is the satellite from the center of the Earth? Applying equation 4.5.1 gives

$$r = \frac{(40{,}000)(1-0.5^2)}{1+0.5\cos 45°} \approx 22{,}163.88 \text{ km.}$$

We can also use equation 4.5.1 to determine how far away the satellite is from the center of the Earth at perigee ($\upsilon = 0°$) and apogee ($\upsilon = 180°$). At perigee the satellite is 20,000 km from the center of the Earth and 60,000 km at apogee.

### 4.5.2 Mean Anomaly

Directly determining the true anomaly for an orbiting object is difficult because the object's orbital speed varies throughout its orbit. But what if an object moved in a circular orbit in which its orbital speed is constant throughout its orbit? That is, assume that the planet in figure 4.9 completes a single orbit around the Sun in $n$ sidereal days. This is called the planet's orbital period. Since there are 360° in a circle, to maintain a constant speed and complete a circular orbit in the same time as the orbital period, the planet must move by $360/n$ degrees per day around that circular orbit.

Figure 4.10 illustrates this very useful idea. We have constructed a fictitious circular orbit, called the mean orbit, so that the center of the mean orbit coincides with the geometric center ($C$) of the planet's true elliptical orbit. The radius of the mean orbit is the length of the semi-major axis $\overline{CA_2}$. $P$ is where the planet is in its true elliptical orbit while $P'$ is where the planet would be at that same instant in time if the planet were following the mean orbit at a constant orbital speed of $360/n$ degrees per day ($n$ is the planet's orbital period). The mean anomaly $M$ is the angle $P'CA_2$. The mean anomaly is analogous to the true anomaly, with the difference being that $M$ is with respect to the mean orbit whereas $\upsilon$ is with respect to the true orbit.

To demonstrate that circular orbits make analysis considerably easier, assume it has been $t$ days since the planet in figure 4.10 was at perihelion, and assume the planet's orbital period is $n$ sidereal days. Then the planet's

# Orbits and Coordinate Systems

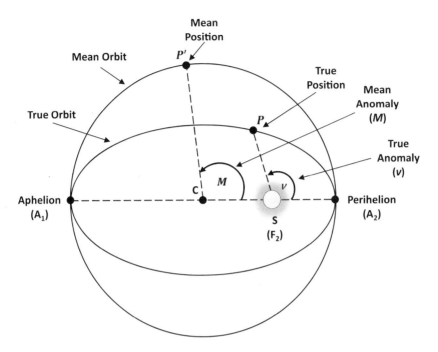

**Figure 4.10** Mean Anomaly
A planet's mean anomaly $M$ is the angle formed by the planet's mean position $P'$, the geometric center $C$ of the mean orbit, and perihelion.

mean position in its mean orbit (its mean anomaly) is given by the very simple equation

$$M = \frac{360°}{n} t. \qquad (4.5.2)$$

As promised, the mean anomaly is indeed easy to calculate!

For example, the orbital period of Mars is 686.97 days. If it has been 300.25 days since Mars was at perihelion, how far has Mars advanced in its mean orbit? We need only apply equation 4.5.2 to compute the mean anomaly:

$$M = \frac{360°}{686.97}(300.25) \approx 157.3431°.$$

How long will it take Mars to go 230° around in its mean orbit from the point of perihelion? Equation 4.5.2 can be written in the equivalent form

$$t = \frac{Mn}{360°}. \qquad (4.5.3)$$

Substituting in the proper values, we have

$$t = \frac{(230°)(686.97)}{360°} = 438.8975 \text{ days}.$$

### 4.5.3 Equation of the Center

Finding the mean anomaly is all well and good, but what we really want is the true anomaly. The mean anomaly and true anomaly are related by the equation of the center, which is the difference between where a planet is in its true elliptical orbit and where it would be assuming a mean orbit (with a constant orbital speed!). The equation of the center is

$$E_c = \upsilon - M. \tag{4.5.4}$$

Note that $E_c$ is exactly 0° at aphelion and perihelion because the planet is halfway through its orbit whether we are referring to its true or mean orbit. If we know the value of $E_c$, equation 4.5.4 can be written in the equivalent form

$$\upsilon = E_c + M \tag{4.5.5}$$

from which it is easy to calculate the true anomaly.

Unfortunately, we most likely will not know $E_c$, so how do we determine its value so that we can calculate the true anomaly? The value for $E_c$ is given by a rather complex equation involving the orbital eccentricity and mean anomaly—namely, the infinite series

$$E_c = \frac{180}{\pi} \left\{ 2e \sin M + \frac{5e^2}{4} \sin(2M) + \frac{e^3}{12} [13 \sin(3M) - 3 \sin M] \cdots \right\}. \tag{4.5.6}$$

This gives $E_c$ in degrees. Without the multiplicative factor $180/\pi$, the result would be in radians. In the reverse case, determining $E_c$ when the true anomaly is known requires a similarly complex infinite series expansion of the orbital eccentricity and true anomaly. The required equation, giving $E_c$ in degrees, is

$$E_c = \frac{180}{\pi} \left\{ 2e \sin \upsilon + \left( \frac{3e^2}{4} + \frac{e^4}{8} \right) \sin(2\upsilon) - \frac{e^3}{3} \sin(3\upsilon) + \frac{5e^4}{32} \sin(4\upsilon) \cdots \right\}. \tag{4.5.7}$$

Rather than evaluating an infinite series, there are various methods for approximating $E_c$, typically by truncating equations 4.5.6 and 4.5.7 when enough terms have been computed to achieve an acceptable degree of accuracy.

# Orbits and Coordinate Systems

We will show how to approximate $E_c$ in later chapters when we compute the position of the Sun, Moon, and planets.

As an example, given that the Earth-Sun orbital eccentricity is 0.0167 and that Earth orbits the Sun in 365.2564 sidereal days, what are Earth's true and mean anomalies when it is 100.25 days past perihelion? First, apply equation 4.5.2 to compute Earth's mean anomaly. That is,

$$M = \frac{360°}{365.2564}(100.25) \approx 98.8073°.$$

Next, apply equation 4.5.6 to compute the equation of the center. Using just the first term in the equation to approximate the equation of the center, we have

$$E_c \approx \frac{180}{\pi}[2(0.0167)\sin(98.8073°)] \approx 1.8911°.$$

Finally, applying equation 4.5.5, we have

$$\upsilon \approx 1.8911° + 98.8073° = 100.6984°.$$

Now repeat this example for when Earth is halfway around in its orbit (i.e., at aphelion, or $\frac{365.2564}{2} = 182.6282$ days after perihelion). By plugging in the appropriate values into the various equations, we obtain

$$M = \frac{360°}{365.2564}(182.6282) = 180.000°$$

$$E_c = \frac{180}{\pi}[2(0.0167)\sin(180.0000°)] = 0.0000°$$

$$\upsilon = 0.0000° + 180.0000° = 180.0000°.$$

These results should not be surprising. We observed earlier that when a planet is exactly halfway around in its orbit (for both its true and mean orbits), the mean position and true position are identical, which means that $M = \upsilon$ at that instant. It is also obvious from equation 4.5.4 that the equation of the center is exactly 0° at that same instant in time.

## 4.5.4 Eccentric Anomaly

The previous 3 subsections showed how to apply Method 1 to determine where an object is in its elliptical orbit. We now turn our attention to the second method, which is more complex but generally provides a more accurate solution than approximating the equation of the center. The second method requires finding the eccentric anomaly and solving Kepler's equation.

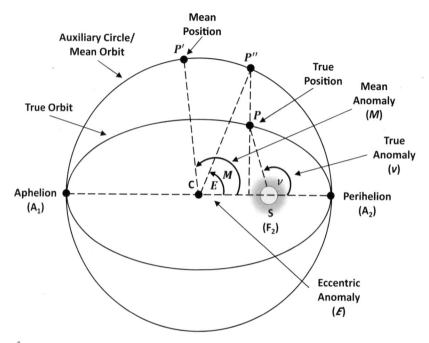

**Figure 4.11** Eccentric Anomaly
A planet's eccentric anomaly $E$ is the angle formed by projecting the planet's true position $P$ onto the auxiliary circle (the point $P''$), the geometric center $C$ of the auxiliary circle, and perihelion.

Just as we did to define the mean anomaly, we start by defining a fictional circular orbit, which is shown as the auxiliary circle in figure 4.11. However, this time we will not assume that a planet's orbital speed is constant as it travels along the auxiliary circle. The auxiliary circle is constructed so that its center coincides with the geometric center of the planet's true elliptical orbit, and so that the auxiliary circle's radius is the length of the elliptical orbit's semi-major axis $\overline{CA_2}$. Obviously, this newly defined fictional orbit is represented by exactly the same circle as for the mean orbit, but the auxiliary circle is *not* a "mean orbit" because the orbital speed varies throughout the auxiliary circle orbit.

In figure 4.11, $P$ is where the planet is in its true elliptical orbit and $P'$ is the planet's mean position. $P''$ is a projection of the planet's position onto the auxiliary circle and is determined by drawing a perpendicular line from the major axis $\overline{A_1 A_2}$ through the planet's true position $P$ until the perpendicular line intersects the auxiliary circle. The eccentric anomaly, designated by $E$, is the angle $P''CA_2$. If we know the eccentric anomaly $E$ and orbital eccentricity

# Orbits and Coordinate Systems

$e$, the true anomaly $v$ can be determined from the equation

$$\tan\left(\frac{v}{2}\right) = \left(\sqrt{\frac{1+e}{1-e}}\right)\tan\left(\frac{E}{2}\right). \tag{4.5.8}$$

An alternative but equivalent equation for relating the true anomaly, mean anomaly, and eccentricity is

$$\cos v = \frac{\cos E - e}{1 - e\cos E}. \tag{4.5.9}$$

For example, assume that the orbital eccentricity for a planet is 0.5 and its eccentric anomaly is 45°. Determine the planet's true anomaly. Applying equation 4.5.8, we have

$$\tan\left(\frac{v}{2}\right) = \left(\sqrt{\frac{1+0.5}{1-0.5}}\right)\tan\left(\frac{45°}{2}\right) = \sqrt{3}\tan 22.5° \approx 0.717439.$$

Taking the inverse tangent of the result and multiplying by 2, we have $v \approx 71.314265°$.

Applying the alternative form in equation 4.5.9, we have

$$\cos v = \frac{\cos 45° - 0.5}{1 - 0.5\cos 45°} \approx \frac{0.207107}{0.646447} \approx 0.320377.$$

Taking the inverse cosine of this result gives us $v \approx 71.314274°$. The difference between the 2 results for the true anomaly is due to round-off errors incurred by using only 6 digits of precision. Even with 6 digits of precision, the 2 results differ by only about 0.03 arcseconds.

If the true anomaly and orbital eccentricity are known, the eccentric anomaly can be found from the equation

$$\tan\left(\frac{E}{2}\right) = \left(\sqrt{\frac{1-e}{1+e}}\right)\tan\left(\frac{v}{2}\right), \tag{4.5.10}$$

which is very similar to equation 4.5.8.

From our previous example, we know if a planet's true anomaly is 71.314265° and its orbital eccentricity is 0.5, the eccentric anomaly is

$$\tan\left(\frac{E}{2}\right) = \left(\sqrt{\frac{1-0.5}{1+0.5}}\right)\tan\left(\frac{71.314265°}{2}\right)$$

$$\approx \sqrt{0.333333}\tan(35.657133°) \approx 0.414213.$$

Taking the inverse tangent and multiplying by 2, we have $E \approx 44.999945°$. The result should, of course, be exactly $45°$. The difference of $\approx 0.20''$ is due to round-off errors when applying equation 4.5.10.

### 4.5.5 Kepler's Equation

Equation 4.5.8, and the equivalent equation 4.5.9, allows us to directly compute the true anomaly if we already know the eccentric anomaly. But how do we determine the eccentric anomaly so that either of these 2 equations can be applied to get the true anomaly?

Kepler discovered that the eccentric and mean anomalies are related by the equation

$$M_r = E_r - e \sin E_r, \qquad (4.5.11)$$

where $e$ is the orbital eccentricity, $M_r$ is the mean anomaly expressed in radians,[7] and $E_r$ is the eccentric anomaly also expressed in radians. This equation is called *Kepler's equation* in his honor. Be very careful when applying equation 4.5.11 to use angles expressed in radians rather than degrees; otherwise, the results will be incorrect!

For example, assume a planet has an orbital eccentricity of 0.5 and an eccentric anomaly of $45°$. What is the planet's mean anomaly in degrees? We will apply equation 4.5.11, but first we must convert the eccentric anomaly from degrees to radians. Doing so, we have

$$E_r = \frac{\pi E}{180°} = \frac{\pi 45°}{180°} \approx 0.785398 \text{ radians}.$$

Inserting this value and the orbital eccentricity into equation 4.5.11, we have

$$M_r \approx 0.785398 - 0.5 \sin(0.785398) \approx 0.431845 \text{ radians}.$$

Converting this value from radians to degrees gives us a mean anomaly of

$$M = \frac{180°(0.431845)}{\pi} \approx 24.742896°.$$

Finding the mean anomaly given the eccentric anomaly is thus straightforward.

If we refer back to section 4.5, we see that step 3 of Method 2 requires determining the true anomaly from the eccentric anomaly. So how do we apply

---

7. An angle $\theta$ expressed in degrees can be converted to radians through the equation $r = \frac{\theta \pi}{180°}$. Conversely, an angle $r$ expressed in radians can be converted to degrees through the equation $\theta = \frac{r 180°}{\pi}$.

# Orbits and Coordinate Systems

equation 4.5.11 to get the eccentric anomaly? It's easy to obtain the mean anomaly from the eccentric anomaly, but the other way around is difficult because equation 4.5.11 is a transcendental equation for which there is no known closed-form solution for $E$ given $M$.

As was true with the equation of the center, there are various methods for approximating Kepler's equation to obtain a value for the eccentric anomaly when given the mean anomaly. If the object's orbital eccentricity is "close" to 0, a reasonable approximation is $E \approx M$. Another approximation that astronomer Meeus gives in *Astronomical Algorithms* is

$$\tan E = \frac{\sin M}{\cos M - e}. \tag{4.5.12}$$

Unfortunately, as with the approximation $E \approx M$, this equation is also unsuitable unless the orbital eccentricity is sufficiently "small."

Rather than attempting to develop an approximation equation that is suitable for "larger" eccentricities, astronomers have devised various numerical algorithms for iteratively solving Kepler's equation. Such methods work for "small" as well as "larger" eccentricities. We will show 1 method here to illustrate how iterative procedures can be applied to solve Kepler's equation.

Let $e$ be the orbital eccentricity and let $M_r$ be an object's mean anomaly expressed in radians. Starting with $E_0 = M_r$, we will iteratively compute

$$E_i = M_r + e \sin E_{i-1} \tag{4.5.13}$$

until the difference between 2 successive estimates of the eccentric anomaly are less than some desired termination criteria. Note that the estimates for the eccentric anomaly in equation 4.5.13 are in radians, not degrees!

As an example, assume that a planet's orbital eccentricity is 0.5 and its mean anomaly is 24.742896°. Use the iterative scheme implied by equation 4.5.13 to determine the eccentric anomaly in degrees. Stop iterating when the difference between 2 successive estimates is less than 0.000002 radians ($\approx 0.4$ arcseconds). Since this is the inverse of the problem we just solved, we should expect to arrive at an eccentric anomaly of 45°.

First, convert the given mean anomaly to radians. Thus,

$$M_r = \frac{\pi M}{180°} = \frac{\pi \, 24.742896°}{180} \approx 0.431845 \text{ radians}.$$

We start our iterative scheme with the estimate $E_0 = M_r$ and use this to compute our next estimate, $E_1$. That is,

$$E_1 = M_r + e \sin E_0 = 0.431845 + 0.5 \sin 0.431845 \approx 0.641119.$$

The difference between this new estimate $E_1$ for the eccentric anomaly and the previous estimate $E_0$ is

$$|E_1 - E_0| = |0.64119 - 0.431845| = 0.209274,$$

which is larger than our desired termination criteria. Our next estimate is given by

$$E_2 = M_r + e \sin E_1 = 0.431845 + 0.5 \sin 0.641119 \approx 0.730891.$$

We again check to see if successive estimates are close enough to stop by computing

$$|E_2 - E_1| = |0.730891 - 0.641119| = 0.089772.$$

This difference is still larger than our termination criteria, so we compute estimate $E_3$ based on estimate $E_2$. The next few iterations and the difference between successive iterations are as follows:

$E_3 \approx 0.765612, \Delta = 0.034721,$

$E_4 \approx 0.778334, \Delta = 0.012722,$

$E_5 \approx 0.782892, \Delta = 0.004558,$

$E_6 \approx 0.784511, \Delta = 0.001619,$

$E_7 \approx 0.785085, \Delta = 0.000573,$

$E_8 \approx 0.785288, \Delta = 0.000203,$

$E_9 \approx 0.785359, \Delta = 0.000072,$

$E_{10} \approx 0.785385, \Delta = 0.000025,$

$E_{11} \approx 0.785394, \Delta = 0.000009,$

$E_{12} \approx 0.785397, \Delta = 0.000003,$

$E_{13} \approx 0.785398, \Delta = 0.000001.$

We can stop at the 13th iteration because the difference between the last 2 estimates ($E_{13}$ and $E_{12}$) is less than our termination criteria of 0.000002 radians. Converting this last estimate to degrees, we have

$$E = \frac{180° E_{13}}{\pi} = \frac{180°(0.785398)}{\pi} \approx 44.999991°,$$

which differs from the expected answer of 45° by only 0.0324 arcseconds.

Although no algorithm is always the "best" for every problem, the simple scheme just presented can be improved to converge more rapidly, especially

# Orbits and Coordinate Systems

for "large" eccentricities. An alternative iterative method that we now describe is based on the Newton/Raphson method for finding the roots of equations. It is also an iterative method, but it uses a different iteration scheme and initial estimate than the previously described method.

For the Newton/Raphson method, we will choose an initial estimate of $E_0 = M_r$ when the orbital eccentricity is 0.75 or less, and $E_0 = \pi$ when the eccentricity is larger than 0.75. By using a different first estimate for the eccentric anomaly for highly eccentric orbits, it is hoped that the algorithm will converge faster for those orbits. The Newton/Raphson method uses the iteration scheme

$$E_i = E_{i-1} - \frac{E_{i-1} - e \sin E_{i-1} - M_r}{1 - e \cos E_{i-1}} \qquad (4.5.14)$$

and stops when the difference between 2 successive estimates is acceptably small. As with the prior iterative method, $E_i$ and $M_r$ must be expressed in radians.

To illustrate the Newton/Raphson approach, let us again solve the previous problem ($e = 0.5$, $M = 24.742896°$). As before, we will iterate until the difference between 2 successive estimates is less than 0.000002 radians ($\approx 0.4$ arcseconds).

As with equation 4.5.13, all calculations must be performed in radians, and hence the first step is to convert the mean anomaly to radians, which we computed previously to be 0.431845 radians. Since the planet's eccentricity is less than 0.75, we will use

$$E_0 = M_r = 0.431845 \text{ radians}$$

as our initial estimate. Applying equation 4.5.14 to this initial estimate, the next few iterations are:

$E_1 \approx 0.815198$, $\Delta = 0.383353$,
$E_2 \approx 0.785642$, $\Delta = 0.029556$,
$E_3 \approx 0.785399$, $\Delta = 0.000244$,
$E_4 \approx 0.785399$, $\Delta = 0.000000$,
$E_5 \approx 0.785399$, $\Delta = 0.000000$.

The Newton/Raphson method does indeed converge rapidly to a solution for this particular problem. We could have stopped after estimate $E_4$ because our termination criterion was already met, but chose to go 1 more iteration.

Within the limits of the computing precision used for this example, no amount of iteration beyond $E_5$ will change the solution because $\Delta$ has reached zero.

Using the estimate $E_4$ and converting it to degrees give us

$$E \approx \frac{180° E_4}{\pi} = \frac{180°(0.785399)}{\pi} \approx 45.000048°,$$

which differs from the expected answer of 45° by 0.173 arcseconds. Interestingly, in this example the Newton/Raphson method converged much faster than the prior method, but it produced a less accurate solution mostly due to round-off errors. In both cases, had we rounded the answer to 4 digits, they would have produced exactly the same result.

When implementing the Newton/Raphson method, or any other such iterative scheme, always use a termination criterion that is larger than zero. Because round-off errors are inherent in digital computers, comparing a decimal number to see when it exactly reaches zero rarely works. In addition to using a termination criterion, set an upper limit on the number of iterations that can be performed in case the termination criterion was set too small, or the algorithm oscillates between values and never converges.

Kepler's equation is a fascinating subject in its own right, and numerous methods have been formulated for solving it (Colwell's book *Solving Kepler's Equation over Three Centuries* is perhaps the most complete treatment of the subject). The simple iteration and Newton/Raphson methods given here are sufficient for our needs.

### 4.5.6 The Ecliptic Plane

Until now we have looked at orbits as essentially being in 2 dimensions; that is, without explicitly saying so we have assumed that an object's orbit, such as a planet's, lies in the same plane as that of Earth and the Sun. The plane containing Earth and the Sun is called the ecliptic plane. Earth's orbit lies entirely within the ecliptic plane, but the orbits of other Solar System objects do not completely lie within the ecliptic plane. In general, a celestial object's orbit will *not* lie in the ecliptic plane, so how an object's orbit is inclined with respect to Earth's orbit is an important consideration for locating that object.

Figure 4.12 illustrates a celestial sphere in which the ecliptic plane is the sphere's "equator." The orbit for planet $P$ is also shown and labeled as the planet's orbital plane. The Sun $S$ is located in the center of the sphere.

# Orbits and Coordinate Systems

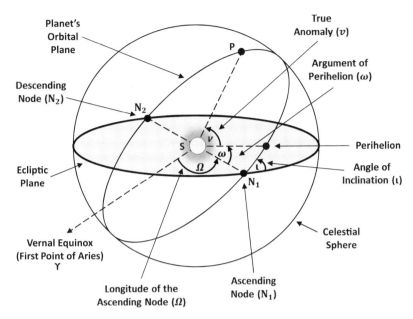

**Figure 4.12** A Planet's Orbital Plane
This illustration shows how a planet's orbital plane is defined. Calculating the location of a planet is complicated by the fact that the planet's orbit does not lie in the Earth-Sun plane.

The point of perihelion, the location of planet $P$, the vernal equinox ($\Upsilon$, the First Point of Aries), and the true anomaly $v$ are also labeled.

When planet $P$ changes in its orbit from being below the ecliptic plane to being above the ecliptic plane, it has passed through the ascending node $N_1$. $N_2$ is the descending node and the point at which the planet transitions from being above the ecliptic plane to being below the ecliptic plane. Angle $\Omega$ in figure 4.12 is measured from the vernal equinox to the ascending node, and it is the longitude of the ascending node. Angle $\omega$, measured from the ascending node to the point of perihelion, is the argument of perihelion. Angle $\iota$ is the angle of inclination and measures the degree to which the planet's orbital plane is inclined with respect to the ecliptic plane.

The angles shown in figure 4.12 can be used to completely describe the orbital characteristics of any object with an elliptical orbit, although the Sun and Earth-Sun plane will have to be replaced with different reference points suitable for the object in question. The chapters ahead will return to these angles to describe the orbits of the Sun, Moon, and planets and will use them to calculate positions for those objects for any given date, time, and location on Earth.

## 4.6 Equatorial Coordinate System

At long last, it is time to turn our attention to the coordinate systems used to locate celestial objects in the sky! We begin with the equatorial coordinate system because it may well be the most widely used coordinate system in astronomy. Star atlases and star catalogs usually give the location of stars, nebulae, quasars, and so forth, in equatorial coordinates whereas astronomy journals, such as *Sky & Telescope* and *Astronomy*, list the location of Solar System objects by equatorial coordinates. A major reason for the popularity of the equatorial coordinate system is that equatorial coordinates are independent of time and an observer's location. (However, carefully note the discussion later in this section of hour angle as a way to make equatorial coordinates vary with time and an observer's location.)

The equatorial coordinate system is also widely used because converting between the various other coordinate systems is simplified by using the equatorial coordinate system as an intermediate step. Figure 4.13 shows the process. For instance, to convert galactic coordinates to horizon coordinates, first convert the galactic coordinates to equatorial coordinates and then convert the equatorial coordinates to horizon coordinates. With this approach we only need to know how to convert a given coordinate system to/from the equatorial coordinate system and not from the given coordinate system to all others.

The equatorial coordinate system is similar in concept to the terrestrial latitude-longitude system for locating a position on Earth. In the equatorial coordinate system, the celestial sphere plays the same role as the Earth in the terrestrial latitude-longitude system, the celestial equator plays the same role as Earth's equator, and the celestial prime meridian (as defined by the First Point of Aries) plays the same role as Greenwich.

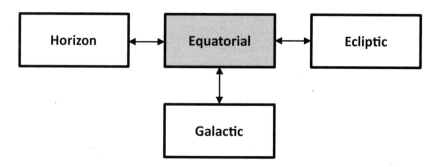

**Figure 4.13** Coordinate System Conversions
Coordinate system conversions are simplified by using the equatorial coordinate system as an intermediate step.

# Orbits and Coordinate Systems

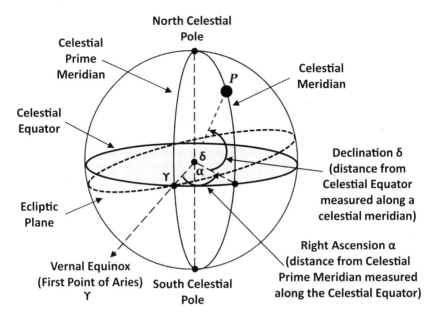

**Figure 4.14** Equatorial Coordinate System
Equatorial coordinates are defined relative to the celestial equator and celestial prime meridian.

Figure 4.14 shows how the equatorial coordinate system is defined. Instead of using the terms longitude and latitude, the equatorial coordinate system uses declination ($\delta$) and right ascension ($\alpha$). Declination is analogous to latitude and indicates how far away an object is from the celestial equator. Declination is measured in degrees and falls within the range $\pm 90°$ with positive angles indicating locations north of the celestial equator and negative angles indicating locations south of the celestial equator. Because declination is measured with respect to the celestial equator, and the celestial equator's location does not vary with time of day or an observer's location, declination for an object is fixed and does not vary with the time of day or an observer's location.

Right ascension is analogous to longitude and indicates how far an object is away from the First Point of Aries, which is the point used to define a celestial prime meridian. Thus, just as longitude measures an object's distance away from the *terrestrial* prime meridian, right ascension measures the distance from an object to the *celestial* prime meridian. As with declination, right ascension does not vary with time of day or an observer's location because it is measured with respect to a fixed location (the First Point of Aries).

Although analogous to terrestrial longitude, right ascension differs from longitude in a very important way. Right ascension is normally expressed in HMS format and is therefore in the range [$0^h$, $24^h$]. Right ascension is expressed in units of time rather than degrees because we normally think of Earth's rotation in terms of time rather than degrees. Right ascension can be easily converted to degrees by multiplying the right ascension by 15. Doing so makes right ascension even more similar to terrestrial longitude.

In our earlier discussion about celestial meridians, we noted that a celestial meridian can be defined relative to an observer rather than the celestial prime meridian. If we use an observer's meridian instead of the celestial prime meridian as a reference point, we have another way to measure "celestial longitude." This frequently used measurement is called the "hour angle" ($H$). While right ascension is really an angular measurement (although expressed in HMS format) of an object's distance from the First Point of Aries, hour angle is very much a time measurement. The hour angle for an object is a measure of how long it has been since the object crossed an observer's meridian.

Because of the way that an hour angle is defined (i.e., relative to an observer's local celestial meridian), it varies both with time of day and an observer's location. Notice also that an hour angle measures time relative to a sidereal day, not a mean solar day. For example, a star will transit an observer's local celestial meridian again after 24 sidereal hours have elapsed, at which time the star's hour angle is precisely $0^h$. Moreover, a star will transit when the local sidereal time (LST) equals the right ascension.

Because both right ascension and hour angle are expressed in HMS format and both express angles in terms of sidereal time rather than degrees, they can be easily confused. To avoid confusion, always check whether right ascension or hour angle is being used. Star catalogs and star atlases typically list objects in right ascension rather than hour angle, but you should always check to be sure. Additionally, remember that right ascension *does not* vary with time of day or an observer's location because it is based on a fixed location (First Point of Aries) whereas the hour angle *does* vary with time of day and an observer's location. Declination does not vary irrespective of whether right ascension or hour angle is being used as the other coordinate.

Hour angle and right ascension are clearly closely related. Converting between them is straightforward because the hour angle is really the difference between the LST and the right ascension. That is,

$$H = LST - \alpha. \qquad (4.6.1)$$

This equation again points out that a star will transit when the LST is equal to the right ascension (i.e., the hour angle is $0^h$ because the star is directly overhead the observer). Recall from chapter 3 that calculating the LST involves

# Orbits and Coordinate Systems

knowing an observer's location. Thus, as expected, the hour angle for a star does indeed vary with an observer's location.

As an example, consider a star whose right ascension is $3^h24^m06^s$. Suppose the *LST* for an observer is $18^h$. Calculate the corresponding hour angle. The required steps are:

1. Convert *LST* to decimal format.
   (Ans: $LST = 18^h$.)
2. Convert the right ascension to decimal format.
   (Ans: $\alpha = 3.401667^h$.)
3. Calculate $H = LST - \alpha$.
   (Ans: $H = 14.598333^h$.)
4. If $H$ is negative, add $24^h$.
   (Ans: no adjustment is necessary.)
5. Convert $H$ to HMS format.
   (Ans: $H = 14^h35^m54^s$.)

This star will culminate when the observer's LST is $3^h24^m06^s$ because, according to equation 4.6.1, that is when the star's hour angle will be $0^h$ for the observer.

The reverse process is no more difficult. Equation 4.6.1 can be rewritten as

$$\alpha = LST - H \qquad (4.6.2)$$

to compute the right ascension ($\alpha$) given the observer's local sidereal time (*LST*) and an object's hour angle ($H$).

Assume a planet's hour angle is $1^h15^m00^s$ and an observer's LST is $21^h$. Find the corresponding right ascension.

1. Convert the LST to decimal format.
   (Ans: $LST = 21.0^h$.)
2. Convert $H$ to decimal format.
   (Ans: $H = 1.25^h$.)
3. Calculate $\alpha = LST - H$.
   (Ans: $\alpha = 19.75^h$.)
4. If $\alpha$ is negative, add $24^h$.
   (Ans: no adjustment is necessary.)
5. Convert $\alpha$ to HMS format.
   (Ans: $\alpha = 19^h45^m00^s$.)

The example program for this chapter performs conversions between hour angle and right ascension. However, the program requires that the UT, date, and observer's longitude be specified rather than just the observer's LST, as was done in these 2 examples.

**4.7  Horizon Coordinate System**

Horizon coordinates are expressed by 2 angles: azimuth and altitude. Consider figure 4.15, which shows an observer and his horizon. $P$ is an object in the sky whose horizon coordinates are desired while $P'$ is the projection of the object onto the plane of the observer. Angle $NOP'$, designated by $A$, is the azimuth and can be easily measured with a compass. Azimuth is in the range of 0° to 360° and indicates how far $P$ is from the north as measured along an observer's horizon. Angle $POP'$ is the altitude, represented by the symbol $h$, and ranges from $-90°$ to $+90°$. Positive altitudes indicate objects above the horizon while negative altitudes indicate objects below the horizon.

Four equations are necessary to convert between the horizon and equatorial coordinate systems. To convert horizon coordinates to equatorial coordinates, apply the following 2 equations ($\phi$ is an observer's latitude, $\delta$ is the declination,

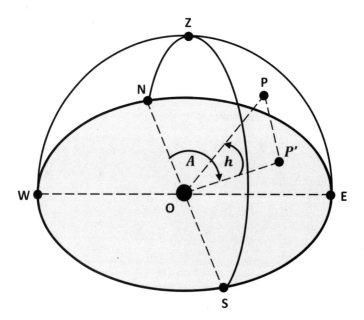

**Figure 4.15**  Horizon Coordinate System
The horizon coordinate system is easy to use, but it varies with location and time of day.

## Orbits and Coordinate Systems

$\alpha$ is the right ascension, and $H$ is the hour angle):

$$\sin\delta = \sin h \sin\phi + \cos h \cos\phi \cos A \qquad (4.7.1)$$

$$\cos H = \frac{\sin h - \sin\phi \sin\delta}{\cos\phi \cos\delta}. \qquad (4.7.2)$$

To convert equatorial to horizon coordinates, apply the following equations:

$$\sin h = \sin\delta \sin\phi + \cos\delta \cos\phi \cos H \qquad (4.7.3)$$

$$\cos A = \frac{\sin\delta - \sin\phi \sin h}{\cos\phi \cos h}. \qquad (4.7.4)$$

There are 2 important items to note about these equations. First, equations 4.7.2 and 4.7.3 use the hour angle ($H$) instead of the right ascension ($\alpha$). This should not be surprising because horizon coordinates change with an observer's location and time of day. Using the hour angle rather than the right ascension in these equations makes the results vary with time of day and an observer's location. Given an observer's LST, equation 4.6.1 can be used to convert between the hour angle and right ascension as required. The second important item is that all coordinates in these equations are expressed in degrees. In particular, be sure to convert $H$ to degrees when applying these equations.

Consider converting altitude 40°, azimuth 115° to equatorial coordinates for an observer at 38° N latitude. We need to apply equations 4.7.1 and 4.7.2.

1. Convert the altitude $H$ to decimal format.
   (Ans: $H = 40.0°$.)
2. Convert the azimuth $A$ to decimal format.
   (Ans: $A = 115.0°$.)
3. Compute $T_0 = \sin h \sin\phi + \cos h \cos\phi \cos A$.
   (Ans: $T_0 = 0.140626$.)
4. $\delta = \sin^{-1}(T_0)$.
   (Ans: $\delta = 8.084044°$.)
5. Compute $T_1 = \sin h - \sin\phi \sin\delta$.
   (Ans: $T_1 = 0.556210$.)
6. Compute $\cos H = \frac{T_1}{\cos\phi \cos\delta}$.
   (Ans: $\cos H = 0.712925$.)
7. Take the arccosine to get $H$.
   (Ans: $H = 44.526604°$.)

8. Compute sin $A$.
   (Ans: sin $A = 0.906308$.)
9. If sin $A$ is positive, then subtract the result of step 7 from $360°$.
   (Ans: $H = 315.473396°$.)
10. Compute $H = \frac{H}{15}$. This converts $H$ to hours to give the hour angle.
    (Ans: $H = 21.031560^h$.)
11. Convert $H$ and $\delta$ to HMS and DMS format.
    (Ans: $H = 21^h 01^m 54^s$, $\delta = 8°05'03''$.)

After step 11 we can use the techniques presented in chapter 3 to determine the observer's LST and apply equation 4.6.1 to convert the hour angle to its corresponding right ascension.

No angle adjustment is needed in step 4 since declination is in the range $\pm 90°$, which matches the range of the arcsine function. However, steps 8 and 9 *are* required since $H$ is in the range $[0^h, 24^h]$ (i.e., $[0°, 360°]$ when expressed as an angle), but step 7 produces an angle in the range $[0°, 180°]$.

Let us now use equations 4.7.3 and 4.7.4 to convert equatorial coordinates to horizon coordinates. Suppose a star is located at $\delta = -0°30'30''$, $H = 16^h 29^m 45^s$. For an observer at $25°$ N latitude, where will the star appear in the sky?

1. Convert $H$ to decimal format.
   (Ans: $H = 16.495833^h$.)
2. Multiply step 1 by 15 to convert to degrees.
   (Ans: $H_{deg} = 247.437500°$.)
3. Convert $\delta$ to decimal format.
   (Ans: $\delta = -0.508333°$.)
4. Compute $T_0 = \sin\delta \sin\phi + \cos\delta \cos\phi \cos H_{deg}$.
   (Ans: $T_0 = -0.351478$.)
5. $h = \sin^{-1}(T_0)$.
   (Ans: $h = -20.577738°$.)
6. $T_1 = \sin\delta - \sin\phi \sin h$.
   (Ans: $T_1 = 0.139669$.)
7. Compute $T_2 = \frac{T_1}{\cos\phi \cos h}$.
   (Ans: $T_2 = 0.164610$.)
8. $A = \cos^{-1}(T_2)$.
   (Ans: $A = 80.525393°$.)

# Orbits and Coordinate Systems

9. Compute $\sin H_{deg}$.
   (Ans: $\sin H_{deg} = -0.923462$.)
10. If the result of step 9 is positive, then $A = 360° - A$.
    (Ans: $A = 80.525393°$.)
11. Convert $A$ and $H$ to DMS format.
    (Ans: $A = 80°31'31''$, $H = -20°34'40''$.)

Since the altitude is negative in this example, the observer will not be able to see the star because it is below his horizon. Notice also that no angle adjustment is needed in step 5. Why? Steps 9 and 10 calculate an azimuth adjustment. Why?

## 4.8 Ecliptic Coordinate System

The equatorial and horizon coordinate systems are not particularly convenient for calculating the location of Solar System objects. Instead, it is generally easier to compute their position in the ecliptic coordinate system and then convert the results to equatorial or horizon coordinates as required. Because objects within the Solar System are constantly moving, their ecliptic coordinates change appreciably over months or even weeks.

Figure 4.16 shows the elements required to describe the ecliptic coordinate system.[8] Shown as a shaded circle, the ecliptic plane is extended until it intersects the celestial sphere to create an "ecliptic equator" as the reference point from which ecliptic latitudes are measured. The ecliptic latitude, denoted by $\beta$, is the angular distance that an object $P$ lies above or below the ecliptic plane and falls within the range $\pm 90°$. Latitudes above the ecliptic plane are positive angles while latitudes below the ecliptic plane are negative angles. An object, such as the Sun, whose orbit lies entirely within the ecliptic plane has an ecliptic latitude of $0°$.

To define an ecliptic longitude, the First Point of Aries (the vernal equinox) again plays a role similar to that of Greenwich, England. Recall that the equinoxes are the 2 points at which the celestial equator intersects the ecliptic plane. The ecliptic longitude, designated by $\lambda$, measures how far away an object is from the First Point of Aries. The ecliptic longitude is in the range $[0°, 360°]$ and measured along the ecliptic toward the First Point of Aries.

---

8. The ecliptic coordinate system described here is a geocentric ecliptic coordinate system because it is defined with respect to the ecliptic plane and uses Earth as the coordinate system's center. A heliocentric ecliptic coordinate system can also be defined that uses the Sun as the coordinate system's center instead of Earth. Heliocentric ecliptic coordinate systems can be useful, but they will not be discussed further.

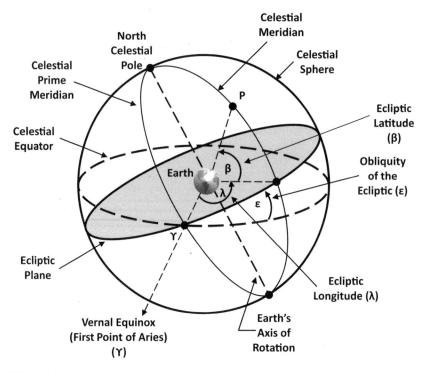

**Figure 4.16** Ecliptic Coordinate System
In the ecliptic coordinate system, the ecliptic plane plays the role of a celestial equator while the point at which the First Point of Aries intersects the ecliptic plane plays the role of a celestial prime meridian.

The ecliptic plane and the plane containing the celestial equator are at an angle to each other, as illustrated in figure 4.16. The angle between the 2 planes is the obliquity of the ecliptic and denoted by $\varepsilon$. Converting between the ecliptic and equatorial coordinate systems requires knowing the obliquity of the ecliptic, whose average value is about 23.4°. For improved accuracy, this angle should be computed since it varies slowly with time. Before showing how to compute the obliquity of the ecliptic, we must digress briefly to discuss the concept of a standard epoch.

Due to an effect called precession of the equinoxes (see section 4.10), right ascension and declination for a given celestial object change very slowly over time. Because of this slow change in coordinates over time, when right ascension and declination are given, it is customary to indicate on what instant of time the measurements are based. That instant of time is called a standard epoch. Star atlases and star catalogs can still be obtained that give

# Orbits and Coordinate Systems 93

equatorial coordinates for the standard epoch 1950.0, although as of the time that this book is being written, most recent atlases and catalogs use the standard epoch 2000.0. The standard epoch 2000.0 is often designated as J2000, which means that the instant of time being referred to is $12^h$ UT on January 1, 2000.

If high accuracy is not needed, the epoch can often be ignored. However, when computations involve the obliquity of the ecliptic, they should be made for the instant of time corresponding to the standard epoch. For example, if the right ascension and declination are based on the standard epoch J2000, then the obliquity of the ecliptic should be calculated for the standard epoch J2000.

Until 1983, an equation developed by the mathematician and astronomer Simon Newcomb was typically used to calculate the obliquity of the ecliptic. His equation is

$$\varepsilon = \varepsilon_0 - \frac{46.845T + 0.0059T^2 - 0.00181T^3}{3,600}, \tag{4.8.1}$$

where $\varepsilon_0 = 23.452294°$ (i.e., $23°27'08.26''$) is the obliquity of the ecliptic at the standard epoch 1900.0 and $T$ is the number of Julian centuries since 1900.0. The fraction on the right-hand side of equation 4.8.1 provides an epoch-dependent correction in degrees for the obliquity of the ecliptic angle.

Starting in 1984, astronomers and publications such as the *Astronomical Almanac* have used an updated equation developed by the Jet Propulsion Laboratory (JPL) to compute the obliquity of the ecliptic. Whereas Newcomb's equation was derived from a massive manual analysis of planetary positions up until about 1895, JPL's equation was derived from a computer analysis of the position of the planets from 1911 through 1979. JPL's equation for computing the obliquity of the ecliptic is given by

$$\varepsilon = \varepsilon_0 - \frac{46.815T + 0.0006T^2 - 0.00181T^3}{3,600}, \tag{4.8.2}$$

where $\varepsilon_0 = 23.439292°$ (i.e., $23°26'21.45''$) is the obliquity of the ecliptic at the standard epoch J2000 and $T$ is the number of Julian centuries since 2000.0.

Using equation 4.8.2, $\varepsilon$ is computed in 4 steps. For the following example, assume that we wish to compute the obliquity of the ecliptic for the standard epoch 2010.0.

1. Calculate the Julian day number for January 0.0 (i.e., midnight) of the desired standard epoch (i.e., 1/0.0/2010).

   (Ans: $JD = 2,455,196.5$.)

2. Compute $T = (JD - 2,451,545.0)/36,525$. This is the number of Julian centuries since 1/0.5/2000.

(Ans: $T = 0.099973$ centuries.)

3. Compute $D_e = 46.815T + 0.0006T^2 - 0.00181T^3$.

(Note: Round-off error can be reduced when implementing this equation by using the equivalent form $D_e = T[46.815 + T(0.0006 - 0.00181T)]$, but this enhancement will be ignored here for the sake of clarity.)

(Ans: $D_e = 4.680222''$.)

4. Compute $\varepsilon = \varepsilon_0 - (D_e/3600)$. This is the desired obliquity of the ecliptic for the standard epoch 2010.0.

(Ans: $\varepsilon_{2010.0} = 23.437992°$.)

The obliquity of the ecliptic $\varepsilon$ could be converted to DMS format in step 4, but it is left in decimal format for ease of use in other procedures.

Using Newcomb's equation for this example instead of JPL's equation yields a result of $\varepsilon_{2010.0} = 23.437979°$, a difference of only $0.000013°$ ($0.0468''$). For the standard epoch 1950.0, Newcomb's equation produces $23.445788°$ as the obliquity of the ecliptic while the JPL equation produces $23.445794°$, a difference of about $0.0216''$. Despite the relatively small differences in the results produced by equations 4.8.1 and 4.8.2, the differences may accumulate to a few arcseconds when computing equatorial coordinates. A few arcseconds may or may not be important depending on whether the instruments being used are sufficiently accurate to even measure arcseconds. All of the programs for this book use JPL's equation instead of Newcomb's equation.

Given the ability to compute the obliquity of the ecliptic, the ecliptic and equatorial coordinate systems are related by the following equations:

$$\sin\delta = \sin\beta\cos\varepsilon + \cos\beta\sin\varepsilon\sin\lambda \qquad (4.8.3)$$

$$\tan\alpha = \frac{\sin\lambda\cos\varepsilon - \tan\beta\sin\varepsilon}{\cos\lambda} \qquad (4.8.4)$$

$$\sin\beta = \sin\delta\cos\varepsilon + \cos\delta\sin\varepsilon\sin\alpha \qquad (4.8.5)$$

$$\tan\lambda = \frac{\sin\alpha\cos\varepsilon - \tan\delta\sin\varepsilon}{\cos\alpha}. \qquad (4.8.6)$$

Equations 4.8.3 and 4.8.4 are used to convert ecliptic coordinates to equatorial coordinates whereas equations 4.8.5 and 4.8.6 are used to convert equatorial coordinates to ecliptic coordinates. $\varepsilon$ in each of these equations is the obliquity of the ecliptic at the standard epoch.

Notice the similarity between these 2 pairs of equations. They are identical when $\delta$ is interchanged with $\beta$ and $\alpha$ is interchanged with $\lambda$. All measurements are expressed as angles, so $\alpha$ must be converted from hours to degrees before using these equations.

Suppose the ecliptic coordinates of Jupiter are ecliptic latitude $1°12'00''$ and ecliptic longitude $184°36'00''$. Use equations 4.8.3 and 4.8.4 to find Jupiter's equatorial coordinates for the standard epoch J2000.

1. Using equation 4.8.2, calculate $\varepsilon$ for the given standard epoch.
   (Ans: $\varepsilon_{2000} = 23.439292°$.)
2. Convert the ecliptic latitude $\beta$ to decimal format.
   (Ans: $\beta = 1.2°$.)
3. Convert the ecliptic longitude $\lambda$ to decimal format.
   (Ans: $\lambda = 184.6°$.)
4. Using equation 4.8.3, compute $T = \sin\beta \cos\varepsilon + \cos\beta \sin\varepsilon \sin\lambda$.
   (Ans: $T = -0.012680$.)
5. $\delta = \sin^{-1}(T)$. This is the declination at the given standard epoch.
   (Ans: $\delta = -0.726531°$.)
6. Compute $y = \sin\lambda \cos\varepsilon - \tan\beta \sin\varepsilon$. This is the numerator of the fraction in equation 4.8.4.
   (Ans: $y = -0.081913$.)
7. Compute $x = \cos\lambda$. This is the denominator of the fraction in equation 4.8.4.
   (Ans: $x = -0.996779$.)
8. Compute $R = \tan^{-1}\left(\frac{y}{x}\right)$.
   (Ans: $R = 4.697898°$.)
9. Compute an angle adjustment for $R$ based on $y$ and $x$ to put the arctangent from step 8 into the correct quadrant. This will be the right ascension in degrees.
   (Ans: Since $x$ and $y$ are both negative, table 4.1 gives a quadrant adjustment factor of $180°$. Applying this adjustment factor gives $\alpha_{deg} = 184.697898°$.)
10. Let $\alpha = (\alpha_{deg}/15)$, which converts the right ascension from degrees to hours.
    (Ans: $\alpha = 12.313193^h$.)
11. Convert $\alpha$ and $\delta$ to HMS and DMS format.
    (Ans: $\alpha_{2000} = 12^h 18^m 47.5^s$, $\delta_{2000} = -0°43'35.5''$.)

Let us illustrate converting right ascension and declination to ecliptic longitude and latitude by doing the previous problem in reverse. Given Jupiter's equatorial coordinates of right ascension $12^h 18^m 47.5^s$, declination

$-0°43'35.5''$, and the standard epoch J2000, compute Jupiter's ecliptic coordinates. This time we need equations 4.8.5 and 4.8.6.

1. Using equation 4.8.2, calculate $\varepsilon$ for the given standard epoch.
   (Ans: $\varepsilon_{2000} = 23.439292°$.)
2. Convert $\alpha_{2000}$ to decimal format.
   (Ans: $\alpha = 12.313194^h$.)
3. Convert $\alpha$ from the previous step to degrees. $\alpha_{deg} = 15\alpha$.
   (Ans: $\alpha_{deg} = 184.697917°$.)
4. Convert $\delta_{2000}$ to decimal format.
   (Ans: $\delta = -0.726528°$.)
5. Using equation 4.8.5, compute $T = \sin\delta\cos\varepsilon + \cos\delta\sin\varepsilon\sin\alpha_{deg}$.
   (Ans: $T = 0.020943$.)
6. $\beta = \sin^{-1}(T)$. This is the ecliptic latitude at the given standard epoch.
   (Ans: $\beta = 1.200010°$.)
7. Compute $y = \sin\alpha_{deg}\cos\varepsilon - \tan\delta\sin\varepsilon$. This is the numerator of the fraction in equation 4.8.6.
   (Ans: $y = -0.080188$.)
8. Compute $x = \cos\alpha_{deg}$. This is the denominator of the fraction in equation 4.8.6.
   (Ans: $x = -0.996640$.)
9. Compute $R = \tan^{-1}\left(\frac{y}{x}\right)$.
   (Ans: $R = 4.600016°$.)
10. Compute an angle adjustment for $R$ based on $y$ and $x$ to put the arctangent result from step 9 into the correct quadrant. Applying the appropriate quadrant adjustment to $R$ gives the ecliptic longitude.
    (Ans: Since $x$ and $y$ are both negative, table 4.1 gives a quadrant adjustment factor of $180°$. Applying this adjustment factor gives $\lambda = 184.600016°$.)
11. Convert the ecliptic latitude and the ecliptic longitude to the DMS format.
    (Ans: $\beta_{2000} = 1°12'00.0''$, $\lambda_{2000} = 184°36'00.0''$.)

## 4.9 Galactic Coordinate System

The last coordinate system we will consider is the galactic coordinate system. Three reference points are needed to define any spherical coordinate system: an origin, an "equator" from which latitudes are measured, and a "prime meridian" from which longitudes are measured. Table 4.2 summarizes these reference points for the coordinate systems discussed so far.

# Orbits and Coordinate Systems

**Table 4.2** Coordinate System Reference Points
Defining a spherical coordinate system requires three reference points.

| System | Origin | "Equator" | "Prime Meridian" |
|---|---|---|---|
| Terrestrial Lat/Lon | Center of Earth | Earth's Equator | Greenwich, England |
| Equatorial | Earth | Celestial Equator | First Point of Aries |
| Horizon | Observer | Observer's Horizon | Observer's Meridian |
| Ecliptic | Earth | Ecliptic Plane | First Point of Aries |

For the galactic coordinate system, the center of our Milky Way is chosen to be the coordinate system's origin. The equatorial coordinates for the center of the Milky Way for the standard epoch 1950.0 are right ascension $17^h42^m$ and declination $-28°45'$. These coordinates were formally defined by the 1958 IAU and were based on the best known estimate for the galactic center at that time. Although modern estimates differ, subsequent measurements place the center of the Milky Way, with respect to the standard epoch J2000, at right ascension $17^h45^m37.22^s$, declination $-28°56'10.23''$.

With a coordinate system origin selected, the plane containing the Sun and the center of the Milky Way is extended until it intersects the celestial sphere to form a "galactic equator." Galactic latitudes, denoted by $b$, are measured with respect to this galactic equator. Galactic latitudes are in the range of $\pm 90°$ with positive angles being north of the galactic plane and negative angles being south of the galactic plane. The galactic equator does not lie in the same plane as the celestial equator, but it is inclined at an angle of approximately $62°$.

Extending a line connecting the Sun and the center of the Milky Way until it intersects the celestial sphere will form 2 points on the celestial sphere. The intersection point farthest away from the Sun that lies in the direction of the constellation Sagittarius is the reference point for measuring galactic longitudes, denoted by the symbol $l$. Galactic longitudes are measured counterclockwise so as to increase from $0°$ to $360°$ in the same direction as increasing right ascension.

The equations that relate equatorial and galactic coordinates require knowing the location of the Galactic North Pole. For the standard epoch 1950.0, the equatorial coordinates of the Galactic North Pole, as agreed to by the 1958 IAU, are exactly right ascension $12^h49^m$, declination $27°24'$. It will be more convenient in the following equations to convert these coordinates to decimal degrees. Thus, the Galactic North Pole's right ascension is $192.25°$ while its declination is $27.4°$. The longitude of the ascending node of the galactic plane is also required, and its value is $33°$. The latitude of the ascending node of the galactic plane is 0°. All of these values are with respect to the standard epoch 1950.0.

The equations relating galactic and equatorial coordinates are

$$\sin\delta = \cos b \cos\delta_0 \sin(l-N_0) + \sin b \sin\delta_0 \qquad (4.9.1)$$

$$\tan\alpha = \frac{\cos b \cos(l-N_0)}{\sin b \cos\delta_0 - \cos b \sin\delta_0 \sin(l-N_0)} + \alpha_0 \qquad (4.9.2)$$

$$\sin b = \cos\delta \cos\delta_0 \cos(\alpha-\alpha_0) + \sin\delta \sin\delta_0 \qquad (4.9.3)$$

$$\tan l = \frac{\sin\delta - \sin b \sin\delta_0}{\cos\delta \sin(\alpha-\alpha_0)\cos\delta_0} + N_0, \qquad (4.9.4)$$

where (for epoch 1950.0) $\alpha_0 = 192.25°$, $\delta_0 = 27.4°$ are the equatorial coordinates of the Galactic North Pole, and $N_0 = 33°$ is the longitude of the ascending node of the galactic plane. Equations 4.9.1 and 4.9.2 convert galactic coordinates to equatorial coordinates while equations 4.9.3 and 4.9.4 convert equatorial coordinates to galactic coordinates.

Strictly speaking, the equatorial coordinates for the Galactic North Pole and the longitude of the ascending node should be adjusted to the same standard epoch as the equatorial coordinates being converted to or being converted from. The next section will demonstrate how to compute precession corrections to adjust the equatorial coordinates given here for the Galactic North Pole. Once the corrected coordinates are known, replace the right ascension (192.25°), declination (27.4°), and longitude of the ascending node (33°) in the previous equations with the corrected coordinates to make all calculations with respect to the new standard epoch. For J2000, the Galactic North Pole is at right ascension $12^h51^m26.36^s$ (192.8598°), declination $27°07'40.90''$ (27.128027°), and the longitude of the ascending node is 32.9319°.

Instead of adjusting equations 4.9.1 through 4.9.4 to a new standard epoch, there is a simpler approach. Equations 4.9.1 and 4.9.2 produce equatorial coordinates with respect to the standard epoch 1950.0. Once calculated, the equatorial coordinates produced can be adjusted for precession to convert them to a different standard epoch. When converting equatorial coordinates to galactic coordinates (equations 4.9.3 and 4.9.4), first adjust the equatorial coordinates to be with respect to epoch 1950.0, and then convert the epoch 1950.0 equatorial coordinates to galactic coordinates.

Galactic coordinates are not used beyond this chapter, so we will ignore the effects of precession and perform all calculations with respect to the epoch 1950.0. This chapter's program allows calculations to be done with respect

# Orbits and Coordinate Systems

to epoch 1950.0 or J2000, although it can be easily modified for any other standard epoch.

Assuming epoch 1950.0, convert galactic latitude $55°20'$, galactic longitude $180°$ to equatorial coordinates. We need equations 4.9.1 and 4.9.2.

1. Convert $b_{1950}$ to decimal format.
   (Ans: $b = 55.333333°$.)

2. Convert $b_{1950}$ to decimal format.
   (Ans: $b = 180.0°$.)

3. Using equation 4.9.1, compute $T = \cos b \cos \delta_0 \sin(l - N_0) + \sin b \sin \delta_0$, where $\delta_0 = 27.4°$ is the right ascension of the Galactic North Pole and $N_0 = 33°$ is the longitude of the ascending node for epoch 1950.0.
   (Ans: $T = 0.653540$.)

4. Compute $\delta = \sin^{-1}(T)$. This is the declination for epoch 1950.0.
   (Ans: $\delta = 40.809063°$.)

5. Compute $y = \cos b \cos(l - N_0)$. This is the numerator in equation 4.9.2.
   (Ans: $y = -0.477037$.)

6. Compute $x = \sin b \cos \delta_0 - \cos b \sin \delta_0 \sin(l - N_0)$. This is the denominator in equation 4.9.2.
   (Ans: $x = 0.587640$.)

7. Compute $R = \tan^{-1}\left(\frac{y}{x}\right)$.
   (Ans: $R = -39.069133°$.)

8. Calculate an adjustment for $R$ based on $y$ and $x$ to put the arctangent in the correct quadrant. Apply the appropriate quadrant adjustment to $R$ and add $N_0$ to get the right ascension in degrees.

   (Ans: Since $x$ is positive and $y$ is negative, table 4.1 gives a quadrant adjustment factor of $360°$. Applying this adjustment factor and adding $N_0$ gives $\alpha_{deg} = 513.180867°$.)

9. If $\alpha_{deg} > 360°$, subtract $360°$.
   (Ans: $\alpha_{deg} = 153.180867°$.)

10. Convert $\alpha_{deg}$ to hours. $\alpha = (\alpha_{deg}/15)$. This is the right ascension for epoch 1950.0.
    (Ans: $\alpha = 10.212058°$.)

11. Convert $\alpha$ and $\delta$ to HMS and DMS format, respectively.
    (Ans: $\alpha_{1950} = 10^h 12^m 43^s$, $\delta_{1950} = 40°48'33''$.)

If done with respect to the standard epoch J2000, the results would be

$\alpha_{2000} = 10^h 15^m 43^s$, $\delta_{2000} = 40°33'35''$.

Again assuming epoch 1950.0, convert right ascension $10^h 12^m 43^s$, declination $40°48'33''$ to galactic coordinates. We need to use equations 4.9.3 and 4.9.4.

1. Convert $\alpha_{1950}$ to decimal format.
   (Ans: $\alpha = 10.211944^h$.)
2. Convert $\alpha$ to degrees. $\alpha_{deg} = 15\alpha$.
   (Ans: $\alpha_{deg} = 153.179167°$.)
3. Convert $\delta_{1950}$ to decimal format.
   (Ans: $\delta = 40.809167°$.)
4. Compute $T_0 = \cos\delta \cos\delta_0 \cos(\alpha - \alpha_0) + \sin\delta \sin\delta_0$.
   (Ans: $T_0 = 0.822462$.)
5. Compute $b = \sin^{-1}(T_0)$. This is the galactic latitude for epoch 1950.0.
   (Ans: $b = 55.332048°$.)
6. Compute $y = \sin\delta - \sin b \sin\delta_0$. This is the numerator of equation 4.9.4.
   (Ans: $y = 0.275045$.)
7. Compute $x = \cos\delta \sin(\alpha - \alpha_0) \cos\delta_0$. This is the denominator of equation 4.9.4.
   (Ans: $x = -0.423535$.)
8. Compute $T_1 = \tan^{-1}\left(\frac{y}{x}\right)$.
   (Ans: $T_1 = -32.999772°$.)
9. Calculate an adjustment for $T_1$ based on $y$ and $x$ to put the arctangent in the correct quadrant. Apply the appropriate quadrant adjustment to $T_1$ and add $N_0$ to get the galactic longitude $l$ for epoch 1950.0.
   (Ans: Since $x$ is negative and $y$ is positive, table 4.1 gives a quadrant adjustment factor of $180°$. Applying this adjustment factor and adding $N_0$ gives $l = 180.000228°$.)
10. If $l > 360°$, subtract $360°$.
    (Ans: $l = 180.000228°$.)
11. Convert $b$ and $l$ to DMS format.
    (Ans: $b_{1950} = 55°19'55''$, $l_{1950} = 180°00'01''$.)

The discrepancy between the original galactic coordinates used in the first example and the galactic coordinates computed here is due to round-off errors.

## 4.10 Precession and Other Corrections

Despite how carefully positional calculations are performed and no matter what coordinate system is used, an object may not appear in the sky in the position that it is calculated to be. This is due to several factors; we will briefly discuss 4 related phenomena: atmospheric refraction, parallax, precession, and nutation. First, consider the impact of atmospheric refraction on the apparent location of a celestial object.

Light rays are bent whenever they pass from one medium, such as air, to another, such as water. This effect is called refraction and is easily observable by filling a glass with water and placing a pencil in the glass. The image of the pencil when viewed through the water is enlarged and appears to be bent with respect to the portion of the pencil that remains outside the water.

Earth is surrounded by an atmosphere, which means that before light from objects outside the Earth reaches our eyes, the light must first pass through Earth's atmosphere. Light rays are bent when they enter the atmosphere so that stars and other celestial objects do not appear to be quite where they are calculated to be. This effect is called atmospheric refraction. The extent to which it distorts the location of an object depends upon air temperature, air pressure, and the object's altitude in the sky. Objects at an observer's zenith are not refracted at all, while those on the horizon may be refracted a great deal. Atmospheric refraction can change the apparent location of an object by as much as $34'$ of arc for objects at the horizon. Because atmospheric refraction is generally greater the closer an object is to the horizon, the higher an object is in the sky the closer its calculated position will come to matching its observed position.

Parallax is an effect in which the apparent position of an object changes when viewed along 2 different lines of sight. The effect can be easily demonstrated. Close your left eye and use your right eye to look at an object a few feet away. Using the index finger on your right hand and with your left eye still closed, point at the center of the object or to some prominent feature on the object. Now without moving your head or your finger, close your right eye and look at the object with your left eye. Relative to your right index finger, the object will appear to have moved to a significantly different position than when viewed through your left eye. This apparent change in position is the parallax effect.

To better understand parallax, consider figure 4.17 in which 2 observers, $A$ and $B$, are looking at the Moon ($M$) from 2 different locations on the surface of the Earth. $S$ is some distant star that both observers are using as a reference point in much the same manner as using our right index finger as a point of

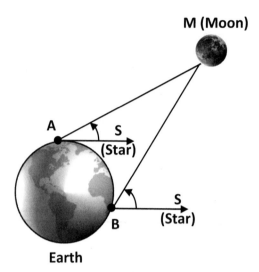

**Figure 4.17** Parallax
Parallax is the result of viewing an object from different locations.

reference in the experiment just described. Since $S$ is so far away relative to the distance between the 2 observers, the distance between the observers is negligible, and both observers will agree on the star's location. However, the 2 observers will not agree on the location of the Moon with respect to the star $S$ because $\angle MAS$ measured by observer $A$ is clearly not the same as $\angle MBS$ measured by observer $B$.

When viewing celestial objects from 2 different locations on Earth, parallax distortion can amount to 1 degree of arc for the Moon but only a few seconds of arc for the planets within our Solar System. Parallax is so small when viewing stars and other distant celestial objects from 2 different locations on Earth that it is virtually unmeasurable. However, the parallax effect can be amplified by increasing the distance between the points at which an object is viewed. When viewing an object from different places along Earth's orbit around the Sun, the effect is referred to as stellar parallax. The greatly increased viewing baseline afforded by Earth's orbit around the Sun, along with some simple trigonometry, provides a way for astronomers to estimate the distance to a star.

The last 2 effects we will describe are precession and nutation. Precession is a gradual shifting of Earth's axis of rotation similar to the axial motion of a rapidly spinning top. When a top spins, its axis of rotation moves around in a circle rather than staying in exactly the same orientation. The circular motion exhibited by the rotational axis of a spinning object is called precession.

# Orbits and Coordinate Systems

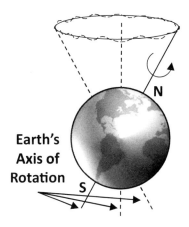

**Figure 4.18** Precession
Earth's rotational axis moves around in a small circle as Earth rotates, similar to the motion of a spinning top. This effect is precession. Earth's axis of rotation also "wobbles" due to the gravitational effects of the Sun and Moon. This effect is nutation.

As Earth rotates, its axis of rotation moves around in a circle so that it gradually sweeps out the surface of a cone. This is shown as a solid circle above the North Pole in figure 4.18. Because of precession, Earth's North Pole completes 1 full trip around this circle about every 25,800 years. Historically, this circular movement of Earth's axis of rotation has been called precession of the equinoxes because the observed effect is that the equinoxes appear to move westward along the ecliptic relative to the fixed stars.

Precession of Earth's axis of rotation is not really a perfect circle as figure 4.18 implies. There is a periodic "wobbling" of Earth's axis due to the combined gravitational pull of the Sun and Moon on Earth's orbit. This wobbling is called nutation and shown in figure 4.18 as a dashed oscillation superimposed on the precession circle. Nutation is a much smaller effect than precession, although nutation may affect an object's apparent location by a few seconds of arc. The nutation oscillation has a period of 18.6 years, which is the same period in which the Moon's orbital plane precesses around Earth, and an amplitude of 9.2 seconds of arc.

Both precession and nutation must be accounted for to obtain highly accurate equatorial coordinates for celestial objects. However, we will present an algorithm only for precession to illustrate how equatorial coordinates can be adjusted to account for various effects.

Calculating precession corrections requires knowing the equatorial coordinates of interest, the reference epoch for which the coordinates are given, and the new standard epoch to which the coordinates are to be converted. The equations required to compute precession corrections for right ascension ($\alpha$) and declination ($\delta$) are

$$\Delta\alpha = (M + N_t \sin\alpha \tan\delta) D \qquad (4.10.1)$$

$$\Delta\delta = (N_d \cos\alpha) D, \qquad (4.10.2)$$

where $D$ is the difference between the new epoch and the reference epoch. $N_d$, $N_t$, and $M$ are corrections calculated from interpolating polynomials that are a function of $T$, the number of Julian centuries from the desired new epoch to the standard epoch 1900.0. The interpolating polynomials are given by the following equations:

$$M = 3.07234^s + 0.00186^s T \qquad (4.10.3)$$

$$N_d = 20.0468'' - 0.0085'' T \qquad (4.10.4)$$

$$N_t = \frac{N_d}{15}. \qquad (4.10.5)$$

Be careful when applying equation 4.10.1. Remember that right ascension ($\alpha$) is normally expressed in hours. It must be converted to degrees (i.e., multiplied by 15) before the sine function can be applied. The result, however, of equation 4.10.1 is a correction to the right ascension in hours, not degrees. Since $N_d$ is a correction in degrees (equation 4.10.4), it must be converted to hours (equation 4.10.5) before being used in equation 4.10.1.

As an example, calculate the equatorial coordinates for the Galactic North Pole with respect to the standard epoch J2000. Recall that the epoch 1950.0 equatorial coordinates for the Galactic North Pole are right ascension $12^h49^m$, declination $27°24'$.

1. Convert $\alpha_{1950}$ to decimal format.
   (Ans: $\alpha_h = 12.816667^h$.)

2. Multiply $\alpha_h$ by 15 to convert it to degrees.
   (Ans: $\alpha_{deg} = 192.250000°$.)

3. Convert $\delta_{1950}$ to decimal format.
   (Ans: $Decl = 27.400000°$.)

4. Let $E_t$ be the epoch to convert to (2000.0 in this example) and compute $T = \frac{E_t - 1900}{100}$.
   (Ans: $T = 1.000000$.)

5. Use equation 4.10.3 to compute $M$ in seconds of time.
   (Ans: $M = 3.074200^s$.)

6. Use equation 4.10.4 to compute $N_d$ in seconds of arc.
   (Ans: $N_d = 20.038300''$.)

# Orbits and Coordinate Systems

7. Compute $N_t = (N_d/15)$. This converts arcseconds to seconds of time.
   (Ans: $N_t = 1.335887^s$.)

8. Let $E_f$ be the epoch to convert from (1950.0 in this example) and compute $D = E_t - E_f$.
   (Ans: $D = 50.0$.)

9. Use equation 4.10.1 to compute $\Delta\alpha$, which is an adjustment factor for the right ascension in seconds of time.
   (Ans: $\Delta\alpha = 146.363795^s$.)

10. Use equation 4.10.2 to compute $\Delta\delta$, which is a correction for the declination in arcseconds.
    (Ans: $\Delta\delta = -979.102504''$.)

11. Divide $\Delta\alpha$ by 3600 to convert it to hours and divide $\Delta\delta$ by 3600 to convert it to degrees.
    (Ans: $\Delta\alpha = 0.040657^h$, $\Delta\delta = -0.271973°$.)

12. Add $\Delta\alpha$ to the right ascension from step 1 and $\Delta\delta$ to the declination from step 3 to obtain the corrected equatorial coordinate for the new standard epoch.
    (Ans: $\alpha = 12.857323^h$, $\delta = 27.128027°$.)

13. Convert $\alpha$ and $\delta$ to HMS and DMS formats, respectively.
    (Ans: $\alpha_{2000} = 12^h51^m26.36^s$, $\delta_{2000} = 27°07'40.90''$.)

Other precession correction methods can be found in the literature, and algorithms can be found to correct for nutation, parallax, and atmospheric refraction. For the purposes of this book, however, it is sufficient to understand that the basic process is to compute the equatorial coordinates for an object, calculate right ascension and declination adjustments, and add the adjustments to arrive at corrected equatorial coordinates.

## 4.11 Program Notes

RunChap4 does all the coordinate system conversions described in this chapter, computes precession corrections, and provides both the simple iterative method and Newton/Raphson method for solving Kepler's equation. The program supports galactic coordinate system conversions for epochs 1950.0 and J2000 only, but can be easily modified to handle other epochs by adding constants with the appropriate galactic coordinates, and then modifying the routines that convert to/from galactic coordinates to use the new epoch.

Modern programming languages provide an arctangent function that takes $y$ and $x$ arguments, but such functions may not return angles in the correct quadrants for use in this book's algorithms. For this reason, this book's programs include routines to properly apply quadrant adjustments for the arctangent function. Additionally, the trigonometric functions in most programming languages assume radians rather than degrees. Thus, this book's programs implement trigonometric functions that accept angles in degrees and inverse trigonometric functions that return results in degrees rather than radians.

The programs for this book allow coordinates expressed in degrees to be entered in decimal or DMS format. Right ascension and hour angle coordinates may likewise be entered in either decimal or HMS format. Be sure to enter a direction of N or S for latitudes and E or W for longitudes when terrestrial coordinates are required.

## 4.12 Exercises

The following example problems should be thoroughly understood before proceeding further. The chapters ahead will frequently perform coordinate system conversions to locate objects, so it is important to have a solid understanding of the material in this chapter.

1. A star with hour angle $15^h 30^m 15^s$ is observed from longitude 64° W. The date is 6/5/1976 and UT is $14^h$. What is the star's right ascension?
   (Ans: $\alpha = 11^h 10^m 13^s$.)

2. A star has right ascension $12^h 32^m 06^s$. An observer is at 40° E longitude, the date is 1/5/2015, and UT is $12^h$. What is the hour angle for this star?
   (Ans: $H = 9^h 06^m 58^s$.)

3. A celestial body is seen in the sky at altitude 10°00′00″, azimuth 200°10′20″. The observer's latitude is 35.6° N. What is the hour angle and declination of the object?
   (Ans: $H = 1^h 46^m 15^s$, $\delta = -40°34′58″$.)

4. The hour angle for a star is $7^h 00^m 00^s$ and its declination is 49°54′20″. If an observer is at latitude 80° S, where will this star appear in the sky?
   (Ans: $h = -51°28′21″$, $A = 267°07′04″$. The star is below the observer's horizon.)

5. Mercury is located at ecliptic latitude 0°00′00″, ecliptic longitude 120°30′30″. What are its equatorial coordinates assuming epoch J2000?
   (Ans: $\alpha_{2000} = 8^h 10^m 50^s$, $\delta_{2000} = 20°02′31″$.)

6. For epoch J2000, assume that a celestial body is at right ascension $11^h 10^m 13^s$, declination 30°05′40″. What are its ecliptic coordinates?
   (Ans: $\beta_{2000} = 22°41′54″$, $\lambda_{2000} = 156°19′09″$.)

# Orbits and Coordinate Systems

7. A star within the Milky Way is at galactic latitude $30°25'40''$, galactic longitude $120°00'00''$ for the standard epoch 1950.0. What are its equatorial coordinates?
   (Ans: $\alpha_{1950} = 15^h29^m53^s$, $\delta_{1950} = 85°59'33''$.)

8. Repeat the previous problem by assuming the coordinates are relative to epoch J2000.
   (Ans: $\alpha_{2000} = 15^h20^m18^s$, $\delta_{2000} = 85°49'05''$.)

9. If a star's location is right ascension $11^h10^m13^s$, declination $30°05'40''$ for epoch 1950.0, what are its galactic coordinates?
   (Ans: $b_{1950} = 68°13'25''$, $l_{1950} = 200°00'15''$.)

10. Repeat the previous problem but assume the coordinates are relative to epoch J2000.
    (Ans: $b_{2000} = 67°38'01''$, $l_{2000} = 199°18'42''$.)

11. If the epoch 1950.0 equatorial coordinates for an object are right ascension $12^h32^m06^s$, declination $30°05'40''$, calculate its epoch J2000 coordinates by correcting for precession.
    (Ans: $\alpha_{2000} = 12^h34^m34^s$, $\delta_{2000} = 29°49'08''$.)

12. If the epoch J2000 equatorial coordinates for an object are right ascension $12^h34^m34^s$, declination $29°49'08''$, calculate the epoch 2015.0 coordinates by correcting for precession.
    (Ans: $\alpha_{2015} = 12^h35^m18^s$, $\delta_{2015} = 29°44'11''$.)

13. Assume that a planet has an orbital eccentricity of 0.00035 and a mean anomaly of $5.498078°$. Use both iterative schemes to solve Kepler's equation for the eccentric anomaly. Use a termination criteria of 0.000002 radians. (The exact answer is $5.5°$.)
    (Ans: For the simple iteration scheme, after 2 iterations $E = 5.500000°$. For the Newton/Raphson method, after 2 iterations $E = 5.500000°$.)

14. Repeat the previous problem assuming an orbital eccentricity of 0.6813025. (The exact answer is $16.744355°$, *not* $5.5°$ because the orbital eccentricity has changed.)
    (Ans: For the simple iteration scheme, after 26 iterations $E = 16.744172°$, or $16°44'39.02''$. For the Newton/Raphson method, after 4 iterations $E = 16.744355°$, or $16°44'39.68''$.)

15. Repeat the previous problem assuming an orbital eccentricity of 0.850000. (The exact answer is $29.422286°$.)
    (Ans: For the simple iteration scheme, after 39 iterations $E = 29.421983°$, or $29°25'19.14''$. For the Newton/Raphson method, after 7 iterations $E = 29.422286°$, or $29°25'20.23''$.)

# 5 Stars in the Nighttime Sky

At this point, the reader may well be a bit confused. Chapter 3 discussed time and presented a maze of calculations for converting from one time system to another. *Why* is all that necessary? Different calendar systems and techniques for dealing with Julian day numbers were also described. *Why* are Julian day numbers necessary?

Chapter 4 may have seemed a little more illuminating when different coordinate systems were presented along with algorithms for converting between them. The horizon and equatorial coordinate systems in particular may have appeared promising and left the reader on the verge of understanding ... something. Still, *how* does time fit in with coordinate system conversions? *How* do orbits and orbital elements relate to time systems and coordinate system conversions? The basic question we're ultimately trying to answer is: *For a given date, time, and location, where will the Sun, Moon, planets, and stars appear in the sky?* The necessary pieces seem to be in place, but how do they fit together?

This chapter will demonstrate *how* and *why* the pieces fit together to locate a specific star, nebula, galaxy, or other deep space object. In fact, this chapter culminates in a sample program that will produce a star chart showing the nighttime sky for any given time, date, latitude, and longitude. Believe it or not, all of this can be done with just the concepts and equations presented for time systems and coordinate system conversions.

There's no need (yet!) to worry about orbital elements because the stars are so far away that we will consider them as stationary in the sense that we will treat them as if they aren't orbiting anything. Of course, that isn't true. Even the most distant celestial objects are in constant motion as they move along their own orbits and as they rotate on their own axes. Nevertheless, we will ignore all that and pretend that they are truly stationary, albeit very distant, celestial objects.

**Figure 5.1** The Mystic Mountain
Located in the Carina Nebula, the Mystic Mountain is an enormous pillar of gas and dust that is 3 light years tall! This image, which is even more stunning in color, was created in 2010 by a team of NASA, European Space Agency, and Space Telescope Science Institute astronomers in celebration of the 20th anniversary of the Hubble Space Telescope. (Image courtesy of NASA/ESA/STScI)

Locating Solar System objects is a different matter. Such objects are clearly not stationary, and so knowing their orbital elements is vital to finding them in the sky. Because they cannot be considered as stationary, calculating the location of objects within the Solar System and understanding how orbital elements fit into the scheme of things must wait a little while longer.

Although this chapter is primarily concerned with deep space objects such as stars, the concepts and methods presented in this chapter are not really limited to "stationary" deep space objects; they can be applied to any celestial object whose equatorial coordinates are known. For example, one could consult an astronomy periodical to get the equatorial coordinates for a planet and then use this chapter's program to locate that planet in horizon coordinates for a given observer's location, date, and time. In later chapters, instead of consulting astronomy periodicals we will directly calculate the equatorial coordinates for the Sun, Moon, and planets for a given point in time. Once we have equatorial coordinates in hand we will then use the results of this chapter to locate such objects in the sky.

Be forewarned that some license is taken in this chapter by omitting or glossing over technical details that might be more confusing than illuminating. This is done to promote conceptual understanding by providing a simple framework in which all the pieces fit. May the purists be forgiving for this simplified treatment! Nonetheless, clean off the table and get some glue ready. The pieces are all assembled. We will now put them together to form a picture of mathematical beauty and elegance. The result should be a much better, albeit still very basic, understanding of how the universe works with regard to orbits.

## 5.1 Locating a Star

Our basic objective in this section is to convert an object's equatorial coordinates (which are fixed because we are assuming the object is stationary) to horizon coordinates for a stated date, time, and location. Once we have an object's horizon coordinates, we then know precisely where to point a telescope to view that object.

To locate a deep space object in horizon coordinates, it is necessary to account for 3 things: date, time of day, and an observer's location. Julian day numbers provide a convenient (well, at least a manageable) method for accounting for dates. Julian day numbers are especially useful for computing elapsed days, and many of the calculations, such as converting UT to GST, do precisely that. Why are elapsed days important?

In order to answer that question, assume Earth is stationary (in the sense that we'll assume it doesn't orbit anything) and that the stars, embedded in the celestial sphere, orbit Earth. If a snapshot is taken of the position of the stars relative to Earth, then in exactly 1 sidereal year the stars will return to precisely the same location in the sky that they were when the snapshot was taken. That is, the stars will have moved through 360° along the celestial sphere. If only half a sidereal year has elapsed, then the stars will have moved only by 180° along the celestial sphere.

The "snapshot" that we have taken is simply the equatorial coordinates of the stars for a stated instant in time (i.e., standard epoch). The amount of elapsed time since we made our snapshot tells us how far along the celestial sphere the stars have moved. Because there are 365.2564 sidereal days in a sidereal year, the stars move along the celestial sphere at a rate of 0.985609° per day. So, if $N$ days have elapsed since we took our snapshot, then the stars will have moved by $0.985609N$ degrees. The date at which we want to know an object's location is thus accounted for by using Julian day numbers to calculate elapsed days, which is precisely what steps 4–7 in converting UT to GST are all about (see section 3.10).

Of course we know that Earth rotates on its axis even if it were not orbiting anything. Because Earth rotates, the time of day affects the apparent motion of the stars in the sky. While the elapsed days tell us how far a star has moved along the celestial sphere, the time of day tells us how far Earth has rotated and must be applied as an adjustment to a star's apparent position in the sky.

Actually, time of day has 3 components that must be considered:

1. Time zone
2. Sidereal time
3. Time of day

First, time zone differences are handled by converting from LCT to UT. Second, our wristwatches measure mean solar time while celestial motion is measured by "star" time. Converting UT to GST accounts for this difference in time units. Third, converting GST to LST accounts for how much Earth has rotated during the fractional part of a day represented by the LCT.

Recall that converting GST to LST (see section 3.12) requires knowing an observer's longitude. An observer's longitude is just another way of expressing how far Earth has rotated from $0^h$ for the calculated UT. The important thing to realize, however, is that the reason for converting from LCT to UT and ultimately to LST is to synchronize clocks so that star time is being used.

Up to this point, calculations have been made to cancel out effects of the date, time of day, and longitude. Only 2 details remain. First, equatorial coordinates are independent of time of day, but they somehow need to be converted to a format that is time (sidereal) dependent. Second, an observer's latitude must be considered. The first detail is resolved by converting an object's right ascension to its equivalent hour angle and using the hour angle to calculate altitude and azimuth. Recall that an hour angle is time varying, which thus provides us with an object's coordinates in a form that is time dependent. Latitude adjustments, the second detail to account for, are made in steps 4–7 of the procedure for converting equatorial coordinates to horizon coordinates (see section 4.7).

Table 5.1 summarizes these steps and explains why each step is needed to locate an object for a given observer's location, local time, and date. The effects of precession, nutation, refraction, and parallax can also be included by applying corrections to the equatorial coordinates before beginning the steps

# Stars in the Nighttime Sky

**Table 5.1** Converting Equatorial to Horizon Coordinates
Converting equatorial to horizon coordinates requires these fundamental steps.

| Calculation | Reason |
|---|---|
| 1. LCT to UT | Account for time zone |
| 2. UT to GST | Synchronize clocks to sidereal time vice solar mean time |
| 3. GST to LST | Account for observer's longitude |
| 4. Date to JD | Calculate elapsed days to account for date |
| 5. Right ascension to hour angle | Make equatorial coordinates time varying |
| 6. Equatorial to horizon | Adjust for observer's latitude |

outlined in the table. Chapters 6, 7, and 8 will demonstrate how to calculate the equatorial coordinates for the Sun, Moon, and planets, which can then be transformed into horizon coordinates using the steps outlined in the table. Once broken down into a series of major steps, the process is not so bewildering after all.

Before working out an example, an important word of warning is in order. Converting UT to GST (step 2 in table 5.1) requires a computation involving the calendar date, as does converting the date to its Julian day number (step 4). It is possible that the calendar date will not be the same for the LCT and UT times. When adding a time zone adjustment to LCT to convert it to UT, the resulting UT might be greater than $24^h$ or less than $0^h$. If the UT computed is greater than $24^h$, then we must subtract $24^h$ from the UT to put it into the proper range. When this occurs, the resulting UT is actually for the next day, so the date must be adjusted accordingly. Likewise, if UT is negative, $24^h$ must be added with the result that the UT is actually for the previous day and the date must be adjusted accordingly. Be sure to watch out for this adjustment in the source code for converting LCT to UT and in the code for converting UT to LCT. As a general rule, time conversions should be done first (as shown in table 5.1) so that the UT date is used for the Julian day number and GST calculations.

Let us proceed with an example and compute the horizon coordinates for Venus. Assume an observer at 38° N latitude, 78° W longitude is in the Eastern Standard Time zone and an astronomy periodical showed that Venus was at right ascension $17^h43^m54^s$, declination $-22°10'00''$ on January 21, 2016. Where did Venus appear at $21^h30^m00^s$ LCT for the observer? Assume that the observer was not on daylight saving time and ignore precession.

1. Convert *LCT* to decimal format.
   (Ans: $LCT = 21.5^h$.)

2. Convert *LCT* to *UT*.
   (Ans: $UT = 2.5^h$, next day so use the date 1/22/2016 for subsequent time-related calculations.)
3. Convert *UT* to *GST*.
   (Ans: $GST = 10.559460^h$.)
4. Convert *GST* to *LST*.
   (Ans: $LST = 5.359460^h$.)
5. Convert $\alpha$ (right ascension) to $H$ (hour angle).
   (Ans: $H = 11.627793^h$.)
6. Convert equatorial coordinates to horizon coordinates.
   (Ans: $h = -73.455227°$, $A = 341.554820°$.)
7. Convert $h$ and $A$ to DMS format.
   (Ans: $h = -73°27'19''$, $A = 341°33'17''$.)

Venus was not visible for the observer at the stated time since its altitude ($h$) was below the observer's horizon. Also notice that the date had to be adjusted in step 2.

Sometimes it is desirable to reverse the process just described and calculate equatorial coordinates for a given set of horizon coordinates. At $21^h45^m00^s$ LCT, the observer from the previous example noticed a bright star at approximately altitude $59°13'$, azimuth $171°05'$. What were the object's equatorial coordinates?

1. Convert *LCT* to decimal format.
   (Ans: $LCT = 21.75^h$.)
2. Convert *LCT* to *UT*.
   (Ans: $UT = 2.75^h$, which is the next day so use the date 1/22/2016 for subsequent time-related calculations.)
3. Convert *UT* to *GST*.
   (Ans: $GST = 10.810145^h$.)
4. Convert *GST* to *LST*.
   (Ans: $LST = 5.610145^h$.)
5. Convert horizon coordinates to equatorial coordinates.
   (Ans: $H = 23.694054^h$, $\delta = 7.498241°$.)
6. Convert $H$ to $\alpha$.
   (Ans: $\alpha = 5.916091^h$.)
7. Convert $\alpha$ and $\delta$ to HMS and DMS format, respectively.
   (Ans: $\alpha = 5^h54^m58^s$, $\delta = 7°29'54''$.)

A star chart shows that Betelgeuse in the constellation Orion (see figure 5.2) is very near that location, so Betelgeuse is most likely what the observer saw.

## 5.2 Star Rising and Setting Times

The preceding section has 1 small problem. The horizon coordinates for a star can be calculated easily enough for a given observer, date, and time, but there is no guarantee that the star will be visible. There are 2 reasons why a star might not be visible. First, it might not be visible because the observer's location is such that the star is never above the horizon, or it may not be above the horizon at the given LCT. For instance, an observer at Earth's South Pole will never be able to see Polaris, the Pole Star, because it is never visible in the Southern Hemisphere. Venus in the example from the previous section was not visible because it was below the observer's horizon at the stated observation time. Another reason a star might not be visible is because it rises and sets during daylight hours. Unless an eclipse occurs, sunlight will prevent one from being able to see a star even if it is above the observer's horizon.

This section will demonstrate how to calculate the rising and setting times for a given set of equatorial coordinates. Once it is known when and if a star will be above the horizon, an observer can choose a convenient viewing time (assuming the star will appear at night!) and then calculate where the star will appear. The following procedure will explicitly tell us when we have chosen a star that will never rise or set for an observer.

Determining star rise and set times is accomplished by first determining a star's LST rising and setting times and then converting the LST times to LCT times. Using the subscript $r$ to denote rising time and the subscript $s$ to denote setting time, the required LST equations are

$$LST_r = 24^h + \alpha - \frac{\cos^{-1}(-\tan\phi \tan\delta)}{15} \qquad (5.2.1)$$

$$LST_s = \alpha + \frac{\cos^{-1}(-\tan\phi \tan\delta)}{15}, \qquad (5.2.2)$$

where $\phi$ is the observer's latitude and $\alpha$ and $\delta$ are the star's equatorial coordinates.

To demonstrate, assume an observer is located at 38° N latitude, 78° W longitude in the Eastern Standard Time zone. What were the rising and setting times for Betelgeuse ($\alpha = 5^h 55^m$, $\delta = 7°30'$) on January 21, 2016? In the following calculations, remember that the subscript $r$ refers to rising times while the subscript $s$ refers to setting times.

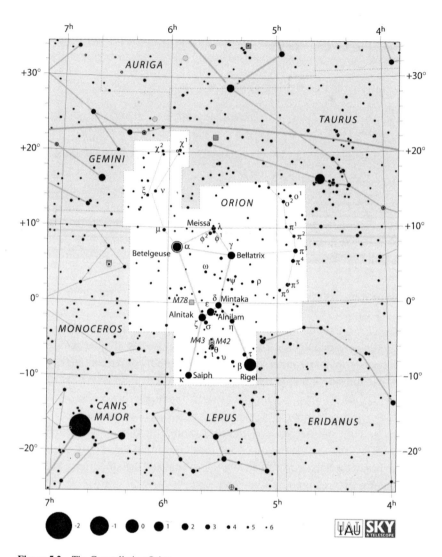

**Figure 5.2** The Constellation Orion
Orion is one of the most easily recognized constellations in the Northern Hemisphere's nighttime sky. The bright star Betelgeuse in Orion's shoulder, which lies some 640 light years away from Earth, is a red supergiant about 20 times more massive than our Sun. (Image courtesy of the IAU and *Sky & Telescope*, Roger Sinnott & Rick Fienberg, released under CC BY 3.0, see http://creativecommons.org/licenses/by/3.0/)

# Stars in the Nighttime Sky

1. Convert $\alpha$ to decimal format.
   (Ans: $\alpha = 5.916667^h$.)
2. Convert $\delta$ to decimal format.
   (Ans: $\delta = 7.5°$.)
3. Convert the observer's latitude to decimal format.
   (Ans: $\phi = 38.000000°$.)
4. Compute the value $A_r = (\sin \delta / \cos \phi)$.
   (Ans: $A_r = 0.165640$.)
5. If $|A_r| > 1$, the star doesn't rise or set.
   (Ans: Star may rise and set since $A_r = 0.165640 < 1$.)
6. $R = \cos^{-1}(A_r)$.
   (Ans: $R = 80.465578°$.)
7. $S = 360° - R$.
   (Ans: $S = 279.534422°$.)
8. $H_1 = \tan \phi \tan \delta$.
   (Ans: $H_1 = 0.102858$.)
9. If $|H_1| > 1$, then the star doesn't rise or set.
   (Ans: Star does rise and set since $H_1 = 0.102858 < 1$.)
10. Calculate $H_2 = \frac{\cos^{-1}(-H_1)}{15}$.
    (Ans: $H_2 = 6.393586^h$.)
11. $LST_r = 24^h + \alpha - H_2$. This is the LST for when the star rises above the observer's horizon.
    (Ans: $LST_r = 23.523081^h$.)
12. If $LST_r > 24^h$, subtract $24^h$.
    (Ans: $LST_r = 23.523081^h$.)
13. $LST_s = \alpha + H_2$. This is the LST for when the star sets below the observer's horizon.
    (Ans: $LST_s = 12.310252^h$.)
14. If $LST_s > 24^h$, then subtract $24^h$.
    (Ans: $LST_s = 12.310252^h$.)
15. Convert $LST_r$ and $LST_s$ to $LCT_r$ and $LCT_s$. These are the star rise and set LCT times.
    (Ans: $LCT_r = 15.679566^h$, $UT_r = 20.679566^h$, $LCT_s = 4.497348^h$, $UT_s = 9.497348^h$.)
16. Convert $LCT_r$ and $LCT_s$ to HMS format.
    (Ans: $LCT_r = 15^h 40^m 46^s$, $LCT_s = 4^h 29^m 50^s$.)

**Figure 5.3** The Horsehead Nebula
The Horsehead Nebula lies just below Alnitak, the easternmost star in Orion's belt. It is far more difficult to view than the Orion Nebula (M42). The Orion Nebula lies south of the middle star in Orion's belt and can be seen with the naked eye whereas the Horsehead Nebula cannot.

In this example, Betelgeuse will rise above the observer's horizon during daylight hours. It will not be visible until the Sun goes down even though it is above the observer's horizon.

There are some important notes to make about this algorithm.

- Steps 4 and 8 are necessary because the arccosine function is not defined for values less than $-1$ or greater than $+1$. The physical interpretation of this fact is that the star never rises or sets for the observer when $A_r$ or $H_1$ is out of range. This does not necessarily mean that the star isn't visible. Polaris, for instance, is always above the horizon for observers in the Northern Hemisphere. This algorithm will indicate that Polaris never rises or sets for our observer, but it will certainly be visible.

- The values $R$ and $S$ appear to be needless calculations. However, if a star does rise and set, $R$ is the star's azimuth when it rises and $S$ is the star's azimuth when it sets. Hence, for this example Betelgeuse will rise at $15^h40^m46^s$ LCT and will appear in the sky at azimuth $80°27'56''$ ($80.465578°$), altitude $0°$. Moreover, Betelgeuse will set at $4^h29^m50^s$ LCT the next day at azimuth $279°32'04''$ ($279.534422°$), altitude $0°$.

- It should be obvious from the calculation for $A_r$ that this algorithm will not work for polar observers (latitude $90°$ N or $90°$ S). This is because $\cos 90° = 0$ so that the value of $A_r$ is undefined. Also note that the algorithm will "blow up" for latitudes close to the poles because as latitude approaches ±90°, the result of the division to produce $A_r$ becomes increasingly large.

This algorithm gives the times when the altitude for a star or other celestial object will be 0°. Because the algorithm does not consider atmospheric refraction or other factors (see section 4.10), the apparent rise and set times will vary from what is calculated. Moreover, objects such as hills may block a view of the horizon.

## 5.3 Creating Star Charts

With the procedures outlined in the previous sections, it is a relatively simple matter to produce star charts for a given date, time, and location. All that is required is to take the equatorial coordinates of stars or other celestial objects, convert them to horizon coordinates, and plot those objects that are above the observer's horizon.

Plotting objects in the sky is a problem that is very similar to making a map of Earth's surface. Both involve converting points on a sphere (3-dimensional space) into points in a plane (2-dimensional space). Unfortunately, mapping points from a sphere onto a plane causes distortion. Techniques for reducing distortion are beyond the scope of this book but can be investigated in texts that deal with cartography.

One way to map points from a 3-dimensional sphere onto a 2-dimensional surface is to project the points on the surface of the sphere into a plane. This can be done by converting each individual point's spherical coordinates to Cartesian coordinates and then simply ignoring the $z$-axis.

To explain the process, consider figure 5.4. Point $P$ is some object located on a sphere of radius $r$. Angle $\theta$, called the azimuthal angle, measures how far around $P$ is in the $xy$ plane while angle $\varphi$, called the polar angle, measures how far down $P$ is from the $z$-axis. Expressed in spherical coordinates, the point $P$ is located at $(r, \theta, \varphi)$.

Now consider figure 5.5, which is the same as figure 5.4 except that the sphere has been removed for clarity and point $P$ is expressed in Cartesian coordinates rather than spherical coordinates. Recall that in the Cartesian coordinate system, an origin is chosen and 3 perpendicular axes are drawn from that origin. We will choose the origin for our Cartesian coordinate system to be the same point as the center of the sphere from figure 5.4. This means that the distance from point $P$ to the origin is $r$, the radius of the sphere from figure 5.4. The location of point $P$ is then expressed as an ordered triple $(x_1, y_1, z_1)$ that indicates how far away point $P$ is from the origin along each of

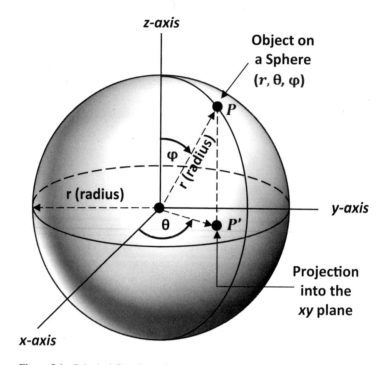

**Figure 5.4** Spherical Coordinate System
The spherical coordinate system locates an object in 3-dimensional space in terms of a radius ($r$), an azimuthal angle ($\theta$), and a polar angle ($\varphi$). The coordinates for point $P$ in this figure are expressed as the ordered triple ($r, \theta, \varphi$).

the 3 axes. $P$ can be projected into the $xy$ plane by simply ignoring the $z$-axis, which is shown as point $P'$ in figure 5.5 and has the 2-dimensional coordinate ($x_1, y_1$).

How do we get the Cartesian coordinates for $P'$ from $P$'s spherical coordinates? We do so by applying 3 equations that relate spherical and Cartesian coordinates:

$$x = r \sin \varphi \cos \theta \quad (5.3.1)$$

$$y = r \sin \varphi \sin \theta \quad (5.3.2)$$

$$z = r \cos \varphi. \quad (5.3.3)$$

Because we are interested in $P'$, the projection of point $P$ into the $xy$ plane, we don't actually need to compute $z$. Now that we can convert spherical coordinates to Cartesian coordinates and project points into a 2-dimensional plane,

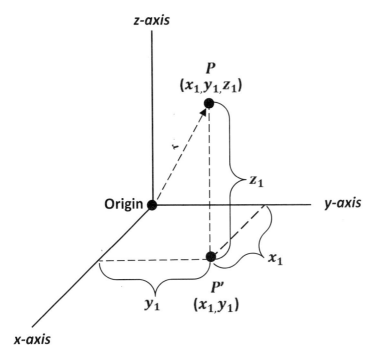

**Figure 5.5** Cartesian Coordinate System
In the Cartesian coordinate system, the location of a point $P$ in 3-dimensional space is expressed as an ordered triple $(x_1, y_1, z_1)$ that indicates how far away from the coordinate system origin $P$ is along each of the 3 axes. $P'$ is the projection of point $P$ into the $xy$ plane.

all that remains is to align our $xy$ plane with the compass directions for a map and relate horizon coordinates to spherical coordinates.

Figure 5.6 shows how the $x$ and $y$ axes from figures 5.4 and 5.5 can be drawn on a flat map with the compass direction north at the top of the map, south at the bottom, west on the left, and east on the right. Our observer is located at the origin. The $x$ and $y$ axes shown without parentheses in figure 5.6 are with respect to the map we are creating with the map's $x$-axis corresponding to east-west and the map's $y$-axis corresponding to north-south. The axes designations in parentheses show how the north-south and east-west directions relate to figures 5.4 and 5.5. Notice that we have chosen to align the $x$-axis of the sphere in figure 5.4 with the $y$-axis (north-south) of figure 5.6, and to align the $y$-axis of the sphere in figure 5.4 with the $x$-axis (east-west) from figure 5.6. This choice is made so that $\theta$ in figure 5.6 is identical to the definition of azimuth in the horizon coordinate system. Furthermore, $\varphi$ and altitude

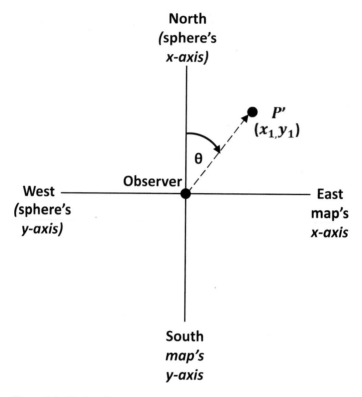

**Figure 5.6** Plotting 3D Points in 2D Space
This figure shows 1 method for plotting 3-dimensional points on the surface of a flat map. After converting spherical coordinates to Cartesian coordinates and performing a projection, the projected points are plotted onto a 2-dimensional map.

are related by the equation

$$\varphi = 90 - h. \tag{5.3.4}$$

For convenience, assume that the radius of our sphere in figures 5.4 and 5.5 is $r = 1$. Then applying equations 5.3.1 and 5.3.2 (and remembering that we have chosen to swap the $x$ and $y$ axes in figure 5.6 from those in figures 5.4 and 5.5), we can derive 2 simple equations for relating an object's horizon coordinates $(h, A)$ to $(x, y)$ coordinates for plotting on a map. The necessary equations are:

$$x = \sin(90° - h) \sin A = \cos h \sin A \tag{5.3.5}$$

$$y = \sin(90° - h) \cos A = \cos h \cos A. \tag{5.3.6}$$

Given these 2 equations, producing star charts is a matter of computing the horizon coordinates, solving the mapping equations, and plotting the resulting $(x, y)$ pairs onto a display. Appropriate scaling factors need to be considered to put each point in its proper place on the display.

The chart produced by this technique will be circular in shape. North will be at the top of the circle, south at the bottom, east to the right, and west to the left. Distortion increases as points get farther away from the center, but the charts produced serve to show the position of constellations relative to each other.

## 5.4 Program Notes

The program **RunChap5** uses the star catalog data files from chapter 1 to produce a star chart for the objects in the selected catalog. Code from previous programs are combined in this chapter's program as the basis for calculating a star's location. In particular, the time conversion routines from chapter 3 are required to convert between LCT, UT, GST, and LST while the coordinate system conversions from chapter 4 are required to convert between equatorial and horizon coordinates. Code is also included to compute precession corrections, although precession corrections are applied only to the specific equatorial coordinates that a user enters. Precession corrections are *not* automatically applied to objects in a star catalog or when creating a star chart. This chapter's program is worth examining since it demonstrates how the previous chapters are tied together.

**RunChap5** uses the methods described in section 5.3 to create a star chart that plots horizon coordinates on a circular plotting surface. The chart created will plot the stars from the currently loaded star catalog as would be seen in the sky for the observer date, time, and location entered into the program. In addition to creating star charts based on horizon coordinates, the program will also create rectangular star charts showing equatorial coordinates. An observer's location is irrelevant to this type of star chart because equatorial coordinates are independent of the date, time, and observer's location.

To avoid entering an observer's latitude, longitude, and time zone each time the program starts, default values can be entered into the DefaultObsLoc.dat data file in the data directory. See the README.TXT file for more details. If the DefaultObsLoc.dat data file is present, it will be read to establish the initial observer latitude, longitude, and time zone each time the program starts.

## 5.5 Exercises

1. An observer is located at latitude 45° N, longitude 100° W in the Pacific Standard Time zone. Assuming the LCT is $9^h00^m00^s$ on December 1, 2015 and the observer is not on daylight saving time, calculate the horizon coordinates for a star at right ascension $6^h00^m00^s$, declination -60°00′00″.
(Ans: $h = -59°41′58″$, $A = 224°15′27″$.)

2. An observer is located at latitude 38.25° N, longitude 78.3° W in the Eastern Standard Time zone. At $21^h00^m00^s$ LCT on June 6, 2015, the observer located an object at altitude 45°00′00″, azimuth 90°00′00″. Assuming this is daylight saving time, what are the object's equatorial coordinates?
(Ans: $\alpha = 16^h14^m42^s$, $\delta = 25°57′41″$.)

3. What are the rising and setting times for the star from problem number 1?
(Ans: Star doesn't rise or set for the observer.)

4. What are the rising and setting times for the star from problem number 2?
(Ans: $LCT_r = 16^h57^m49^s$, $LCT_s = 7^h59^m51^s$.)

# 6 The Sun

Of all the objects in the sky, the Sun is the most important to life on Earth. The Sun is responsible for giving us heat and light without which life as we know it would be impossible. We are able to see the Moon because it reflects light from the Sun. Moreover, we are able to see the planets and other objects in the Solar System because they too reflect light from the Sun.

In this chapter, we will consider the Sun in more detail. Calculations will be presented for determining the Sun's ecliptic coordinates from which we can apply the results of the preceding chapter to locate it in the sky for any observer. Of course, on a clear day the Sun's location is readily apparent, but being able to calculate the Sun's position is important for locating the planets and predicting details about the Moon, such as when eclipses will occur. Other calculations will be presented for determining sunrise and sunset, the equinoxes and solstices, and the distance from Earth to the Sun.

## 6.1 Some Notes about the Sun

Before diving into the mathematics, let's consider some basic facts about the Sun. Our Sun is a star much like the stars we see in the nighttime sky—a fact that is sometimes overlooked because the Sun is so much closer to Earth than any other star. Our Sun belongs to a class of stars called yellow dwarfs, which appear to be a very common type of star in the universe. Astronomers study the Sun, among other reasons, because understanding the processes and physical laws that govern how our Sun operates may help explain how other stars in the universe behave.

The Sun is not a stationary object by any means. It rotates on its axis every 25–30 days. Interestingly, the Sun rotates faster at its equator than at its polar regions, which is a consequence of the Sun being a massive ball of gas instead of a solid object as Earth is. Additionally, the Sun moves in an elliptical orbit around the center of the Milky Way Galaxy, cruising through our galaxy

**Figure 6.1**  Solar Flare
This extraordinary picture taken on August 31, 2012, shows both a solar flare (bright area on the top left) and an enormous solar prominence extending outward from the Sun's surface (bottom left). The prominence shown here erupted, sending hot plasma and electrically charged particles out into space at over 900 miles per second. Although the eruption did not directly strike Earth, it struck a glancing blow that caused an aurora to appear in northern skies on September 3. (Image courtesy of NASA/SDO/AIA/Goddard Space Flight Center)

at 782,000 km/hour (486,000 miles/hour). It takes the Sun 225–250 million years to complete 1 orbit. By comparison, Earth rotates at a speed of 1770 km/hour (1,100 miles/hour) and revolves around the Sun in 365.25 days at 108,000 km/hour (67,000 miles/hour). Lying roughly 93 million miles from Earth, the Sun is nearly perfectly spherical in shape as opposed to the ellipsoid shape of the Earth. Although estimates of its size and mass vary considerably, the Sun's diameter is about 1,391,000 km (864,000 miles), which is 109 times larger than Earth's diameter.

The first person known to have estimated the mass of the Sun was Sir Isaac Newton, who in the third edition of his *Principia Mathematica* estimated the Sun to be 169,282 times more massive than Earth. A more modern estimate is that the Sun's mass is $1.9885 \times 10^{30}$ kg ($2.1919 \times 10^{27}$ tons), making the Sun 333,000 times more massive than Earth and by far the most massive object in our Solar System. In fact, the Sun alone accounts for about 99.86 percent of the total mass in our Solar System. Applying Cecilia Payne-Gaposchkin's groundbreaking research in stellar astronomy,[1] astronomers did a spectral

---

1. In 1925, Payne-Gaposchkin showed that a star's temperature and light spectrum are related. She was the first person to receive a PhD in astronomy from Harvard University and the first woman to chair Harvard's Astronomy Department.

# The Sun

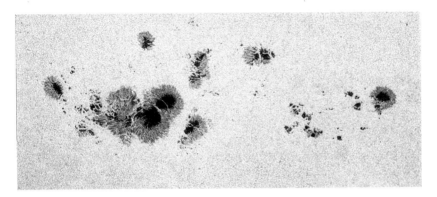

**Figure 6.2** Sunspots
This group of sunspots was photographed in July 2012 and designated as AR11520. These sunspots stretched some 200,000 miles across the Sun's surface. The large sunspot at the bottom left is 11 times larger than Earth. (Image courtesy of NASA Goddard and Alan Friedman, released under CC BY 2.0, see https://creativecommons.org/licenses/by/2.0/)

analysis of sunlight and discovered that the Sun is composed primarily of hydrogen and helium, with hydrogen accounting for three-fourths of the Sun's total mass. Because the Sun is made up mostly of these lighter elements, its mean density is only 1.4 times that of water whereas Earth has a mean density of 5.5 times that of water.

The chemical composition of the Sun is the key to why it gives off heat and why it shines. Our Sun is like a gigantic nuclear furnace with thermonuclear reactions constantly occurring inside it at incredible rates. It is estimated that the Sun converts about 600 million tons of hydrogen into helium each second through the process of nuclear fusion. As a by-product of this ongoing nuclear fusion, the Sun loses about 4 million tons of mass per second as that mass is released into space in the form of light and heat energy. Despite such unimaginable losses in mass each second, it is estimated that the Sun will not run out of fuel for another 5–7 billion years. The heat generated by the Sun's thermonuclear reactions is estimated to be up to 27 million °F at the Sun's core. The Sun's surface is significantly cooler, but it is still a scorching 10,000 °F.

The enormous heat generated by the burning Sun makes it very bright. In terms of visual magnitude, the Sun is magnitude −26.7. By contrast, a Full Moon is magnitude −12.7 while the faintest star visible to the naked eye is about magnitude 6.5. It is very hazardous to look directly at the Sun and even more dangerous to directly view the Sun through a telescope. The only safe way to view the Sun, even during an eclipse, is to project its image onto a screen or use specially designed filters.

When an image of the Sun is projected onto a screen, its surface features become more apparent. Sunspots, which Galileo wrote about in 1610 after viewing them through his telescope,[2] appear on the surface of the Sun as dark blotches. These spots appear dark only because they are cooler than their surrounding environment. Estimated to have temperatures in the range 6,000–7,000 °F, sunspots are indeed substantially cooler than the Sun's surface. Sunspots are typically enormous in size relative to Earth. The largest sunspot ever recorded was in March 1947 and was 40 times larger than Earth!

Sunspots increase and decrease in frequency over a cycle that averages about 11 years, although the reasons why sunspots form and tend to occur over an 11-year cycle are not completely understood. Astronomers do know that sunspots are a result of magnetic storms on the surface of the Sun, and that sunspots usually occur in pairs.

In addition to sunspot activity, bright areas are often observed on the surface of the Sun that may last for a few minutes or for several hours. These bright spots are solar flares and are the largest-known explosions in our Solar System. Solar flares periodically erupt and shoot out from the Sun's surface with great speed, sending out showers of proton particles that reach Earth in a matter of hours. Such showers of protons may cause disruptions in Earth's magnetic field, which can in turn disrupt radio and other forms of electromagnetic communications. Solar flares may also affect Earth's climate.

Landing on the surface of the Sun is impossible since the Sun is not a solid body and because of the Sun's extreme temperatures. If a probe *could* somehow land on the surface of the Sun, it would have to travel at 618 km/second (384 miles/second) to escape the Sun's gravitational pull. By comparison, Earth's escape velocity is a mere 11.19 km/second (7 miles/second).

Although getting close to the Sun's surface is impractical, spacecraft can be sent to within a few million miles of the Sun. Such probes can greatly increase our understanding of the Sun and hence our understanding of the stars. A number of probes have been launched to study the Sun, starting in the 1960s with the Pioneer probes, whose missions were to observe solar flares and other solar-related phenomena. More recently, in 2006 NASA launched the Solar TErrestrial RElations Observatory (STEREO) probes, which produced stunning ultraviolet images of the far side of the Sun that were beamed back to Earth in July 2015. The images returned by STEREO A and B have been combined to produce 3-dimensional views of the Sun, Earth's nearest star.

---

2. No, Galileo did not go blind by looking at the Sun through a telescope! His blindness was caused by cataracts and glaucoma. Still, looking directly at the Sun with or without a telescope is a *very* bad idea!

# The Sun

**Figure 6.3** Total Solar Eclipse
A solar eclipse occurs when the Moon comes between Earth and the Sun. In this photograph of a total solar eclipse that occurred in 1999, the Sun's corona can be clearly seen as an irregularly shaped halo encircling the Sun. (Image courtesy of Oregon State University, released under CC BY-SA 2.0, see http://creativecommons.org/licenses/by-sa/2.0/)

NASA launched the Parker Solar Probe in 2018 to make close-range observations of the Sun and even fly into the Sun's corona, which is the bright halo of light around the Sun that extends for millions of miles into space. The Sun's corona can be most easily observed during a total solar eclipse. Solar Probe Plus will be a truly historic milestone in mankind's exploration of space because it will be our very first visit to a star.

As impressive as the Sun is, there are much larger and more impressive stars in the universe. For example, the red hypergiant VY Canis Majoris in the constellation Canis Major is 17 times more massive than the Sun. As one of the largest stars in our galaxy, VY Canis Majoris is 3,900 light years away from

Earth and has an estimated diameter of 1.2 billion miles, making it 1,420 times larger in diameter than the Sun. If placed at the center of our Solar System, the surface of this massive star would extend beyond Jupiter!

As of 2015, hundreds of stars with diameters significantly larger than that of our Sun have been cataloged. The largest one presently known is UY Scuti in the constellation Scutum. This red supergiant, which is 7–10 times more massive than our Sun, has a diameter of over 1.5 billion miles, making it 1708 times larger than the Sun. If UY Scuti were placed at the center of our Solar System, its surface would extend into the orbit of the planet Saturn.

The most massive star known as of the year 2015 is 265 times more massive than the Sun. It is the star R136a1 in the Tarantula Nebula, which is in the Large Magellanic Cloud Galaxy. The star with the hottest known surface temperature (377, 540 °F) is the star WR 102 in the constellation Cygnus. Compared to stars such as these, our Sun in many respects is an average and somewhat unremarkable star indeed.

One last fun fact: we've been taught since grade school that Earth and the planets revolve around the Sun, but to be technical and precise about the matter, this is not true! Earth and the planets actually revolve around the Solar System's center of mass (its barycenter), and *not* the center of the Sun. To understand the concept of a barycenter, imagine trying to balance a ruler on the end of your finger. The ruler will balance on your finger when there is an equal amount of the ruler's mass on either side of your finger. The position of your finger relative to the ends of the ruler is the barycenter.

To determine the location of the Solar System's barycenter, and hence the real point around which Earth and the planets revolve, we must theoretically know the position of the Sun and every planet, asteroid, comet, and interstellar speck of dust in our Solar System. Because all these objects are in constant motion along their respective orbits, the location of the Solar System's barycenter is constantly changing.

Imagine that Earth and the planets are all arranged in a straight line on the same side of the Sun (say, at perihelion). Then instead of Earth revolving around the Sun's center, Earth would revolve around a point 800,000 km (500,000 miles) above the surface of the Sun! The Solar System's barycenter is frequently located above the surface of the Sun, but it is never beyond the Sun's corona.

Astronomers cannot always ignore the barycenter of a system, particularly when studying deep space objects. There are many instances in which 2 objects in relatively close proximity are sufficiently close to having the same mass that their common center of mass is between them. This means neither object orbits the other; they both orbit a point that lies somewhere between them. When 2 relatively close objects orbit a barycenter located between them, they form what astronomers call a binary system. Binary systems are very common. It is estimated that 50 percent of all stars are part of a binary system and that 80 percent of all stars are part of a multistar system comprised of at least 2 stars.

Situated about 4 light years away, Alpha Centauri in the Centaurus constellation, is the closest star to Earth, other than the Sun. It is the third brightest star in the nighttime sky, with Sirius being the brightest star and Canopus being the second brightest. What appears to be a single star to the naked eye is actually 2 stars, Alpha Centauri A and Alpha Centauri B, that form a binary star system. It may also turn out that another very faint red dwarf star, Proxima Centauri, is gravitationally bound to Alpha Centauri A and B, and if true, they form a nearby 3-star system revolving around a common barycenter that lies between them.

We don't have to look into deep space to find a binary system. Pluto and its moon Charon orbit around a point situated between the 2 of them, thus forming a binary system in our very own Solar System.

We will ignore the Solar System's barycenter and smugly state that Earth and the planets revolve around the Sun. We can safely say so because the Sun is so much more massive than anything else in the Solar System and the distances involved are so great that except for extreme accuracy (such as to detect relativistic effects), the difference between the Solar System's barycenter and the Sun's center is negligible. We will also ignore the barycenter for objects orbiting Earth. Earth is so much more massive than the Moon that the Earth-Moon barycenter is always located below the surface of the Earth.

## 6.2 Locating the Sun

Calculating the position of the Sun, Moon, or a planet may seem a daunting task. Conceptually, however, the process is really quite simple:

- Take a snapshot of the ecliptic coordinates for the object of interest at some convenient instant in time.
- Calculate how many days ($D$), including fractional parts of a day, have elapsed since the snapshot was taken.
- Calculate how far the object has moved along its orbit in $D$ days.
- If necessary, apply corrections, such as precession, to account for irregularities in the object's orbit.
- Convert the corrected ecliptic coordinates to horizon coordinates.

In the Sun's case, this process is further simplified because only the ecliptic longitude ($\lambda$) needs to be calculated. Recall from chapter 4 that the ecliptic

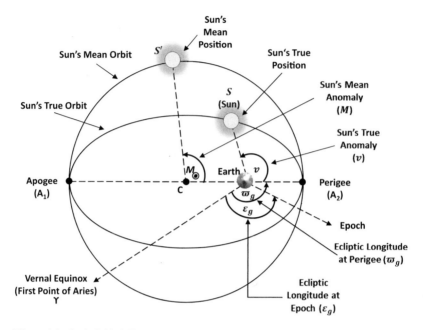

**Figure 6.4** Sun's Orbital Elements
This illustration shows the Sun's orbital elements. Assuming a circular geocentric orbit simplifies calculating the Sun's location.

latitude ($\beta$) of the Sun is 0° because the Earth-Sun orbit lies completely within the ecliptic plane.

Figure 6.4 shows the orbital elements we will use to locate the Sun. For convenience, a geocentric model is chosen in which the Sun ($S$) revolves around Earth in an elliptical orbit, labeled as the *Sun's True Orbit*, with Earth at the occupied focus of that orbit. An imaginary mean Sun ($S'$) is defined that moves in a constant-speed circular orbit, labeled as the *Sun's Mean Orbit*, around Earth. The mean orbit is defined so that its center coincides with the geometric center ($C$) of the true elliptical orbit and its radius is the length of the elliptical orbit's semi-major axis $\overline{CA_2}$. The figure is exaggerated because the true Earth-Sun orbit has an eccentricity of about 0.0167, making the Sun's true orbit much more circular than the figure suggests.

Figure 6.4 should look very familiar because of its close similarity to figure 4.10. However, there are 2 significant differences between the 2 figures. First, figure 6.4 is a geocentric model instead of a heliocentric model and therefore uses terminology appropriate for a geocentric Earth-Sun orbit (such as apogee/perigee rather than aphelion/perihelion, Earth at the occupied focus rather than the Sun). Second, figure 6.4 shows the position of

# The Sun

the vernal equinox and a line labeled *Epoch*, which are used to define 2 new angles, $\varepsilon_g$ and $\varpi_g$. Both of these angles will be discussed later in this section. Since figures 6.4 and 4.10 are so clearly similar, the reader may wish to review section 4.5 to remember why we define a mean anomaly and how it is used.

As shown in figure 6.4, the Sun's mean anomaly $M$ is measured from the moment of perigee. Since Earth completes an orbit of 360° around the Sun in 365.242191 mean days, in a geocentric model the mean Sun moves along its circular orbit by

$$\frac{360°}{365.242191 \text{ days}} \approx 0.985647° \text{ per day.}$$

The mean anomaly is easy to calculate from this relationship because it is simply how far the mean Sun has gone around its mean orbit since the moment of perigee. The mean anomaly is thus given by

$$M = \frac{360° D_p}{365.242191 \text{ days}}, \qquad (6.2.1)$$

where $D_p$ is the number of days (including fractional portions of a day) that have elapsed since the moment of perigee. The subscript $p$ emphasizes that the moment of perigee is being used as a reference point. This equation should also look familiar because it is the same as equation 4.5.2 with the Sun's orbital period substituted for $n$ and the number of elapsed days $D_p$ substituted for $t$.

Take a moment to compare figure 6.4 with figure 4.16, which we used to describe the ecliptic coordinate system. The Sun plays the role of point $P$ (the point whose ecliptic coordinates are desired) in figure 4.16. As should be clear from figure 4.16, the Sun's ecliptic latitude is $\beta = 0°$ because the Sun's orbital plane coincides with the ecliptic plane.

There are 2 problems with using equation 6.2.1 to calculate the Sun's mean anomaly and then applying section 4.5 to derive the Sun's true anomaly. First, the mean anomaly is defined with respect to the moment of perigee, which occurs in early January but does not occur at the same time of day or even on the same day from year to year. How do we relate the varying date and time at which perigee occurs to the date and time at which we want to know the Sun's position? Selecting a specific place in the Sun's orbit (perigee) as a reference is really an arbitrary choice. Instead of using a specific *place* in the Sun's orbit, we could just as easily use a specific *instant in time* (e.g., $12^h$ UT on January 1, 2000) as a reference point. This is precisely what we will do to resolve this first problem.

The second problem is that although we can determine the Sun's true anomaly from its mean anomaly, how do we relate the Sun's true anomaly to its ecliptic longitude? The true anomaly is measured with respect to perigee, but the ecliptic longitude is measured with respect to the First Point of Aries. We will resolve this second difficulty by defining an angle relative to the First Point of Aries that we can combine with the true anomaly to get the Sun's ecliptic longitude. Let's now see how both of these difficulties are resolved.

Figure 6.4 shows the position of a standard epoch, which we are free to choose to be whatever instant in time we wish, and defines a new angle $\varepsilon_g$ as the angular distance from the First Point of Aries to the standard epoch. The subscript $g$ is used to emphasize that we are using a geocentric model. Defined this way, $\varepsilon_g$ is the Sun's ecliptic longitude at the instant in which the Sun is located at the standard epoch. This resolves our first problem because it allows us to compute the Sun's mean anomaly relative to a fixed epoch instead of a varying moment of perigee.

Figure 6.4 defines another angle, $\varpi_g$, which is the ecliptic longitude of the Sun at the moment of perigee for the standard epoch. This resolves our second problem because we can now easily relate the true anomaly and the ecliptic longitude. We merely add $\varpi_g$ to the true anomaly $\upsilon$ to get the ecliptic longitude.

Given $\varepsilon_g$ and $\varpi_g$, we first adjust the mean anomaly so that it is with respect to the moment of perigee rather than the standard epoch. That is, if $M$ is the mean anomaly measured from the standard epoch we have chosen, then the mean anomaly $M_\odot$ measured from the moment of perigee is

$$M_\odot = M + \varepsilon_g - \varpi_g. \tag{6.2.2}$$

The subscript $\odot$ is used to emphasize that we are describing elements of the Sun's orbit.[3]

We can combine equations 6.2.1 and 6.2.2, which gives

$$M_\odot = \frac{360° D_e}{365.242191 \text{ days}} + \varepsilon_g - \varpi_g, \tag{6.2.3}$$

where $D_e$ is the number of days since the standard epoch. Once we have the Sun's mean anomaly, we can solve either the equation of the center or Kepler's equation to obtain the Sun's true anomaly. For the purposes of this chapter, we

---

3. The symbol $\odot$ is frequently used in astronomy to refer to the Sun.

# The Sun

**Table 6.1** Sun's Orbital Elements
This is a geocentric snapshot of the Sun's orbital elements at the standard epoch J2000 as given by *The Astronomical Almanac 2000*.

| Orbital Element | Value |
| --- | --- |
| $e$, eccentricity of the Earth-Sun orbit | 0.016708 |
| $a_0$, length of the Earth-Sun orbital semi-major axis | $1.495985E08$ km |
| $\theta_0$, Sun's angular diameter when a distance of $a_0$ from Earth | 0.533128° |
| $\varepsilon_g$, Sun's ecliptic longitude at the epoch | 280.466069° |
| $\varpi_g$, Sun's ecliptic longitude at perigee at the epoch | 282.938346° |

will approximate the equation of the center by

$$E_c \approx \frac{360°}{\pi} e \sin M_\odot. \tag{6.2.4}$$

The true anomaly is then

$$\upsilon_\odot = M_\odot + E_c. \tag{6.2.5}$$

Once the true anomaly is known, determining the Sun's ecliptic coordinates is a simple matter. The ecliptic latitude is always 0° while the ecliptic longitude, as can be seen from figure 6.4, is simply the true anomaly adjusted by $\varpi_g$. That is,

$$\lambda_\odot = \upsilon_\odot + \varpi_g. \tag{6.2.6}$$

Table 6.1 shows some of the Sun's orbital elements with respect to the standard epoch J2000. *The Astronomical Almanac 2000* provides some useful interpolation equations that allow the Sun's orbital elements to be referenced to another epoch, such as 2010. In *Practical Astronomy with your Calculator or Spreadsheet*, Duffett-Smith also provides interpolation equations that allow $e$, $\varepsilon_g$, and $\varpi_g$ to be determined for a different standard epoch. ($a_0$ and $\theta_0$ do not need to be adjusted because their values are independent of the epoch.) If $JD_e$ is the Julian day number for the desired standard epoch, Duffett-Smith's equations for adjusting $e$, $\varepsilon_g$, and $\varpi_g$ to the standard epoch are:

$$T = \frac{JD_e - 2{,}415{,}020.0}{36{,}525} \tag{6.2.7}$$

$$e = 0.01675104 - 0.0000418T - 0.000000126T^2 \tag{6.2.8}$$

$$\varepsilon_g = 279.6966778 + 36{,}000.76892T + 0.0003025T^2 \tag{6.2.9}$$

$$\varpi_g = 281.2208444 + 1.719175T + 0.000452778T^2. \tag{6.2.10}$$

Note that equation 6.2.7 calculates the number of Julian centuries the standard epoch is from $12^h$ UT (noon) on January 0, 1900.

Using the equation of the center to find the true anomaly, let's now work through a complete example to find the Sun's location on February 5, 2015, at $12^h$ LCT. Assume that an observer is in the Eastern Standard Time zone at 78° W longitude, 38° N latitude and is not on daylight saving time.

1. Use the equations from chapter 3 to perform all needed time conversions. That is, compute UT, GST, and LST from the given LCT. Adjust the date if necessary.

    (Ans: $UT = 17.0^h$, $GST = 2.035013^h$, $LST = 20.835013^h$, Date = 2/5/2015.)

2. Compute the Julian day number $JD_e$ for the standard epoch. Be sure to include the fractional part of the day.

    (Ans: $JD_e = 2,451,545.0$ for J2000.)

3. Compute the Julian day number $JD$ for the desired date. Use the Greenwich date and UT from step 1, *not* the LCT time and date, and be sure to include the fractional part of the day. From step 1, we need the Julian day number for $17.0^h$ UT on 2/5/2015.

    (Ans: $JD = 2,457,059.20833$.)

4. Compute $D_e$, the total number of elapsed days since the standard epoch, by subtracting $JD_e$ from $JD$.

    (Ans: $D_e = 5514.20833$ days.)

5. Use equation 6.2.3 to compute $M_\odot$.

    (Ans: $M_\odot = 5432.592589.°$)

6. Add or subtract multiples of 360° to adjust $M_\odot$ to the range of 0° to 360°. (Hint: use $M_\odot$ mod 360°.)

    (Ans: $M_\odot = 32.592589°$.)

7. Use equation 6.2.4 to approximate the equation of the center.

    (Ans: $E_c = 1.031320°$.)

8. Add $E_c$ to $M_\odot$ to get the true anomaly.

    (Ans: $\upsilon_\odot = 33.623909°$.)

9. Add or subtract multiples of 360° to adjust $\upsilon_\odot$ to the range of 0° to 360°.
    (Ans: no adjustment is necessary.)

10. Add $\upsilon_\odot$ and $\varpi_g$ to get the ecliptic longitude.
    (Ans: $\lambda_\odot = 316.562255°$.)

11. If $\lambda_\odot > 360°$, subtract 360°.
    (Ans: no adjustment is necessary.)

# The Sun

At this point, the Sun's ecliptic coordinates are $\lambda_\odot$ ecliptic longitude, $0°$ ecliptic latitude.

12. Convert the ecliptic coordinates from the prior step to equatorial coordinates.
    (Ans: $\alpha = 21.267801^h$, $\delta = -15.872529°$.)
13. Finally, convert the equatorial coordinates to horizon coordinates. (Remember that converting equatorial to horizon coordinates requires that the right ascension $\alpha$ from the previous step be converted to an hour angle.)
    (Ans: $h = 35°47'01''$, $A = 172°17'46''$.)

As we pointed out earlier, we can solve either the equation of the center or Kepler's equation to obtain the true anomaly from the mean anomaly. Let's repeat the previous example to determine the Sun's ecliptic coordinates by using the simple iterative scheme from subsection 4.5.5 to solve Kepler's equation. Steps 1–6 of the algorithm above are identical whether solving the equation of the center or solving Kepler's equation. Thus, our objective is to solve Kepler's equation to determine the true anomaly from the mean anomaly found in step 6 ($M_\odot = 32.592589°$).

Recall from subsection 4.5.5 that we will iteratively compute the equation

$$E_i = M_r + e \sin E_{i-1},$$

where both the eccentric anomaly and mean anomaly are expressed in *radians*, not *degrees*. Thus, we must first convert $M_\odot$ from degrees to radians, giving us

$$M_r = \frac{M_\odot \pi}{180°} = \frac{32.592589° \pi}{180°} \approx 0.568848 \text{ radians}.$$

This gives us a first estimate of the eccentric anomaly as $E_0 = M_r = 0.568848$ radians. The next few iterations follow, where $\Delta$ is the difference between successive approximations to the eccentric anomaly.

$E_1 \approx 0.577848$, $\Delta = 0.009000$,

$E_2 \approx 0.577974$, $\Delta = 0.000126$,

$E_3 \approx 0.577976$, $\Delta = 0.000002$.

Only 3 iterations are necessary to obtain a sufficiently accurate approximation for the eccentric anomaly (0.577976 radians, which is $\approx 33.115588°$). The rapid convergence to a solution is achieved because the Earth-Sun orbital eccentricity is so small.

Given the eccentric anomaly, we can obtain the true anomaly by applying equation 4.5.8, which is

$$\tan\left(\frac{v_\odot}{2}\right) = \left(\sqrt{\frac{1+e}{1-e}}\right)\tan\left(\frac{E}{2}\right).$$

Doing so gives $v_\odot \approx 33.642307°$. This differs from the result in step 8 by about $0.018°$, or by a little over 1 arcminute.

Solving Kepler's equation replaces steps 7 and 8, after which we merely continue with steps 9–13 to determine the Sun's horizon coordinates. Using the true anomaly obtained by solving Kepler's equation instead of the value from the equation of the center, we obtain

$\lambda_\odot = 316.580653°,$

$\alpha = 21.269018^h, \delta = -15.867003°,$

$h = 35°47'13'', A = 172°16'25''.$

The method presented here for determining the position of the Sun, whether by solving the equation of the center or Kepler's equation, is usually accurate to within about $1'$, which is sufficient for the purposes of this book. For the best accuracy, however, one should use a standard epoch as close as possible to the date for which the Sun's position is desired, or use equations 6.2.7 through 6.2.10 to adjust the Sun's orbital elements to the desired date.

More accurate algorithms can be found from other sources, such as *The Astronomical Almanac* or Meeus's *Astronomical Algorithms*. Although more accurate algorithms exist, the method presented here is sufficient to demonstrate how orbital elements, the anomalies, and the various coordinate systems are interrelated.

## 6.3 Sunrise and Sunset

As with locating a star in horizon coordinates, there is no guarantee that the Sun will be visible in the sky at the given date, time, and location. For example, let's use the algorithm from the previous section (and solve the equation of the center) to compute the Sun's location at $20^h$ LCT on February 5, 2015, for an observer in the Eastern Standard Time zone at 78° W longitude, 38° N latitude. Assume the observer was not on daylight saving time. For that observer, date, and time, the Sun's horizon coordinates were $h = -28°01'04''$, $A = 271°26'04''$, which was well below the observer's horizon. This should not be a surprise because it is easy to know when the Sun will or will not

# The Sun

be visible. Quite simply, the Sun cannot be seen during the LCT evening hours.

Determining the times for sunrise and sunset is not as easy as determining when a star will rise or set. The reason is that the coordinates calculated for the Sun are good only for the specific time chosen. For stars we applied 2 equations (5.2.1 and 5.2.2) to a star's fixed equatorial coordinates to determine its rising and setting LST times. However, since the Sun moves rather rapidly across the sky, its equatorial coordinates at sunrise can differ substantially from its coordinates at sunset. So, the simple approach we used for stars based on a single set of fixed coordinates will not work.

Our strategy for handling the Sun's rapidly changing coordinates is to first calculate its position at midnight for the date in question and then again at midnight the next day. Given the Sun's ecliptic longitude $\lambda_1$, its ecliptic longitude $24^h$ later is

$$\lambda_2 = \lambda_1 + 0.985647° \qquad (6.3.1)$$

because the Sun advances in its orbit by $0.985647°$ per day. Using these 2 locations $\lambda_1$ and $\lambda_2$, we will next apply the procedure from chapter 5 to compute 2 sets of rising and setting times, *ST1* and *ST2*. Finally, we will interpolate these 2 sets of rising and setting times to arrive at an interpolated sunrise and sunset time. The interpolation equation required is

$$T = \frac{24.07 ST1}{24.07 + ST1 - ST2}. \qquad (6.3.2)$$

Let's work through an example by calculating sunrise and sunset on February 5, 2015, for an observer at 78° W longitude, 38° N latitude in the Eastern Standard Time zone. Assume the observer is not on daylight saving time.

1. Calculate the Sun's ecliptic location at midnight ($UT = 0^h$) for the date in question.
   (Ans: $\lambda_1 = 315.844356°$, $\beta_1 = 0°$.)

2. Convert the Sun's ecliptic coordinates to equatorial coordinates.
   (Ans: $\alpha_1 = 21.220290^h$, $\delta_1 = -16.086932°$.)

3. Using the equatorial coordinates from the previous step, treat those coordinates as if they were a star and apply equations 5.2.1 and 5.2.2 to compute $ST1_r$ and $ST1_s$, the Sun's LST rising and setting times for the first set of coordinates.
   (Ans: $ST1_r = 16.088377^h$, $ST1_s = 2.352202^h$.)

4. Use equation 6.3.1 to calculate the Sun's ecliptic coordinates $24^h$ later.
   (Ans: $\lambda_2 = 316.830003°$, $\beta_2 = 0°$.)

5. If $\lambda_2 > 360°$, subtract $360°$.

   (Ans: no adjustment is required.)

6. Convert the ecliptic coordinates from the previous step to equatorial coordinates.

   (Ans: $\alpha_2 = 21.285495^h$, $\delta_2 = -15.791967°$.)

7. Using the equatorial coordinates from the previous step, calculate $ST2_r$ and $ST2_s$, the Sun's LST rising and setting times for the second set of coordinates.

   (Ans: $ST2_r = 16.136536^h$, $ST2_s = 2.434454^h$.)

8. Use equation 6.3.2 to interpolate the 2 sets of LST rising times.

   (Ans: $T_r = 16.120631^h$.)

9. Interpolate the 2 sets of LST setting times.

   (Ans: $T_s = 2.360268^h$.)

10. Convert the LST times to their corresponding LCT times.

    (Ans: $LCT_r = 7.298498^h$, $LCT_s = 17.510180^h$.)

11. Convert the LCT times to HMS format.

    (Ans: $LCT_r = 7^h 18^m$, $LCT_s = 17^h 31^m$.)

The calculated sunrise and sunset times may not match what is observed for the same reasons that the star rise and set times from chapter 5 may not match actual observations. Additionally, atmospheric conditions create another effect that is particularly observable with the Sun. This effect is what we call dawn in the morning and twilight in the evening. When the Sun is less than 18° below the horizon at sunrise, light is reflected by the atmosphere so that we see sunlight before the Sun actually rises. The same situation occurs at sundown when the Sun is no more than 18° below the horizon.

Calculating the LCT times of sunrise and sunset over a period of several months illustrates very clearly that days as defined by the apparent Sun are not equal in length. An interesting experiment to perform is to plot the length of daylight hours throughout the year and observe when the days are the shortest and when they are the longest.

## 6.4 Equinoxes and Solstices

As we noted earlier, the ecliptic plane is inclined with respect to the celestial equator (see figure 6.5 and compare it to figure 4.16) by approximately 23°26'. Because the Earth-Sun orbit is inclined with respect to the celestial equator, we can use the geometry of that fact to identify 4 specific points in the Earth-Sun orbit: 2 equinoxes and 2 solstices.

# The Sun

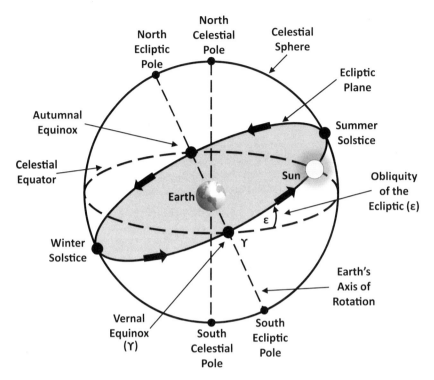

**Figure 6.5** Equinoxes and Solstices
Equinoxes and solstices mark the beginning and end of the seasons.

Referring to the geocentric Earth-Sun model in figure 6.5, it is clear that the Sun crosses the celestial equator twice a year in its orbit around Earth. The 2 times at which those crossings occur are the equinoxes, which take place at about March 21 and September 22. The March equinox, known as the vernal or spring equinox, is the point at which the Sun transitions from being south of the celestial equator to north of the celestial equator. The September equinox is called the autumnal or fall equinox, and it is 180° around the ecliptic plane from the vernal equinox. The autumnal equinox is the point at which the Sun transitions from being north of the celestial equator to south of the celestial equator. There are 2 interesting facts to note about the equinoxes. First, the Sun's right ascension is $0^h$ at the vernal equinox while its right ascension is $12^h$ at the autumnal equinox. Second, at the times of the equinoxes, the lengths of day and night are very close to the same. This can be shown quite easily by comparing the sunrise and sunset times around March 21 and September 22.

The solstices occur when the Sun has no apparent northward or southward motion. That is, the solstices occur when the Sun is at its highest point above the plane of the celestial equator and when it is at its lowest point below the celestial equator. Those 2 times occur at about June 22 (summer solstice) and December 22 (winter solstice). The summer solstice is the longest day of the year while the winter solstice is the shortest day of the year. The word "day" is being used here in the sense of the number of daylight hours.

Both the solstices and equinoxes play a practical role in our calendar system because passage through a solstice or equinox defines the beginning of a season. The March equinox marks the beginning of spring, the June solstice marks the beginning of summer, the September equinox marks the beginning of fall, and the December solstice marks the beginning of winter.

Although it may not be readily apparent from figure 6.5, the Sun's ecliptic longitude at the equinoxes and solstices is always an integer multiple of 90° (0°, 90°, 180°, or 270°). This fact could theoretically be exploited to calculate the precise moment at which the equinoxes and solstices occur by working backward from the Sun's ecliptic coordinates to get the true anomaly, then the mean anomaly, and from there a Julian date. However, taking such an approach requires a great deal of mathematical expertise.

Instead of attempting to solve such a difficult mathematical problem, an easier approach is to collect several years of equinox/solstice data and apply curve fitting to create a function, based on historical data, that approximates when an equinox or solstice will occur. The curve-fitting functions we will use are derived from those presented by Meeus in *Astronomical Formulae for Calculators*, 3rd edition. His 4 equations are:

March Equinox:

$$JD_M = 1{,}721{,}139.2855 + 365.2421376Y + 0.0679190T^2 - 0.0027879T^3$$

(6.4.1)

June Solstice:

$$JD_J = 1{,}721{,}233.2486 + 365.2417284Y - 0.0530180T^2 + 0.0093320T^3$$

(6.4.2)

September Equinox:

$$JD_S = 1{,}721{,}325.6978 + 365.2425055Y - 0.126689T^2 + 0.0019401T^3$$

(6.4.3)

# The Sun

**Table 6.2** Equinoxes and Solstices
This table shows the 2004 equinoxes and solstices as calculated by the equations in this book and by the more accurate equations in *Astronomical Algorithms*.

|                    | This Book                    | Meeus                        | Δ       |
|--------------------|------------------------------|------------------------------|---------|
| March Equinox      | 3/20 at $6^h42^m36^s$        | 3/20 at $6^h49^m42^s$        | $7^m$   |
| June Solstice      | 6/21 at $0^h49^m41^s$        | 6/21 at $0^h57^m57^s$        | $8^m$   |
| September Equinox  | 9/22 at $16^h27^m20^s$       | 9/22 at $16^h30^m54^s$       | $4^m$   |
| December Equinox   | 12/21 at $12^h44^m22^s$      | 12/21 at $12^h42^m40^s$      | $2^m$   |

*Note*: Meeus' results have been converted to UT.

December Solstice:

$$JD_D = 1{,}721{,}414.3920 + 365.2428898Y - 0.0109650T^2 - 0.0084885T^3,$$

(6.4.4)

where $Y$ is the year in question and $T$ is the year divided by 1000. The result of each equation is a Julian day number.

Let us calculate the solstices and equinoxes for the year 2010. The previous 4 equations are accurate only to about 15–20 minutes, although the intermediate calculations that follow below will be given to the nearest second.

1. Calculate $T = (Year/1000)$.
   (Ans: $T = 2.01$.)
2. Using equations 6.13 through 6.16, compute the Julian day numbers.
   (Ans: March $JD_M = 2{,}455{,}276.23384$, June $JD_J = 2{,}455{,}368.98427$, September $JD_S = 2{,}455{,}462.63777$, December $JD_D = 2{,}455{,}552.48727$.)
3. Convert the Julian day numbers from the previous step to calendar dates.
   (Ans: $JD_M$ gives 3/20.733836/2010, $JD_J$ gives 6/21.484267/2010, $JD_S$ gives 9/23.137774/2010, $JD_D$ gives 12/21.987267/2010.)
4. Convert the day part of the dates to a day and decimal format UT.
   (Ans: March: day 20, $17.612067^h$; June: day 21, $11.622418^h$; September: day 23, $3.306565^h$; December: day 21, $23.694398^h$.)
5. Convert the UT results to HMS format.
   (Ans: March equinox occurred 3/20/2010 at $17^h37^m$, June solstice occurred 6/21/2010 at $11^h37^m$, September equinox occurred 9/23/2010 at $3^h18^m$, December solstice occurred 12/21/2010 at $23^h42^m$.)

Meeus gives a more accurate set of curve-fitting equations in *Astronomical Algorithms*, but those equations require considerably more computation than the ones presented here. For example, table 6.2 shows the equinoxes and solstices for the year 2004 as calculated by equations 6.4.1–6.4.4 and as calculated by Meeus's more accurate equations. For the purposes of this book, the additional calculations required for greater accuracy are not warranted, especially when the gain in accuracy is only a few minutes.

## 6.5 Solar Distance and Angular Diameter

The Sun is about 93 million miles away, but its distance from Earth varies throughout the year because Earth's orbit is elliptical in shape. If Earth's orbit were a perfect circle, the Earth-Sun distance would be constant. At first glance, it might seem that the varying Earth-Sun distance is the reason for our seasons. It is true that distance from the Sun is an important factor in determining how much heat reaches a planet's surface. For example, Mercury's surface is very hot because its orbit is so close to the Sun whereas Pluto's surface is very cold since its orbit is so far away from the Sun. However, Earth's seasons are more attributable to the inclination of the axis of rotation with respect to Earth's orbit than to the varying Earth-Sun distance.

The heliocentric model in figure 6.6 shows Earth orbiting the Sun in an elliptical orbit. The figure is highly exaggerated to make it easier to see the relationship of Earth's rotational axis, shown as a line extending through the Earth, to Earth's orbit. The axis of rotation is not perpendicular to the plane of Earth's orbit, but is inclined by approximately 23°26′. As we have noted before, this angle is the obliquity of the ecliptic.

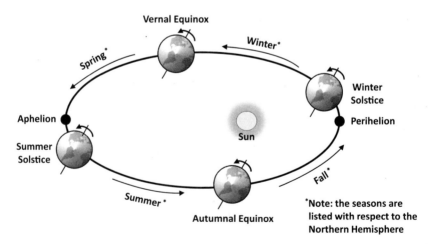

**Figure 6.6** Earth's Seasons
Our seasons are more attributable to the tilt of Earth's axis than the varying Earth-Sun distance.

# The Sun

Figure 6.6 also shows where Earth is in its orbit for the equinoxes and solstices. It may seem surprising that the solstices do not coincide with perihelion and aphelion, or that the equinoxes do not lie on the orbit's semi-major axis. This is because the solstices and equinoxes are defined relative to when Earth crosses the plane of the celestial equator, not with respect to the apsides.

The seasons labeled in figure 6.6 are for the Northern Hemisphere. Since Earth's axis of rotation is not perpendicular to Earth's orbital plane, the Sun shines more directly on the Northern Hemisphere during 1 portion of the year (summer) than 6 months earlier or later (winter). The result is that more sunlight reaches the Northern Hemisphere during the summer, which results in warmer weather than in winter even though Earth is actually farther away from the Sun during summer than it is in winter.

Figure 6.6 also explains why seasons in the Southern Hemisphere occur at opposite times during the year from when they occur in the Northern Hemisphere. For example, it is clear from the figure that the Sun shines more directly on the Southern Hemisphere when it is winter in the Northern Hemisphere and vice versa. Perhaps less commonly known is that summer in the Northern Hemisphere lasts 2–3 days longer than in the Southern Hemisphere. This is a direct consequence of Kepler's second law, which in essence states that Earth moves faster along its orbit near perihelion than when it is near aphelion.

Because the distance to the Sun varies throughout the year, the Sun's apparent size varies too. As figure 6.7 illustrates, apparent size, which is synonymous with angular diameter and angular size, is an angular measurement of how large an object appears to be when viewed from some distance away. If an object of diameter $d$ is viewed from a distance $D$, then the object's angular diameter is given by the equation

$$\theta = 2\tan^{-1}\left(\frac{d}{2D}\right), \tag{6.5.1}$$

where $d$ and $D$ must be expressed in the same units (e.g., km).

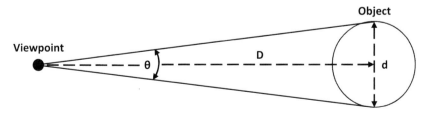

**Figure 6.7** Defining Angular Diameter
The angular diameter, or apparent size, is expressed in degrees and measures how large an object appears to be when viewed from some distance away.

For example, the Sun's diameter is approximately 1,391,000 km. From table 6.1, the Earth-Sun orbital semi-major axis is $1.495985E08$ km in length. So, applying equation 6.5.1, when the Sun is that distance away from Earth, its apparent size is $0.532745°$. This differs from the value given in table 6.1 because the astronomers who provided the value in the table used more precise measurements for the angular diameter and distances involved.

Equation 6.5.1 can be rewritten as

$$d = 2D \tan\left(\frac{\theta}{2}\right), \qquad (6.5.2)$$

which allows one to find an object's diameter when its angular diameter and distance away have been measured or otherwise determined. Thus, using equation 6.5.2 with the angular diameter ($\theta_0$) and distance ($a_0$) from table 6.1, one can calculate that the Sun's diameter is approximately 1,392,000 km.

Note that equation 6.5.1 requires that we know both an object's diameter and its distance away. Although the Sun's diameter can be considered as constant, its distance varies significantly throughout the year so that equation 6.5.1 is not very convenient for determining the Sun's angular diameter. However, once we know the Sun's true anomaly so that we know where it is in its orbit, the distance to the Sun and its angular diameter can be calculated as by-products from calculating the Sun's position. The equations required are

$$Dist_\odot = \frac{a_0(1-e^2)}{1+e\cos v_\odot} \qquad (6.5.3)$$

and

$$\theta_\odot = \frac{\theta_0\left(1+e\cos v_\odot\right)}{1-e^2}, \qquad (6.5.4)$$

where $a_0$ is the length of the semi-major axis of the Earth-Sun orbit, $\theta_0$ is the angular diameter of the Sun when it is $a_0$ distance away, and $e$ is the orbital eccentricity. $a_0$, $\theta_0$, and $e$ are listed in table 6.1. If we let

$$F = \frac{1+e\cos v_\odot}{1-e^2}, \qquad (6.5.5)$$

then equations 6.5.3 and 6.5.4 can be rewritten in the equivalent forms

$$Dist_\odot = \frac{a_0}{F} \qquad (6.5.6)$$

and

$$\theta_\odot = \theta_0 F. \qquad (6.5.7)$$

# The Sun

Equation 6.5.3 should look familiar. We saw it earlier in subsection 4.5.1 when the true anomaly concept was first introduced. The equation holds for any object in an elliptical orbit and expresses the relationship between an orbiting object's true anomaly and the characteristics of its orbit (specifically, the orbital eccentricity and the semi-latus rectum).

To illustrate, let us find the distance to the Sun and its angular diameter on February 15, 2015.

1. Using the steps in section 6.2, compute the Sun's true anomaly for the given date at $0^h$ UT. For this example, we'll solve the equation of the center rather than Kepler's equation and assume that the observer is at $0°$ latitude, $0°$ longitude.
   (Ans: $\upsilon_\odot = 43.025813°$.)
2. Compute $F$ from equation 6.5.5.
   (Ans: $F = 1.012497$.)
3. Use the values from table 6.1 and equation 6.5.6 to compute the distance to the Sun.
   (Ans: $Dist_\odot = 1.478 E08$ km.)
4. Use equation 6.5.7 to compute the Sun's angular diameter.
   (Ans: $\theta_\odot = 0.539790°$.)
5. Convert distance to miles and angular diameter to DMS format if desired.
   (Ans: $Dist_\odot = 9.181 E07$ miles, $\theta_\odot = 0°32'$.)

## 6.6 Equation of Time

Recall from section 3.1 that the apparent motion of the Sun can be used to define a solar day whose time can be measured by a sundial. However, in a geocentric model the motion of the Sun as it orbits Earth is irregular throughout the year. For this reason, the motion of a mean Sun is used to define a mean solar day. The amount by which time for a mean solar day and time for an apparent solar day differ varies throughout the year. The difference between these 2 times is the equation of time. Mathematically, the relationship between apparent solar time ($T_\odot$), mean solar time ($T_{M_\odot}$), and the equation of time ($\Delta T$) is

$$\Delta T = T_{M_\odot} - T_\odot. \tag{6.6.1}$$

Finding the equation of time is straightforward. By definition, the mean Sun transits an observer's horizon at exactly $12^h$, which gives us a value for $T_{M_\odot}$ at a specific instant in time. All that is needed now is to find the equivalent solar time at that same instant. This can be done by recalling that the right ascension and sidereal time are the same at transit. Therefore, finding the Sun's right ascension when it transits an observer's horizon will give the GST from which the UT can be calculated. The equation of time is the difference between $12^h$ and the calculated UT.

An alternative way to find the equation of time is to compute it directly from the Sun's ecliptic longitude $\lambda_\odot$ and mean anomaly $M_\odot$. Adapted from W. M. Smart's equation in *Textbook on Spherical Astronomy*, $\Delta T$ is given by

$$-\Delta T = y \sin(2\lambda_\odot) - 2e \sin M_\odot + 4ey \sin M_\odot \cos(2\lambda_\odot)$$
$$- \frac{y^2}{2} \sin(4\lambda_\odot) - \frac{5e^2}{4} \sin(2M_\odot), \tag{6.6.2}$$

where $y = \tan^2\left(\frac{\varepsilon}{2}\right)$ and the resulting $\Delta T$ is in radians.

Table 6.1 gives Earth's orbital eccentricity $e$ while an acceptable value for the obliquity of the ecliptic $\varepsilon$ is 23.439292°, which was its value at the standard epoch J2000. For better accuracy, one can use JPL's equation presented in section 4.8 to compute $\varepsilon$ for any particular date. It is important to note that equation 6.6.2 gives the equation of time in radians, which must be converted to degrees and then to hours.

To illustrate, calculate the equation of time for May 5, 2016.

1. Compute the Sun's ecliptic longitude and mean anomaly for the given date. For this example, we will solve the equation of the center rather than Kepler's equation.
   (Ans: $\lambda_\odot = 44.954290°$, $M_\odot = 120.363970°$.)

2. Use equation 4.8.2 to compute the obliquity of the ecliptic for the given year. (An acceptable degree of accuracy can be obtained by using $\varepsilon = 23.439292°$ rather than computing it.)
   (Ans: $\varepsilon = 23.437212°$.)

3. Compute $y = \tan^2\left(\frac{\varepsilon}{2}\right)$.
   (Ans: $y = 0.043027$.)

4. Using equation 6.6.2, compute $\Delta T$ in radians.
   (Ans: $\Delta T = 0.014500$ radians.)

5. Multiply by $-1$ to account for the minus sign on the left side of equation 6.6.2.
   (Ans: $\Delta T = -0.014500$ radians.)

6. Convert radians to degrees.
   (Ans: $\Delta T = -0.830768°$.)

7. Convert degrees to hours by dividing by 15.
  (Ans: $\Delta T = -0.055385^h$.)
8. Convert the equation of time to HMS format if desired.
  (Ans: $\Delta T = -0^h03^m19^s$.)

For this example, $-0^h03^m19^s$ must be subtracted from "sundial" time to get the corresponding "wristwatch" time.

## 6.7 Program Notes

The program for this chapter contains all the algorithms described in this chapter for calculating the Sun's location, sunrise and sunset, the equinoxes and solstices, and the equation of time. The program also includes an option to select whether the equation of the center or Kepler's equation will be used to calculate the Sun's position.

## 6.8 Exercises

For the following problems, use the Sun's orbital elements for the standard epoch J2000.

1. An observer is located at 95° W longitude, 30° N latitude within the Central Standard Time zone. The date is August 9, 2000, and the observer is on daylight saving time. If the LCT is $12^h$, what are the Sun's ecliptic coordinates? What are the Sun's equatorial coordinates? What are the Sun's horizon coordinates? Solve the equation of the center for this problem.
  (Ans: $\beta_\odot = 0°$, $\lambda_\odot = 137.386004°$, $\alpha = 9.322193^h$, $\delta = 15.623648°$, $h = 65°43'$, $A = 121°34'$.)

2. What are the LCT sunrise and sunset times for the previous observer?
  (Ans: $LCT_r = 6^h46^m$, $LCT_s = 20^h05^m$.)

3. Another observer is located at 30° W longitude, 20° S latitude within the Eastern Standard Time zone. Assume the observer is on daylight saving time. If the date is May 6, 2015, and the LCT is $14^h30^m00^s$, what are the Sun's ecliptic, equatorial, and horizon coordinates? Use the Newton/Raphson method to solve Kepler's equation for this problem.
  (Ans: $\beta_\odot = 0°$, $\lambda_\odot = 45.917857°$, $\alpha = 2.896770^h$, $\delta = 16.603118°$, $h = 13°34'$, $A = 293°37'$.)

4. What are the LCT sunrise and sunset times for the previous observer?
   (Ans: $LCT_r = 4^h24^m$, $LCT_s = 15^h30^m$.)

For the remaining problems, when computing the Sun's location, solve the equation of the center rather than Kepler's equation.

5. For the year 2010, what were the UT times for the solstices and equinoxes?
   (Ans: March equinox: March 20 at $17^h37^m$; June solstice: June 21 at $11^h37^m$; September equinox: September 23 at $3^h18^m$; December solstice: December 21 at $23^h42^m$.)

6. On August 9, 2015, how far away was the Sun? What was its angular diameter?
   (Ans: $Dist_\odot = 1.517E08$ km or $9.425E07$ miles, $\theta_\odot = 0°32'$.)

7. On May 6, 2010, how far away was the Sun? What was its angular diameter?
   (Ans: $Dist_\odot = 1.509E08$ km or $9.377E07$ miles, $\theta_\odot = 0°32'$.)

8. On August 9, 2015, what was the equation of time?
   (Ans: $\Delta T = 0^h05^m37^s$.)

9. On May 6, 2010, what was the equation of time?
   (Ans: $\Delta T = -0^h03^m21^s$.)

10. On January 1, 2020, what will be the estimated equation of time?
    (Ans: $\Delta T = 0^h03^m07^s$.)

# 7 The Moon

This chapter is devoted to Earth's only natural satellite—the Moon. Our Moon is by far the most visible celestial object in the night sky. Unlike most other objects that grace the nighttime sky, it is quite often visible during daytime hours. The Moon's domination of the nighttime sky and its quiet beauty have caused it to be a source of wonder throughout the ages. Even when viewed with the naked eye, the Moon is an interesting object to behold. Its splendor is greatly enhanced by a low-powered telescope or a good pair of binoculars.

This chapter will demonstrate how to calculate the position of the Moon. This may seem to be more trouble than it's worth since the Moon's location can be readily determined by simply looking up into the sky. However, calculating the Moon's position is required for predicting eclipses and determining when the Moon will rise and set. This chapter will show how to calculate the Moon's rising and setting times, but it will only describe predicting eclipses in general terms. This chapter will also show how to calculate the distance to the Moon and its angular diameter.

The historical definition of a month is closely tied to the Moon. The reason a month is divided into roughly 4 weeks is related to the 4 phases of the Moon. This chapter will discuss different ways of describing the Moon's phases and show how to predict the lunar phases. It will conclude with a general discussion about eclipses and some of the rules that govern when eclipses must and cannot occur.

## 7.1 Some Notes about the Moon

Although the Moon is the only natural satellite orbiting Earth, it is hardly the only moon in our Solar System. Pluto and all full-fledged planets in our Solar System, except Mercury and Venus, also have moons. Observing the motion of

152                                                                                      Chapter 7

**Figure 7.1** The Moon from Apollo 11
This photograph was taken from Apollo 11 during its journey homeward when the spacecraft was about 10,000 nautical miles away from the Moon. (Image courtesy of NASA)

the moons of Jupiter is part of what convinced Galileo that the Sun is the center of our Solar System. However, as Isaac Asimov once observed, our Moon may have been a contributing factor for the delay in the widespread acceptance of a heliocentric Solar System. Since the Moon obviously orbits Earth, it could be argued, then so must all the other objects within the Solar System and the universe.

The Moon is roughly 240,000 miles away from Earth. At perigee, it is 225,700 miles away while at apogee it is 252,000 miles away. With a diameter of nearly 2,160 miles, the Moon is one-fourth the size of Earth and has a mass 0.0123 times that of Earth. Because it has such a low mass, the Moon's gravitational field is too weak to hold much of an atmosphere. Hence, the Moon is a place with no sound because there is no atmosphere to carry sound waves. Shadows on the Moon are very dark and sharply defined because there is no atmosphere to scatter light rays. There is no weather to alter the Moon's landscape, so its surface remains unchanged over eons of time except for the impact of meteoroids and the occasional Moonquake, or the infrequent visits by humans and our machines.

The absence of an atmosphere is largely responsible for the very hot lunar days and very cold lunar nights. With no atmosphere to shield the Moon from the intense solar rays that reach it, surface temperatures during the day may soar to 250 °F while with no atmosphere to hold in heat, nighttime surface temperatures may drop to −300 °F. In 2009 NASA's Lunar Reconnaissance Orbiter (LRO) measured the temperature at the bottom of some of the craters at the Moon's North and South Poles to be −413 °F, which is colder than the surface temperature of Pluto and the lowest naturally occurring temperature ever recorded anywhere in our Solar System.

Lunar days with their searing heat last for a little over 13.5 Earth days while the Moon's bitterly cold nights also last 13.5 days. Despite the Moon's extreme surface temperatures, establishing a lunar base is feasible, especially if part of the base is built under the Moon's surface. Underground homes on Earth are a rarity, but they could well be the norm on the Moon when colonizing the Moon becomes a reality.

Fascination with the Moon inspired early science fiction writers to create fanciful tales about lunar creatures and daring space flights. Perhaps most notable among these writers was the French author Jules Verne, who wrote an imaginary tale of a trip to the Moon with surprising scientific accuracy. His approach to sending a man to the Moon is not all that far from what modern day rockets do. Verne envisioned a spacecraft being shot out of a gigantic cannon, which may be a fair description of how astronauts feel when they are blasted into space. His rather accurate description of the state of weightlessness on such an imaginary flight shows how prophetic Verne turned out to be.

The Moon has a more direct effect on our planet than simply fueling our imaginations and providing romantic Moonlight. Tides are caused by the gravitational pull of the Sun and Moon upon the oceans and are readily visible effects to anyone who lives near an ocean. Moreover, the gravitational pull of the Moon affects Earth's orbit while the gravitational pull of Earth in turn affects the Moon's orbit. In an effect called tidal braking, the Moon's gravitational pull is slowing down Earth's rate of rotation and thereby increasing the length of our days by 2 milliseconds per century. The rotational energy that

**Figure 7.2** Tides at the Bay of Fundy
The author took these photographs at Hopewell Cape's *The Rocks* in the Bay of Fundy. The photograph on the top was taken at high tide; the one on the bottom was taken 16.5 hours later at low tide. The people in the bottom photograph are quite literally walking on the ocean's floor! Spring tides at Hopewell Cape average 50 feet while the highest tide ever recorded there was 71 feet. The Bay of Fundy's geography contributes to the extreme tides found there, but it is the Moon's gravitational pull that sets this awesome power into motion.

Earth loses is transferred to the Moon with the result that the Moon's orbit is moving farther away from Earth at a rate of about 4 cm a year.

A curious fact about the Moon is that only 1 of its sides is ever visible from Earth. The Moon rotates on its axis at 10 miles per hour, which means that a lunar day with respect to the stars is 27.322 Earth days. This is the same amount of time that it takes the Moon to complete 1 orbit around Earth with respect to the stars. It is because the Moon's rotational and orbital periods are the same that Earthbound observers can only ever see 1 side of the Moon. However, over time we can actually view almost 60 percent of the Moon's surface because of the effect of lunar libration, a slow oscillation in the Moon's orbit and axis of rotation. The net result of this lunar wobbling is that slightly different parts of the lunar surface are viewable from Earth over time.

The unseen side of the Moon, erroneously called the "dark side," was first photographed in 1959 by the Soviet Union's Luna 3 space probe and was later seen by human eyes during the Apollo Moon missions. There is in reality no "dark side of the Moon" because, except for the bottoms of some of the deepest craters at the Moon's poles, all areas of the Moon's surface are exposed to sunlight.

Features on the Moon's surface enhance its romantic attraction and splendor. Its silvery color, with its contrasting dark patches scattered here and there, have caused some observers to look at the full Moon and visualize the face of a man. Others deny there is a "man in the Moon" and have seen the profile of a woman instead. Still others have imagined great oceans on the Moon, as evidenced by the dark areas on its surface. For this reason, the Moon's dark areas are called *maria*, which is Latin for seas. The lighter colored areas are called *terrae*, which is Latin for earth or dry land. Satellite photographs and samples returned from manned and robotic lunar expeditions confirm that the great lunar maria were never seas at all; they are in fact relatively flat plains formed by ancient volcanic eruptions. Lunar terrae are highlands or mountainous regions that are at a higher elevation than maria.

Samples returned from lunar missions confirm that although the Moon's surface appears bright white in color, it is actually quite dark and reflects only about 7–12 percent of the light that strikes the surface. By contrast, grass reflects 15 percent of the light that strikes it whereas snow reflects 40–85 percent, depending on how directly the light strikes the snow.

Satellite photographs reveal surprising differences between the Moon's far side and near side (the side that always faces Earth). While about 31 percent of the near side of the Moon is covered in maria, less than 2 percent of the far side has any maria at all. Almost 90 percent of the Moon's far side is covered in craterous regions while only 60 percent of the side visible from Earth is

**Figure 7.3** *Eagle* Leaving the Moon
In this photograph, Apollo 11's Lunar Excursion Module *Eagle* is about to dock with the Command Module *Columbia* in preparation for returning home. Earth is in the background above the lunar horizon while an extensive maria area can be seen on the surface. (Image courtesy of NASA)

covered in craters. Astronomers continue to debate why there are such striking differences between the 2 sides of the Moon.

Modern studies have dispelled many of the romantic notions once held about the Moon. Lunar features are easily recognizable with even a moderately sized telescope, and because of this, much of the Moon's mystery and romance are gone, or at least greatly diminished. This is certainly true of the craters seen in great profusion across the lunar surface, denoting the Moon's violent past.

# The Moon

Scientists have long debated whether lunar craters were a result of volcanic eruptions or the impact of objects such as meteoroids or asteroids striking the Moon.[1] Evidence collected from the Apollo missions proved conclusively that lunar craters were formed by the impact of objects striking the Moon's surface.

Easily visible to the naked eye and located in the southern part of the Moon, the crater Tycho gives mute evidence of the tremendous size of the objects that must have struck the Moon in ages past. Tycho is a sharply defined crater approximately 53 miles in diameter and 3 miles deep. It is of considerable scientific interest because it appears to be the most recently formed of the ancient lunar craters. Tycho's clearly discernible rays emanating from around the crater's rim offer important clues to the Moon's age and how lunar craters were formed.

Located slightly northwest of the center of the Moon's near side, the giant impact crater Copernicus is 58 miles wide and 2.4 miles deep. Copernicus is a relatively young crater, although not as young as Tycho. Apollo 12 astronauts brought back samples from the Copernicus crater region that have yielded a wealth of geological information about the Moon, how it was formed, and the timescale on which the lunar features were formed.

Tycho and Copernicus are not the largest lunar craters. They are actually rather average in size compared to other craters. The largest impact crater is the Aitken Basin, which is located at the Moon's South Pole and is mostly on the far side of the Moon so that it cannot be observed from Earth. The Aitken Basin is 1,550 miles in diameter and over 5 miles deep in some places. By comparison, the puny Meteor Crater in Arizona is only 3,900 feet in diameter and 560 feet deep. The largest impact crater on Earth is the Vredefort crater in South Africa, which is estimated to have been originally 185 miles in diameter but has decreased in diameter over time due to the effects of erosion.

In 2005, NASA initiated a program to actively monitor the Moon to determine how often it is struck by objects more massive in size than a few ounces. Understanding how often objects bombard the Moon is important because of the dangers such objects would pose to spacecraft orbiting the Moon and to any colony established on the Moon. Although meteor showers regularly pepper Earth, we are protected by Earth's atmosphere, so most "shooting stars" burn up long before they reach the surface. However, the Moon is not so fortunate because it essentially has no atmosphere. In fact, NASA has detected hundreds of impacts on the lunar surface since their formal monitoring efforts began.

---

1. Meteoroids and asteroids are both rocky objects in our Solar System. Asteroids are generally considered to be the same as meteoroids, except that an asteroid is much larger in size ("planet or moon sized" versus "rock sized") than a meteoroid. When a meteoroid enters Earth's atmosphere, it is called a meteor or shooting star. If a meteor does not vaporize in Earth's atmosphere but reaches the surface, it is called a meteorite.

The most phenomenal impact detected to date occurred on September 11, 2013, when a meteoroid, estimated to have a mass of 880 pounds and a width of 2–4.5 feet, struck the Moon at a speed of nearly 40,000 miles per hour. The impact gouged a 130-foot crater in the lunar surface and generated a flash of light that was bright enough to be visible from Earth with the naked eye. Had such an object struck a lunar colony or an orbiting spacecraft, the results would have been disastrous.

Another striking feature of the lunar surface is that it contains a number of large mountain ranges. The most mountainous region visible from Earth is near the lunar South Pole where the Leibnitz Mountains have peaks that reach 32,800 feet. These mountains are actually part of the Aitken Basin impact crater's rim, and they are the highest mountain range on the Moon. Other mountainous regions include the lunar Caucasian Mountains, which tower to 19,600 feet, and the lunar Alps, which reach a height of 12,000 feet. By comparison, Earth's Mount Everest is 29,000 feet high while Mount Kilimanjaro is 19,300 feet high.

Surface water cannot last long on the Moon because direct exposure to the Sun's rays would cause it to vaporize and quickly be lost into space. Despite this physical fact and the fact that the lunar maria are known not to be ancient sea beds, scientists debated whether water was ever on the Moon. The long debate is now over. In October 2008, India launched the Chandrayaan-1 spacecraft whose mission included attempting to detect water on the Moon. The Chandrayaan-1 sent a probe to the Moon's surface that detected evidence of water in the thin atmosphere just above the Moon's surface.

NASA's Lunar Crater Observation and Sensing Satellite (LCROSS) arrived at the Moon in 2009 and confirmed the presence of ice in the permanently shadowed crater Cabeus located at the Moon's South Pole. Even more conclusively, in 2010 sensors aboard the still-functioning Chandrayaan-1 discovered over 40 permanently shadowed craters near the North Pole that may contain as much as 600 million tons of water in the form of ice. Although no liquid water exists on the Moon's surface, it is now believed that water in the form of ice is abundant. With the prospect of available sources of water, can the establishment of a lunar colony be much further in our future?

## 7.2 Lunar Exploration

It would be grossly untrue to claim that significant exploration of the Moon did not begin until the advent of the Space Age. Nevertheless, it is perhaps fair to say that the most exciting phase of lunar exploration began when

Americans and Russians first sent spacecraft to the Moon. The first space probe to successfully reach the Moon was the Soviet's Luna 2, which was intentionally crashed onto the Moon in September 1959. In October of that same year, the Luna 3 sent back humankind's first photographs of the far side of the Moon before it, too, was crashed onto the lunar surface.

The United States, which had been trailing behind the successes of the Soviet Luna Program, launched a series of Ranger space probes in the early 1960s to explore the lunar surface. The Ranger probes were designed to send back lunar photographs before crashing into the Moon. Between 1966 and 1968, the United States launched a series of Surveyor space probes intended to actually land on the Moon and send back important scientific data in preparation for a manned mission. Once again, however, the Soviets were first with Luna 9, which in February 1966 was the first space probe to successfully soft land on a celestial object. Luna 9 predated Apollo 11's lunar landing by 3 years and sent back the first close-up pictures of the lunar surface, which were also the first pictures ever to be sent back from the surface of another world.

Despite Luna being a successful exploration program that contributed greatly to our understanding of the Moon and space travel in general, Soviet achievements were largely overshadowed by the manned American Apollo missions. The last spacecraft in the Luna Program was Luna 24, which was launched in 1976. Luna 24 successfully landed on the Moon and returned 6 ounces of soil that Soviet scientists analyzed and discovered evidence of water on the Moon. Other scientists within the international community dismissed the Soviets' findings, disagreed with them, or were simply unaware of them.

The pinnacle of the American space program was reached when Apollo 11 crew members Neil Armstrong, Edwin "Buzz" Aldrin, and Michael Collins succeeded in accomplishing a feat that has been dreamed about since time immemorial. On July 20, 1969, Neil Armstrong stepped onto the lunar surface while the entire world watched in awe and wonder. It is doubtful that the excitement of that moment will be realized again until a manned mission succeeds in landing on a planet within the Solar System.

Apollo 17 was launched in December 1972 and remains to date the last manned mission to the Moon. Before leaving the Moon, Apollo 17 astronauts Eugene Cernan and Harrison Schmitt unveiled a plaque on their Lunar Excursion Module (LEM) that would remain behind as they returned to Earth. The plaque read, "Here man completed his first exploration of the Moon, December 1972 AD. May the spirit of peace in which we came be reflected in the lives of all mankind." To date only 12 humans, all of whom were astronauts from the Apollo Program, have walked on the surface of the Moon. All manned

**Figure 7.4** Crew of Apollo 11
From left to right, the crew of Apollo 11: Commander Neil A. Armstrong, Command Module Pilot Michael Collins, and Lunar Module Pilot Edwin E. Aldrin Jr. On July 20, 1969, the Lunar Module *Eagle* landed at the Sea of Tranquility. Shortly thereafter, Neil Armstrong became the first person in history to set foot on another world. (Image courtesy of NASA)

landings took place on the near side of the Moon because of the difficulty of communicating with a spacecraft on the far side of the Moon.

Fortunately, exploration of the Moon did not end with the manned Apollo missions. After a hiatus of over 20 years, NASA returned to the Moon with the Clementine mission. Launched in January 1994, the Clementine spacecraft orbited the Moon and spent 2 months extensively mapping the lunar surface. NASA returned again in 1998 with the Lunar Prospector to continue mapping the Moon's surface, and again in 2011 with the Lunar Reconnaissance Orbiter (LRO) spacecraft. As a result of these missions, we now have extraordinarily detailed and precise maps of the entire Moon. The LRO was the spacecraft that carried the previously mentioned LCROSS probe, whose mission was to look for evidence of water on the Moon. The LRO is still operational, doing its part to add to our growing knowledge about the Moon.

Lunar exploration is presently a truly international effort. The European Space Agency (ESA) launched the Small Missions for Advanced Research in Technology-1 (SMART-1) spacecraft in 2003, whose primary mission was to capture 3-dimensional X-ray and infrared images of the lunar surface. In

# The Moon

2007 the Japan Aerospace Exploration Agency (JAXA), Japan's equivalent of NASA, launched the SELENE (also called Kaguya) lunar orbit explorer to investigate the origin and evolution of the Moon. Also in 2007, the Chinese National Space Agency (CNSA) launched the first of their Chang'e lunar orbiters. Chang'e 3, which soft-landed on the Moon in December 2013, was the first unmanned spacecraft to successfully do so since the Soviet Luna 24 spacecraft in 1976. The Chang'e 3 deployed a robotic rover that moves around the lunar surface as it performs various scientific experiments.

In 2008, India's national space agency, the Indian Space Research Organization (ISRO), launched the lunar orbiter Chandrayaan-1 that played such a vital role in confirming the presence of water on the Moon. ISRO is currently engaged in an ambitious effort to create a 3-dimensional atlas of the entire Moon and conduct a thorough mineralogical mapping of the lunar surface.

All the previously mentioned lunar missions were directed and funded by various governmental agencies. However, commercial industries have now entered the modern "race" to the Moon. In 2007, Google announced that it would award $30 million to the first privately funded team to successfully land a robot on the Moon that can travel at least 0.3 miles on the lunar surface and send back still images and video to Earth. More recently, in October 2014 the German company LuxSpace, whose spacecraft was carried onboard the Chang'e 5 test spacecraft, made a successful lunar flyby as a memorial to the German space pioneer Manfred Fuchs.

These exploration efforts and others make it clear that humans will someday return to the Moon to explore it, take advantage of its resources, and live there in what will surely be the first of many colonies formed throughout our Solar System. As but 1 example of the current race to return to the Moon, JAXA has announced plans for a manned lunar landing around 2020 and the establishment of a manned lunar base by 2030. It is indeed an exciting time in the history of the exploration of the Moon!

## 7.3 Locating the Moon

Calculating the Moon's position is the most ambitious and complex calculation we have presented so far. The complexity is because—in the parlance of astrophysics—determining the Moon's position requires solving a 3-body problem. That is, we must consider 3 bodies (Earth, Sun, and Moon) in our calculations because the masses of the Earth and Sun are so much greater than the Moon's mass and they are so close to the Moon that they significantly

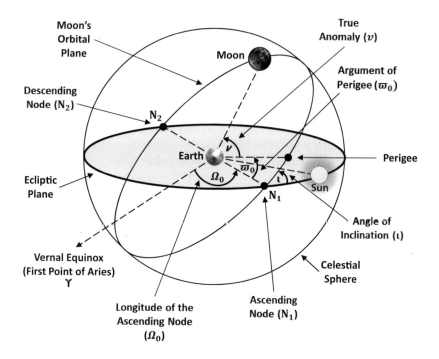

**Figure 7.5** The Moon's Orbital Elements
This illustration shows the Moon's orbital plane and orbital elements. Calculating the Moon's position requires knowing the Sun's position as well.

affect the Moon's orbit. In contrast, determining the position of the Sun can be accomplished to an acceptable degree of accuracy by solving only a 2-body (Earth and Sun) problem. Although all masses in the Solar System (planets, asteroids, comets, etc.) affect the Earth-Sun orbit, their effects are negligible for most purposes because the collective mass of those objects, coupled with their relatively large distances, is greatly overwhelmed by the combined masses of the Earth and Sun.

Figure 7.5 shows the angles required for locating the Moon. This figure is similar to figure 4.12 (section 4.5), except that figure 7.5 is a geocentric model and uses the nomenclature from table 7.1 to explicitly show the Moon's orbital elements. The Sun's position is also shown because it has such a significant impact on calculating the Moon's position.

To determine the Moon's position, we will proceed conceptually in much the same manner as for determining the Sun's location. The basic idea is to calculate the Moon's true anomaly from which the Moon's position can be determined. However, we will not attempt to solve Kepler's equation to find

# The Moon

**Table 7.1** Moon's Orbital Elements
These geocentric orbital elements for the Moon are for the standard epoch J2000, as given in *The Astronomical Almanac 2000*.

| Orbital Element | Value |
| --- | --- |
| $e$, eccentricity of the Moon's orbit | 0.0549 |
| $a_0$, length of the Moon's orbital semi-major axis | 384,400 km |
| $\theta_0$, Moon's angular diameter when a distance of $a_0$ from Earth | 0.5181° |
| $\iota$, inclination of the Moon's orbit with respect to the ecliptic | 5.1453964° |
| $\lambda_0$, Moon's ecliptic longitude at the epoch | 218.316433° |
| $\varpi_0$, Moon's ecliptic longitude at perigee at the epoch | 83.353451° |
| $\Omega_0$, Moon's ecliptic longitude of the ascending node at the epoch | 125.044522° |

the Moon's true anomaly but will instead rely on solving the equation of the center. This will considerably simplify our calculations.

Besides being a 3-body problem, calculating the Moon's position is more complicated than calculating the Sun's position in 2 ways. First, locating the Moon is a 3-dimensional problem because the Moon's orbit does not lie within the ecliptic plane. Recall that the Sun's orbit *does* lie completely within the ecliptic plane, so only the Sun's ecliptic longitude must be calculated to determine the Sun's position (the Sun's ecliptic latitude is 0°). The Moon's ecliptic latitude and longitude must both be calculated.

Second, several adjustments must be made to account for various effects. We will limit our considerations to 4 effects:

- Annual equation correction ($A_e$)
- Variation correction ($V$)
- Evection correction ($E_v$)
- Mean anomaly correction ($C_a$)

The annual equation and variation corrections account for the gravitational pull of the Sun, which varies depending on where the Moon and Earth are relative to the Sun. The annual equation accounts for the fact that the Earth-Sun distance varies throughout the year and hence affects the gravitational pull of the Sun on the Moon. The Sun's gravitational pull on the Moon is greater as Earth approaches perihelion and less as Earth approaches aphelion. The variation correction is an adjustment that varies with where the Moon is in its orbit around Earth. When the Moon is between the Earth and Sun, the Sun's gravitational attraction is stronger. When Earth is between the Moon and Sun, the Sun's gravitational attraction is weaker. As for the remaining effects, the evection correction accounts for changes in the Moon's orbital eccentricity, which is also affected by the Sun's gravitational pull. The mean anomaly correction is applied to obtain a better estimate of the Moon's true anomaly when solving the equation of the center.

Professional astronomers make many more corrections than these to calculate the Moon's position. In fact, to achieve high accuracy, hundreds of corrections are necessary to account for the gravitational effects of the planets, parallax, variations in the longitude of the ascending node, variations in the position of perigee, precession, and other effects! Fortunately, sufficient accuracy can be obtained for our purposes with just these 4 corrections. The accuracy achieved in doing so is better than a quarter of a degree,[2] which is sufficient for this book since our primary goal is to convey a conceptual understanding of the factors involved rather than achieving arcsecond accuracy.

The approach for calculating the Moon's position can be summarized as follows:

- Calculate the Sun's ecliptic longitude $\lambda_\odot$ (section 6.2). Better accuracy will usually be obtained in finding the Sun's location by solving Kepler's equation instead of the equation of the center, but the additional effort may not be warranted.
- Compute the Moon's mean ecliptic longitude ($\lambda$) with respect to the standard epoch.
- Compute the Moon's mean ecliptic longitude for the ascending node ($\Omega$).
- Compute the Moon's mean anomaly ($M_m$) with respect to the standard epoch.
- Compute corrections $A_e$, $E_v$, and $C_a$ and use $C_a$ to solve the equation of the center to get the Moon's true anomaly ($v_m$).
- Compute the variation correction $V$.
- Apply corrections to $\lambda$ and $\Omega$ from which the Moon's true ecliptic coordinates will be determined.

Several equations are required for this rather lengthy process. In order to avoid confusion, we will use the subscript $m$ when referring to the Moon and the subscript $\odot$ when referring to the Sun. You may wish to refer back to section 6.2 to see how the approach for calculating the Moon's position parallels the approach for calculating the Sun's position.

---

2. Duffett-Smith provides an accuracy of one-fifth of a degree in *Practical Astronomy with Your Calculator or Spreadsheet*. The algorithm Meeus presents in *Astronomical Algorithms* is accurate to within $10''$. For greater accuracy than these sources, refer to the resources cited in sections 10.6 and 10.7.

# The Moon

Since the Moon orbits Earth with respect to the stars in 27.3217 days, the Moon moves along its mean orbit by

$$\frac{360°}{27.3217 \text{ days}} \approx 13.176339686° \text{ per day}.$$

Thus, the Moon's mean ecliptic longitude before any corrections are made is

$$\lambda = 13.176339686 D_e + \lambda_0, \tag{7.3.1}$$

where $D_e$ is the number of days (including fractional days) since the standard epoch, and $\lambda_0$ is the Moon's ecliptic longitude at the epoch (see table 7.1).

The Moon's (uncorrected) mean ecliptic longitude of the ascending node is

$$\Omega = \Omega_0 - 0.0529539 D_e, \tag{7.3.2}$$

where $\Omega_0$ is the Moon's ecliptic longitude of the ascending node at the standard epoch (see table 7.1). Recall that the ascending node is the point in an orbit at which an object transitions from being below its orbital plane to being above it.

The Moon's (uncorrected) mean anomaly is

$$M_m = \lambda - 0.1114041 D_e - \varpi_0, \tag{7.3.3}$$

where $\varpi_0$ is the Moon's ecliptic longitude at the moment of perigee for the standard epoch.

The annual equation, evection, and mean anomaly corrections are given by

$$A_e = 0.1858 \sin M_\odot \tag{7.3.4}$$

$$E_v = 1.2739 \sin \left[2 \left(\lambda - \lambda_\odot\right) - M_m\right] \tag{7.3.5}$$

$$C_a = M_m + E_v - A_e - 0.37 \sin M_\odot. \tag{7.3.6}$$

Given the mean anomaly correction, we will use the following approximation to solve the equation of the center to get the Moon's true anomaly:

$$\upsilon_m = 6.2886 \sin C_a + 0.214 \sin (2 C_a). \tag{7.3.7}$$

The variation correction is given by

$$V = 0.6583 \sin \left[2 \left(\lambda' - \lambda_\odot\right)\right], \tag{7.3.8}$$

where $\lambda'$ is a corrected ecliptic longitude that includes the true anomaly $\upsilon_m$ plus the $A_e$ and $E_v$ corrections. That is,

$$\lambda' = \lambda + E_v + \upsilon_m - A_e. \tag{7.3.9}$$

Applying the variation correction $V$ to $\lambda'$ gives the Moon's true ecliptic longitude, $\lambda_t$.

$$\lambda_t = \lambda' + V. \qquad (7.3.10)$$

The corrected ecliptic longitude of the ascending node is given by

$$\Omega' = \Omega - 0.16 \sin M_\odot. \qquad (7.3.11)$$

Finally, once $\Omega'$, $\lambda_t$, and the Moon's orbital inclination $\iota$ (see table 7.1) are known, the Moon's ecliptic longitude ($\lambda_m$) and latitude ($\beta_m$) are given by:

$$\lambda_m = \Omega' + \tan^{-1}\left[\frac{\sin(\lambda_t - \Omega')\cos \iota}{\cos(\lambda_t - \Omega')}\right] \qquad (7.3.12)$$

$$\beta_m = \sin^{-1}\left[\sin(\lambda_t - \Omega')\sin \iota\right]. \qquad (7.3.13)$$

The resulting ecliptic coordinates can then be converted to the equatorial or horizon coordinate systems as desired.

For all the preceding chapters, we blissfully ignored the difference between TT and UT. (TT was briefly discussed in chapter 3.) For the Moon, however, the difference between TT and UT can be significant and should be considered for greater accuracy because of the distance that the Moon travels in its orbit in a mere few minutes. The difference between TT and UT varies year to year and day to day. Nevertheless, we will approximate it as $63.8^s$, which was the difference between TT and UT at the beginning of the year 2000. We will assume that

$$TT = UT + 63.8^s \qquad (7.3.14)$$

for all calculations in this chapter, irrespective of the date and actual time difference. One can consult the US Naval Observatory website, *The Astronomical Almanac*, or any of a number of other readily available sources to get a more accurate time difference and therefore improve the accuracy of predicting the Moon's position.

To illustrate the lengthy process required, let's compute the Moon's position for January 1, 2015, at $22^h$ LCT. Assume an observer is at 38° N latitude, 78° W longitude within the Eastern Standard Time zone, and that the observer is not on daylight saving time.

1. Convert LCT to UT, GST, and LST times and adjust the date if needed.
   (Ans: $UT = 3.00^h$, $GST = 9.762547^h$, $LST = 4.562547^h$, Date = 1/2/2015.)
2. Use equation 7.3.14 to compute TT.
   (Ans: $TT = 3.017722^h$.)

# The Moon

3. Compute the Julian day number for the standard epoch.
   (Ans: Epoch: 2000.0, $JD_e = 2,451,545.00$.)

4. Compute the Julian day number for the desired date using the Greenwich date and TT from steps 1 and 2, and include the fractional part of the day.
   (Ans: $JD = 2,457,024.62574$.)

5. Compute the total number of elapsed days, including fractional days, since the standard epoch (i.e., $JD - JD_e$).
   (Ans: $D_e = 5479.625738$ days.)

6. Use the algorithm from section 6.2 to calculate the Sun's ecliptic longitude and mean anomaly for the given UT date and time. (For this example, we will solve the equation of the center when computing the Sun's position.)
   (Ans: $\lambda_\odot = 281.394034°$, $M_\odot = 358.505618°$.)

7. Apply equation 7.3.1 to calculate the Moon's (uncorrected) mean ecliptic longitude.
   (Ans: $\lambda = 72,419.726515°$.)

8. If necessary, use the MOD function to put $\lambda$ into the range $[0°, 360°]$.
   (Ans: $\lambda = 59.726515°$.)

9. Apply equation 7.3.2 to compute the Moon's (uncorrected) mean ecliptic longitude of the ascending node.
   (Ans: $\Omega = -165.123031°$.)

10. If necessary, adjust $\Omega$ to be in the range $[0°, 360°]$ (i.e., $\Omega$ MOD $360°$).
    (Ans: $\Omega = 194.876969°$.)

11. Apply equation 7.3.3 to compute the Moon's (uncorrected) mean anomaly.
    (Ans: $M_m = -634.079710°$.)

12. Adjust $M_m$ if necessary to be in the range $[0°, 360°]$.
    (Ans: $M_m = 85.920290°$.)

13. Use equation 7.3.4 to compute the annual equation correction.
    (Ans: $A_e = -0.004845°$.)

14. Use equation 7.3.5 to compute the evection correction.
    (Ans: $E_v = -0.237497°$.)

15. Use equation 7.3.6 to compute the mean anomaly correction.
    (Ans: $C_a = 85.697288°$.)

16. Use equation 7.3.7 to compute the Moon's true anomaly.
    (Ans: $v_m = 6.302897°$.)

17. Use equation 7.3.9 to apply all of the applicable corrections and the true anomaly to arrive at a corrected mean ecliptic longitude.
    (Ans: $\lambda' = 65.796760°$.)

18. Use equation 7.3.8 to compute the variation correction.
    (Ans: $V = -0.623159°$.)

19. Apply equation 7.3.10 to calculate the Moon's true ecliptic longitude.
    (Ans: $\lambda_t = 65.173601°$.)

20. Apply equation 7.3.11 to compute a corrected ecliptic longitude of the ascending node.
    (Ans: $\Omega' = 194.881141°$.)

21. Compute $y = \sin(\lambda_t - \Omega')\cos\iota$ where $\iota$ is the inclination of the Moon's orbit with respect to the ecliptic (see table 7.1). This is the numerator of the fraction in equation 7.3.12.
    (Ans: $y = -0.766215$.)

22. Compute $x = \cos(\lambda_t - \Omega')$. This is the denominator of the fraction in equation 7.3.12.
    (Ans: $x = -0.638869$.)

23. Compute $T = \tan^{-1}\left(\frac{y}{x}\right)$.
    (Ans: $T = 50.178711°$.)

24. Using the algebraic signs of $y$ and $x$, determine a quadrant adjustment for $T$ to remove the angle ambiguity. This value will be used to calculate the Moon's true ecliptic longitude, which must be in the range $[0°, 360°]$.
    (Ans: $Adjustment = 180°$, so $T = 230.178711°$.)

25. Calculate $\lambda_m = \Omega' + T$. This uses the temporary results computed in the preceding steps to apply equation 7.3.12.
    (Ans: $\lambda_m = 425.059853°$.)

26. If $\lambda_m > 360°$, then subtract $360°$.
    (Ans: $\lambda_m = 65.059853°$.)

27. Use equation 7.3.13 to compute the Moon's ecliptic latitude.
    (Ans: $\beta_m = -3.956258°$.)

28. Convert the Moon's ecliptic latitude ($\beta_m$) and longitude ($\lambda_m$) to their corresponding equatorial coordinates.
    (Ans: $\alpha_m = 4.257714^h$, $\delta_m = 17.248880°$.)

29. Convert the equatorial coordinates to horizon coordinates for the observer stated at the beginning of these calculations.
    (Ans: $h_m = 68°52'$, $A_m = 192°11'$.)

The calculations just presented are long indeed. Care must be taken at each step to avoid making mistakes in such a lengthy process!

Instead of using the equation of the center to compute $\lambda_\odot$ and $M_\odot$ in step 7, we could have solved Kepler's equation. Had we done so with the simple

# The Moon

iterative method, for this example the result would be

$\lambda_\odot = 281.392970°, M_\odot = 358.505618°.$

Using these values and carrying through the calculations, steps 27 and 29 would yield

$\lambda_m = 65.059814°, \beta_m = -3.956255°,$

which then leads to horizon coordinates of

$h_m = 68°52', A_m = 192°11'$

for the stated observer. The additional effort required to solve Kepler's equation is not worth it in this example.

## 7.4 Moonrise and Moonset

In section 6.3, we noted that the Sun's equatorial coordinates change rapidly and that fact makes it more difficult to determine when the Sun will rise/set than to determine when a star will rise/set. So, we employed a strategy in which we computed the Sun's position at midnight on the date for which we want to calculate sunrise and sunset and then computed the Sun's position $24^h$ later. We used those 2 locations to compute 2 sets of rising and setting times and then interpolated to estimate sunrise and sunset for the date of interest.

The Moon's equatorial coordinates also change rapidly with time. Hence, to compute moonrise and moonset, we will employ the same strategy, except that we will calculate positions for the Moon that are $12^h$ apart rather than $24^h$ apart. Specifically, we will first determine the Moon's position at midnight ($UT = 0^h$) on the date in question. Using the equatorial coordinates obtained, we will compute an initial set of rising and setting times. Next, we will compute the Moon's position $12^h$ later to arrive at a second set of rising and setting times and then interpolate the 2 sets of rising and setting times to arrive at an estimate for moonrise and moonset.

Determining the Moon's position at midnight requires the calculations presented in section 7.3. Although we can repeat those calculations to determine the Moon's position $12^h$ later, there is a much easier way that is sufficiently accurate for our purposes. Assume that the Moon's position at some point in time is ecliptic latitude $\beta_1$, longitude $\lambda_1$. Then the Moon's coordinates $t$ hours later are given by:

$$\beta_2 = \beta_1 + 0.05 \cos\left(\lambda_{t_1} - \Omega'_1\right) t \qquad (7.4.1)$$

$$\lambda_2 = \lambda_1 + \left[0.55 + 0.06 \cos\left(C_{a_1}\right)\right] t. \qquad (7.4.2)$$

$\lambda_{t_1}$ is the Moon's true ecliptic longitude (equation 7.3.10), $\Omega'_1$ is the corrected ecliptic longitude of the ascending node (equation 7.3.11), and $C_{a_1}$ is the mean anomaly correction (equation 7.3.6) that result from calculating the Moon's position at midnight (i.e., when computing $\beta_1$ and $\lambda_1$).

The equation we will use to interpolate the 2 sets of rising and setting times is

$$T = \frac{12.03 ST1}{12.03 + ST1 - ST2}. \qquad (7.4.3)$$

As an example, compute the rising and setting times for the observer from the previous section. That is, compute moonrise and moonset on January 1, 2015, for an observer at 38° N latitude, 78° W longitude within the Eastern Standard Time zone who is not on daylight saving time.

1. Calculate the Moon's ecliptic coordinates for midnight ($UT = 0^h$) for the stated date. Save the values of $\lambda_{t_1}$, $\Omega'_1$, and $C_{a_1}$ for use in a later step.
   (Ans: $\beta_1 = -2.981288°$, $\lambda_1 = 50.279952°$, $\lambda_{t_1} = 50.389191°$, $\Omega'_1 = 194.943809°$, $C_{a_1} = 71.289991°$.)

2. Compute $\alpha_1$ and $\delta_1$, the equatorial coordinates for the ecliptic coordinates calculated in the previous step.
   (Ans: $\alpha_1 = 3.244116^h$, $\delta_1 = 14.941252°$.)

3. Using $\alpha_1$ and $\delta_1$, compute $ST1_r$ and $ST1_s$, which will be the LST rising and setting times for these equatorial coordinates. See equations 5.2.1 and 5.2.2 to calculate rising and setting times.
   (Ans: $ST1_r = 20.441871^h$, $ST1_s = 10.046360^h$.)

4. Use $\lambda_{t_1}$, $\Omega'_1$, $C_{a_1}$, and the Moon's ecliptic coordinates from step 1 with equation 7.4.1 to compute the Moon's ecliptic latitude 12 hours (i.e., $t = 12$) later.
   (Ans: $\beta_2 = -3.470089°$.)

5. Use equation 7.4.2 to compute the Moon's ecliptic longitude $12^h$ later.
   (Ans: $\lambda_2 = 57.110913°$.)

6. If $\lambda_2 > 360$, subtract 360 degrees.
   (Ans: $\lambda_2 = 57.110913°$.)

7. Compute $\alpha_2$ and $\delta_2$, the equatorial coordinates for the ecliptic coordinates calculated in the previous steps.
   (Ans: $\alpha_2 = 3.170028^h$, $\delta_2 = 16.133568°$.)

8. Using $\alpha_2$ and $\delta_2$, compute $ST2_r$ and $ST2_s$, which are also rising and setting times.
   (Ans: $ST2_r = 20.839239^h$, $ST2_s = 10.580817^h$.)

# The Moon

9. Use the interpolation formula given in equation 7.4.3 with $ST1_r$ and $ST2_r$ to compute the LST for moonrise.
   (Ans: $T_r = 21.140161^h$.)
10. Use equation 7.4.3 with $ST1_s$ and $ST2_s$ to compute the LST for moonset.
    (Ans: $T_s = 10.513441^h$.)
11. Convert the LST times $T_r$ and $T_s$ from the previous 2 steps into their corresponding LCT times for moonrise and moonset.
    (Ans: $LCT_r = 14.597889^h$, $LCT_s = 4.000179^h$.)
12. If desired, convert the LCT times to HMS format.
    (Ans: $LCT_r = 14^h 36^m$, $LCT_s = 4^h 00^m$.)

This straightforward algorithm will not always work because converting an LST setting time to LCT may give a calculated setting time that is on the day *before* the calculated rising time. Should this happen, use the algorithm to determine the LCT rise/set times for a date *prior* to the date of interest and for a date *after* the date of interest that do not result in a setting time that is on the day before the rising time. Use those 2 sets of LCT rise/set times to calculate how much the rise/set times are changing over those dates, then use that information to estimate when the Moon will rise and set on the date of interest.

To illustrate, compute the Moon's rising and setting times for the previous observer for January 22, 2015. Converting the LST rise/set times to LCT gives a setting time that is the day before the rising time. So, try the algorithm again for January 21, which gives

$$LCT_{rp} = 7.929372^h, LCT_{sp} = 17.978205^h.$$

Now try January 23. This date also gives a setting time on the day before the rising time when converting LST times to LCT times. The first date after January 22 that does not have this problem is January 27. For that date, we obtain

$$LCT_{ra} = 11.932303^h, LCT_{sa} = 0.944280^h.$$

Given these 2 sets of rising/setting times around the date of interest, take the 2 rising times and divide their difference by the number of days between January 21 and January 27. This gives

$$avg_r = \frac{LCT_{ra} - LCT_{rp}}{6} = 0.667155^h.$$

This means that on average, the Moon's rising time is increasing by $0.667155^h$ per day. So, the estimated rise time for January 22 is

$$LCT_{rp} + 0.667155^h = 8.596527^h.$$

The estimated rising time for January 23, 2 days later, would be

$$LCT_{rp} + 2 * 0.667155^h = 9.263682^h,$$

and so on, for any other date in the interval January 21 through January 27. Performing the same analysis for the setting times gives

$$avg_s = \frac{LCT_{sa} - LCT_{sp}}{6} = -2.838988^h,$$

which means that on average, the Moon's setting time is *decreasing* by $2.838988^h$ per day. So, the estimated setting time for January 22 is

$$LCT_{sp} + (-2.838988^h) = 15.139217^h.$$

This chapter's program detects when converting a setting time to LCT creates a problem. However, the program does not perform the steps just outlined to estimate new rise/set times. That extra bit of work is left as an easy exercise for the reader!

## 7.5 Lunar Distance and Angular Diameter

Because the Moon's orbit is an ellipse instead of a circle, the Earth-Moon distance varies throughout the month by as much as 50,000 km. Moreover, the apparent size of the Moon changes as the distance between Earth and the Moon changes. When the Moon is at perigee, it appears to be 14 percent larger and 30 percent brighter than when it is at apogee. When either a Full Moon or a New Moon occurs at perigee, it is called a supermoon. Conversely, when a Full Moon or New Moon occurs at apogee, it is called a micromoon. Three to 6 supermoons occur each year, and 3 to 6 micromoons occur each year.

Calculating the distance to the Moon and the Moon's angular diameter is a very simple task once the Moon's true anomaly $v_m$ is known. The distance is given by the equation

$$Dist_m = \frac{a_0(1-e^2)}{1+e\cos v_m} \tag{7.5.1}$$

while the angular diameter is given by

$$\theta_m = \frac{\theta_0(1+e\cos v_m)}{1-e^2}. \tag{7.5.2}$$

$a_0$ is the length of the semi-major axis of the Moon's orbit, $\theta_0$ is the angular diameter of the Moon when it is $a_0$ distance away from Earth, and $e$ is the

# The Moon

Moon's orbital eccentricity (see table 7.1). These 2 equations are identical to those in section 6.5 for determining the Sun's distance and angular diameter, except obviously that data for the Moon must be used in the equations instead of data for the Sun. As we did for the Sun, let

$$F = \frac{1 + e \cos \upsilon_m}{1 - e^2}, \qquad (7.5.3)$$

so that equations 7.5.1 and 7.5.2 can be rewritten as

$$Dist_m = \frac{a_0}{F} \qquad (7.5.4)$$

and

$$\theta_m = \theta_0 F. \qquad (7.5.5)$$

For example, calculate the distance to the Moon and the Moon's angular diameter on January 1, 2015, for an observer at 38° N latitude, 78° W longitude within the Eastern Standard Time zone who is not on daylight saving time.

1. Using the steps in section 7.3, calculate the Moon's true anomaly at $0^h$ UT for the given date.
   (Ans: $\upsilon_m = 6.086312°$.)
2. Use equation 7.5.3 to compute $F$.
   (Ans: $F = 1.057779$.)
3. Use the values from table 7.1 and equation 7.5.4 to compute the distance to the Moon.
   (Ans: $Dist_m = 363{,}403$ km.)
4. Use equation 7.5.5 to compute the Moon's angular diameter.
   (Ans: $\theta_m = 0.548035°$.)
5. Convert distance to miles and angular diameter to DMS format if desired.
   (Ans: $Dist_m = 225{,}808$ miles, $\theta_m = 0°33'$.)

## 7.6 Phases of the Moon

When viewed from Earth, the amount of the Moon's surface that we see as illuminated changes as the Moon orbits Earth. The phase of the Moon, age of the Moon, and percentage of illumination all describe how much of the Moon will appear to be illuminated to an Earthbound observer. Phase of the Moon refers to the shape of the portion of the Moon illuminated by the Sun as seen by an observer on Earth. Figure 7.6 shows the phases of the Moon, which cycle

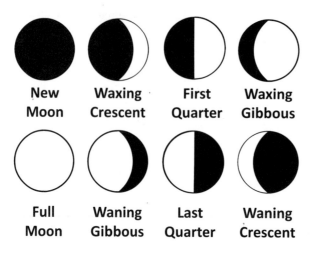

**Figure 7.6** Phases of the Moon
The phases of the Moon, as seen from Earth, continuously cycle from no illumination (New Moon) to fully illuminated (Full Moon).

from no illumination (New Moon) to being fully illuminated (Full Moon) by the Sun's rays.

Figure 7.7 shows the phases of the Moon in the context of where the Moon is in its orbit with respect to Earth and the Sun. As the figure demonstrates, half of the Moon is always illuminated by the Sun (except, of course, during an eclipse) regardless of where the Moon is in its orbit. Likewise, half of the Moon is always facing Earth regardless of where the Moon is in its orbit. The amount of the Moon's surface that appears illuminated as seen from Earth constantly changes because the angle $A$ between the Moon and the light source (the Sun) continually changes.

Before proceeding, there are 2 points to make about figure 7.7. First, apogee and perigee are not shown because they are irrelevant to the phase of the Moon. Therefore, do *not* interpret the figure as meaning that a Full Moon can occur only at apogee simply because earlier figures showed apogee/aphelion on the left side of a figure depicting an orbit. Any phase of the Moon can occur at any point in the Moon's orbit. Second, in section 7.3 we pointed out that the plane containing the Earth-Moon orbit is not in the same plane as the Earth-Sun orbit (see figure 7.5). In fact, the two orbital planes are inclined with respect to each other by about 5° (see table 7.1). Hence, figure 7.7 must be understood as a 3-dimensional representation, *not* as a 2-dimensional one. If the Earth-Moon orbit did lie in the same plane as the Earth-Sun orbit, then a solar eclipse would

# The Moon

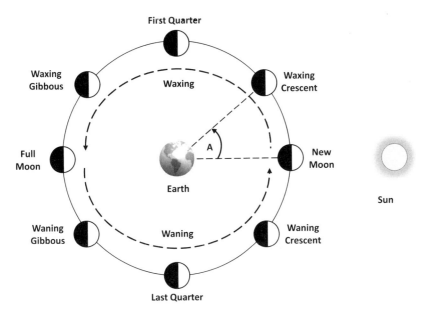

**Figure 7.7** Cycle of Lunar Phases
The Moon's phases were an early basis for defining the months. In modern times, the phase cycle defines the synodic month (29.5306 days).

occur at each New Moon while a lunar eclipse would occur at each Full Moon, which clearly is untrue.

Age of the Moon is another way to express how much of the Moon appears to be illuminated. In this context, age of the Moon has nothing to do with how old the Moon is, but it is instead an angle that measures how much of the Moon's orbit has been completed with respect to some reference. The age of the Moon, $\angle A$ in figure 7.7, is measured from the position where the Moon is not illuminated at all (as viewed from Earth) to the Moon's current position. The position of no illumination is New Moon, as shown in the figure.

The equation for computing the Moon's age $A$ (in degrees) is

$$A = \lambda_t - \lambda_\odot, \tag{7.6.1}$$

where $\lambda_t$ is the Moon's true ecliptic longitude and $\lambda_\odot$ is the Sun's ecliptic longitude. Instead of degrees, the age of the Moon is often expressed in days. Since the Moon orbits Earth in 29.5306 days, it moves by

$$\frac{360°}{29.5306 \text{ days}} \approx 12.1907° \text{ per day.}$$

So, the Moon's age can be expressed in days by

$$A_{days} = \frac{A}{12.1907°}, \quad (7.6.2)$$

or, by combining with equation 7.6.1, as

$$A_{days} = \frac{\lambda_t - \lambda_\odot}{12.1907°}. \quad (7.6.3)$$

When equations 7.6.2 and 7.6.3 are used, be sure that the numerator ($A$ or $\lambda_t - \lambda_\odot$) is adjusted to be in the range [0°, 360°] before the division is done.

At New Moon, the age of the Moon is 0° (0 days). When the Moon's age is 90° (7.4 days), half of the visible portion of the Moon is illuminated and it is a First Quarter Moon. At age 180° (14.8 days), the Moon is at maximum illumination and is a Full Moon. Ninety degrees later, when the Moon's age is 270° (22.1 days), the visible portion of the Moon is again half illuminated and is a Last Quarter Moon. First Quarter and Last Quarter Moon are equally illuminated, but the geographic area of the Moon that appears illuminated to an Earthbound observer is different for First and Last Quarter.

When the age of the Moon is expressed in days, the Moon transitions between major phases (New Moon, First Quarter Moon, Full Moon, Last Quarter Moon) about every 7 days. Alternatively, another way to express this idea is that the amount of the visible Moon that appears to be illuminated changes by 50 percent about every 7 days. Also, when the age of the Moon is 360°, the Moon has completed a full orbit from New Moon to New Moon. Using equation 7.6.2 with $A = 360°$, we see that the Moon completes an orbit about every 29.5 days. Thus, the phase of the Moon provides an astronomical event for measuring both a week (transition between major phases of the Moon) and a month (transition from New Moon back to New Moon).

Instead of arbitrarily limiting the phase of the Moon to the 8 phases shown in figure 7.6, the phase of the Moon can be precisely defined mathematically in terms of the age of the Moon. The required equation is

$$F = \frac{1 - \cos A}{2}, \quad (7.6.4)$$

where the Moon's age $A$ is expressed in degrees. This equation always returns a result in the range [0.0, 1.0], which is intuitively appealing because it ranges from a value of 0.0 when the visible portion of the Moon is not illuminated at all (New Moon) to 1.0 when the visible portion of the Moon is fully illuminated (Full Moon).

Figure 7.7 shows that the terms "waxing" and "waning" are used to describe the Moon as it transitions from 1 phase to the next. As the Moon transitions

# The Moon

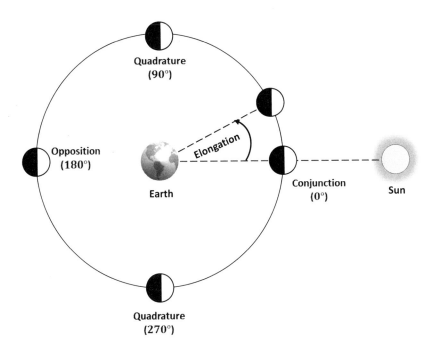

**Figure 7.8** Lunar Elongation
Elongation is the angle between the Sun and a celestial object when Earth is a reference point.

from New Moon to Full Moon, it is waxing because the amount of illumination (as seen from Earth) increases and correspondingly, the age of the Moon increases from 0° to 180°. When the Moon proceeds from Full Moon back to New Moon, it is waning because the amount of illumination is steadily decreasing. When the Moon is waning, the age of the Moon is increasing from 180° to 360°. The term "crescent" refers to the phases of the Moon in which the visible portion of the Moon is less than half illuminated. "Gibbous" means that more than half of the visible portion of the Moon is illuminated, but the Moon is less than fully illuminated.

Astronomers are often interested in the angle between a celestial object, such as the Moon, and the Sun when Earth is used as a reference point. This angle, shown in figure 7.8, is called an object's "elongation." When comparing figure 7.8 to figure 7.7, it should be clear that the Moon's elongation and age refer to the same angle.

At Full Moon, the Moon's elongation (and age) is 180° and the Moon is said to be at the point of opposition with the Sun. Two objects are at opposition when they are on opposite sides of the Earth and have ecliptic longitudes

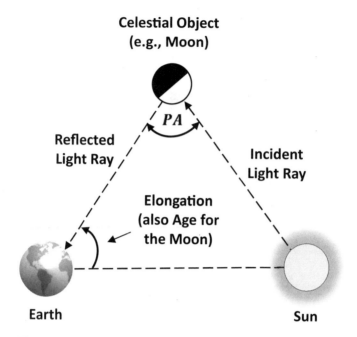

**Figure 7.9** Phase Angle
The phase angle is the angle made between a light ray striking an object and the light ray reflected back from the object to an observer.

that are 180° apart. At New Moon, the Moon's elongation is 0° and it is said to be in conjunction with the Sun. Two objects are in conjunction when they are on the same side of the Earth and have the same ecliptic longitude. Knowing when lunar conjunctions and oppositions occur is required to determine when an eclipse will occur. When the Moon is at First or Last Quarter, its elongation is 90° and 270°, respectively. In both cases, the Moon is said to be in "quadrature," a term astronomers use to denote that when viewed from Earth, a celestial object is at a right angle with respect to the Sun.

The amount of the visible Moon that appears illuminated to an Earth-bound observer can also be expressed as a percentage. Carefully note that this refers to how much of the visible portion of the Moon appears to be illuminated and ranges from 0 percent to 100 percent. It does *not* refer to how much of the Moon is illuminated by the Sun, which is always 50 percent. Calculating the percent illumination requires knowing an object's phase angle, which is angle $\angle PA$ in figure 7.9. The phase angle is the angle made between an incident light ray striking an object and the light ray reflected back from the object. In other words, the phase angle is the angle formed by the Earth-Object-Sun.

The phase angle is related to, but it is not the same as, an object's elongation. An object's elongation is the angle Object-Earth-Sun, which in the case of the Moon is the same as the Moon's age.

Three equations are required to calculate percent illumination. First,

$$d = \cos^{-1}\left[\cos\left(\lambda_m - \lambda_\odot\right)\cos\beta_m\right], \quad (7.6.5)$$

where $\lambda_m$ and $\beta_m$ are the Moon's ecliptic coordinates and $\lambda_\odot$ is the Sun's ecliptic longitude. $d$ is actually the same as the Moon's age $A$, but we will use a different variable here to keep it distinct from equation 7.6.1 in which we arrived at the Moon's age by a different approximation. Second, the equation

$$PA = 180 - d - 0.1468\left[\frac{1 - 0.0549\sin M_m}{1 - 0.0167\sin M_m}\right]\sin d \quad (7.6.6)$$

approximates the Moon's phase angle from the Moon's mean anomaly. Finally, percent illumination is given by

$$K_\% = 100\left[\frac{1 + \cos(PA)}{2}\right]. \quad (7.6.7)$$

It should be clear that phase, age, and percent illumination are simply different ways of expressing how much of the Moon is illuminated by the Sun's rays as seen from Earth. Table 7.2 summarizes the relationship between age, phase, and percent illumination for each of the 8 phases of the Moon shown in figure 7.6.

Table 7.2 Age, Phase, and Percent Illuminated
All three measures in this table are ways of expressing how much of the Moon is illuminated as seen from Earth.

| Phase | Age | | | |
| --- | --- | --- | --- | --- |
| | Degrees | Days | F | % |
| New Moon | 0 | 0.0 | 0.00 | 0 |
| Waxing Crescent | 45 | 3.7 | 0.15 | 15 |
| First Quarter | 90 | 7.4 | 0.50 | 50 |
| Waxing Gibbous | 135 | 11.1 | 0.85 | 85 |
| Full Moon | 180 | 14.8 | 1.00 | 100 |
| Waning Gibbous | 225 | 18.5 | 0.85 | 85 |
| Last Quarter | 270 | 22.1 | 0.50 | 50 |
| Waning Crescent | 315 | 25.8 | 0.15 | 15 |
| New Moon | 360 | 29.5 | 0.00 | 0 |

Let us work through an example by calculating the age of the Moon (in degrees and days) and its phase on January 1, 2015, for the observer from the previous section.

1. Compute the Sun's ecliptic longitude. (We will solve the equation of the center for this example to compute the Sun's ecliptic longitude.)
   (Ans: $\lambda_\odot = 280.248151°$.)
2. Compute the Moon's true longitude for the given date.
   (Ans: $\lambda_t = 50.389181°$.)
3. Use equation 7.6.1 to calculate the age of the Moon in degrees.
   (Ans: $A = -229.858970°$.)
4. If necessary, adjust $A$ to be in the range $[0°, 360°]$.
   (Ans: $A = 130.141030°$, or $A_{days} = 10.7$ days.)
5. Apply equation 7.6.4 to compute the Moon's phase.
   (Ans: $F = 0.82$.)

Since we calculated the phase to be about 130°, this is closest to a Waxing Gibbous Moon (see table 7.2).

Let us continue with this example by also computing the percent illumination.

1. Compute the Sun's ecliptic longitude. (We will solve the equation of the center for this example to compute the Sun's ecliptic longitude.)
   (Ans: $\lambda_\odot = 280.248151°$.)
2. Calculate the Moon's mean anomaly and ecliptic coordinates.
   (Ans: $M_m = 71.222237°$, $\beta_m = -2.981288°$, $\lambda_m = 50.279952°$.)
3. Apply equation 7.6.5 to compute the Moon's age in degrees. (Note how this value compares to the Moon's age obtained earlier when equation 7.6.1 was used.)
   (Ans: $d = 129.966690°$.)
4. Use equation 7.6.6 to calculate the Moon's phase angle.
   (Ans: $PA = 49.924934°$.)
5. Use equation 7.6.7 to calculate the percent illumination.
   (Ans: $K_\% = 82\%$.)

Again, be sure to note that percent illumination (82 percent in this case) refers only to how much of the visible portion of the Moon is illuminated. Also, note that if the phase $F$ obtained for this example (0.82) is multiplied by 100, it gives the Moon's illumination as a percentage. The values obtained for $F$ and $K_\%$ may differ slightly because they are obtained by different approximations,

# The Moon

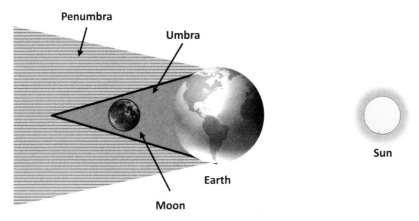

**Figure 7.10** Lunar Eclipse
In a lunar eclipse, Earth passes between the Moon and Sun. (This figure is not to scale.)

in particular by how the Moon's age is calculated. However, they should be the same value to at least 1 decimal point of accuracy.

## 7.7 Eclipses

An eclipse is one of the most exciting astronomical events to observe. An eclipse occurs when one celestial object, such as the Moon, moves into the shadow of another, such as Earth. Similar to but distinct from an eclipse is an occultation. An occultation occurs when one object is hidden or partially obscured because another object is in the line of sight. For example, when the Moon passes in front of a star or a planet, the Moon occults the star or planet. One could hardly say that the star or planet is in the Moon's shadow; hence, this type of astronomical event is an occultation rather than an eclipse. We will not consider occultations any further.

There are 2 types of eclipses: lunar and solar. Both are caused when the Moon or the Earth passes into a shadow, thus blocking out the Sun's light. Figure 7.10 shows a total lunar eclipse. In this case, the Moon enters Earth's shadow. The darkest portion of the shadow is the umbra whereas the lighter portion of the shadow is the penumbra. A total lunar eclipse occurs when the Moon is totally within the umbra. A partial lunar eclipse occurs when the Moon is partly in the umbra and partly in the penumbra.

A lunar eclipse can be seen from anywhere on the nighttime part of the Earth. Determining whether a particular observer can see an eclipse is easy. Once the time for a lunar eclipse is known, compute the moonrise and moonset

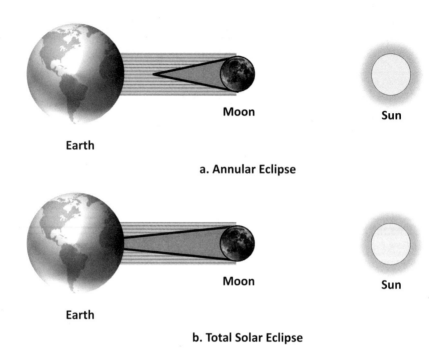

**Figure 7.11** Solar Eclipse
In a solar eclipse, the Moon passes between Earth and the Sun. (This figure is not to scale.)

time for the observer. If the time at which the eclipse will occur falls within the moonrise and moonset times for the observer, then that observer can see the eclipse.

Solar eclipses occur when Earth moves into the shadow cast by the Moon. Refer to figure 7.11. Because the Moon is a smaller object, the shadow cast by the Moon is not as long or as large as that cast by Earth. Therefore, as shown in figure 7.11a, the umbra formed during a solar eclipse does not always reach Earth. This type of eclipse is an annular solar eclipse and the result, as seen from Earth, is that the Moon appears as a smaller disk on top of the Sun.

When the orbital geometry is just right, the umbra in a solar eclipse will reach Earth, as shown in figure 7.11b. This type of eclipse is a total solar eclipse. During a total solar eclipse only a small part of Earth is in the umbra of the Moon's shadow, yet a larger portion of Earth is in the penumbra. Thus, solar eclipses are more difficult to predict because an observer's location on Earth is important. Observers who fall within the umbra will see the Sun as totally obscured by the Moon while observers who fall within the penumbra will still see an eclipse, but they will only see the Sun as partially obscured. Observers outside the penumbra will not see an eclipse at all.

Through careful observation, astronomers have devised a set of rules that can be used to determine if an eclipse will occur. Although we will not actually compute the occurrence of an eclipse, some of the rules can easily be used to determine whether an eclipse is likely. Determining exactly when and where an eclipse will occur is much more difficult and beyond the scope of this book.

After a moment's reflection, it should be obvious that eclipses can occur only when the Moon is near opposition to, or conjunction with, the Sun. That is, an eclipse can occur only near a Full Moon or a New Moon. At a Full Moon, Earth is between the Sun and Moon so a lunar eclipse is possible only at a Full Moon. At a New Moon, the Moon is between the Earth and Sun so a solar eclipse is possible only at a New Moon. However, eclipses do not occur each month because the position of the Moon must be such that it is in, or very close to, the ecliptic plane.

Astronomers have observed that eclipses follow a pattern called the Saros cycle, which, although not exact, is 18 years, 11 days, 8 hours in length. There are at least 2 solar eclipses, even if they are only partial solar eclipses, but not more than 5 solar eclipses each year. While at least 2 solar eclipses will occur each year, there may not be any lunar eclipses. If there are any lunar eclipses, there will be at most 3 in a single year. However, the total number of eclipses (solar and lunar) for a given year will not exceed 7. A lunar eclipse is often preceded by a solar eclipse that occurred 2 weeks earlier, or it is followed by a solar eclipse that will occur 2 weeks later.

In the algorithm presented in section 7.3 for determining the Moon's position, the Moon's true orbital longitude ($\lambda_t$) and corrected ecliptic longitude of the ascending node ($\Omega'$) are computed. Those values can be used to determine when an eclipse *cannot* occur as well as when an eclipse *must* occur.

With regards to lunar eclipses, if

$$|\lambda_t - \Omega'| > 12°15' \qquad (7.7.1)$$

or

$$|\lambda_t - \Omega' - 180°| > 12°15', \qquad (7.7.2)$$

then a lunar eclipse *cannot* occur. However, if

$$|\lambda_t - \Omega'| < 9°30' \qquad (7.7.3)$$

or

$$|\lambda_t - \Omega' - 180°| < 9°30', \qquad (7.7.4)$$

then a lunar eclipse *must* occur. These seemingly arbitrary bounding angles (12°15′, 9°30′, and the 2 bounding angles presented in the following equations for solar eclipses) are arrived at by considering the Moon's angular size and how close the Moon must be to the ecliptic plane for an eclipse to occur. When a lunar eclipse occurs, the maximum time that it can be seen, including the time that the Moon is in Earth's penumbra, is $3^h 40^m$. The maximum time for the umbral phase of a lunar eclipse is $1^h 40^m$.

With regards to solar eclipses, if

$$|\lambda_t - \Omega'| > 18°31' \qquad (7.7.5)$$

or

$$|\lambda_t - \Omega' - 180°| > 18°31', \qquad (7.7.6)$$

then a solar eclipse *cannot* occur. However, if

$$|\lambda_t - \Omega'| < 15°31' \qquad (7.7.7)$$

or

$$|\lambda_t - \Omega' - 180°| < 15°31', \qquad (7.7.8)$$

then a solar eclipse *must* occur. The maximum time that an annular solar eclipse will last for a given place on the Earth's surface is $12^m 24^s$, while a total solar eclipse will last for a maximum of $7^m 40^s$.

In applying the various rules for determining the occurrence of an eclipse, it is important to remember that an eclipse can occur only during Full Moon or New Moon. Thus, for example, there is no point in looking at $\lambda_t$ and $\Omega'$ for a First Quarter Moon to determine if an eclipse will occur.

## 7.8 Program Notes

The program for this chapter builds upon many of the routines from prior chapters, such as the routines for performing time and coordinate system conversions, and calculating the position of the Sun. When calculating the Sun's position, the program for this chapter allows choosing whether to solve the equation of the center or Kepler's equation. However, the program only solves the equation of the center to determine the Moon's position.

# The Moon

## 7.9 Exercises

For these sample problems, solve the equation of the center when it is necessary to calculate the Sun's position.

1. An observer is at 95° W longitude, 30° N latitude. The date is August 9, 2000, and the observer is in the Central Standard Time zone on daylight saving time. If the LCT is $12^h$, what are the Moon's ecliptic, equatorial, and horizon coordinates?
   (Ans: $\beta_m = 3.044500°$, $\lambda_m = 257.219940°$, $\alpha_m = 17.094802^h$, $\delta_m = -19.794427°$, $h_m = -50°44'$, $A_m = 84°56'$.)

2. For the observer in the last problem, at what time will the Moon rise and set?
   (Ans: $LCT_r = 15^h47^m$, $LCT_s = 2^h35^m$ the next day.)

3. Another observer is located at 30° W longitude, 20° S latitude within the Eastern Standard Time zone. For the date May 15, 2010, assume that the observer is on daylight saving time. If the LCT is $14^h30^m$, what are the ecliptic, equatorial, and horizon coordinates for the Moon?
   (Ans: $\beta_m = 2.417166°$, $\lambda_m = 76.416359°$, $\alpha_m = 4.998364^h$, $\delta = 25.150750°$, $h_m = 26°32'$, $A_m = 313°24'$.)

4. At what time will the Moon rise and set for the previous observer?
   (Ans: $LCT_r = 6^h21^m$, $LCT_s = 16^h25^m$.)

5. At $12^h$ UT on August 9, 2005, for an observer at 0° latitude, 0° longitude, how far away was the Moon? What was its angular diameter?
   (Ans: $Dist_m = 363{,}361$ km or $225{,}782$ miles, $\theta_m = 0°33'$.)

6. At $14^h30^m$ UT on May 6, 2005, for an observer at 0° latitude, 0° longitude, how far away was the Moon? What was its angular diameter?
   (Ans: $Dist_m = 363{,}402$ km or $225{,}807$ miles, $\theta_m = 0°33'$.)

7. On August 9, 2005, at $12^h$ UT, what was the Moon's age? What was the phase of the Moon? What percentage of the visible portion of the Moon was illuminated?
   (Ans: $A = 42°$, $A_{days} = 3.5$ days, $F = 0.13$, $K_\% = 13\%$, which is closest to a Waxing Crescent Moon.)

8. On May 6, 2005, at $14^h30^m$ UT, what was the Moon's age? What was the phase of the Moon? What percentage of the visible portion of the Moon was illuminated?
   (Ans: $A = 331°$, $A_{days} = 27.1$ days, $F = 0.06$, $K_\% = 6\%$, which is closest to a Waning Crescent Moon.)

# 8 Our Solar System

The structure of our Solar System as the ancients understood it was simple and straightforward: it consisted of the Sun, Earth, Moon, and a handful of planets. Modern astronomers, however, recognize that our Solar System has a complex structure with a rich variety of objects held captive by the Sun's gravitational pull. We will briefly examine our Solar System's structure and some of its objects before we proceed to algorithms for locating the planets. Building upon the foundations laid in preceding chapters, this chapter provides a valuable set of additional mathematical tools for exploring the nighttime sky.

Broadly speaking, astronomers describe our Solar System's structure in terms of regions of space. Objects within those regions orbit the Sun or are locked in orbit around another object that in turn orbits the Sun (e.g., Earth and the Moon). Proceeding outward from the Sun, the structure[1] can be categorized as:

• Inner Planets: This region of space contains the 4 planets closest to the Sun (Mercury, Venus, Earth, and Mars).

• Asteroid Belt: Situated between Mars and Jupiter, this region of space is a massive debris field containing the dwarf planet Ceres and tens of thousands of asteroids.

• Outer Planets: This region lies beyond the Asteroid Belt and contains the Solar System's largest planets (Jupiter, Saturn, Uranus, and Neptune). While the inner planets are rocky objects, the outer planets are enormous gaseous objects.

• Kuiper Belt: This region of space is a large debris field beyond Neptune that extends to about 50 AUs (4.65 billion miles) from the Sun. Three dwarf planets (Pluto, Haumea, and Makemake) lie within the Kuiper Belt.

---

1. Astronomers do not universally agree that these groupings are the best way to describe the Solar System's structure. Moreover, the regions may change as our knowledge grows, and regions may overlap because their boundaries are not sharply defined.

- Scattered Disc: Icy objects with highly eccentric orbits that may extend 100 AUs (9.29 billion miles) or more from the Sun characterize this sparsely populated region of space. The dwarf planet Eris lies within this region.[2]
- Oort Cloud: This theoretical region of space marks the outermost boundary of our Solar System. It begins beyond the Scattered Disc region and extends 200,000 AUs (18.5 trillion miles) from the Sun.

Astronomers categorize Solar System objects in a variety of ways, including by the region of space in which they reside or originate, by their physical size, whether they are massive enough to have cleared their neighborhood, and by various orbital characteristics. Trojans and centaurs are 2 interesting classes of objects. A trojan shares the same orbit as a planet or a larger moon and is in a stable orbit in which it remains in the same position relative to that larger object. For technical reasons beyond the scope of this book, a trojan's position in the shared orbit is 60° ahead of or behind the larger object.[3] To stay in a stable position relative to the object it leads or follows, a trojan must have roughly the same orbital period as the object whose orbit it shares.

More than 6,000 trojans have been discovered in Jupiter's orbit and there may be millions more that are larger than 1 km in size. Astronomers have confirmed the presence of 7 trojans in Mars's orbit, 12 in Neptune's orbit, 1 in Uranus's orbit, and a temporary one in Venus's orbit. Trojans have been discovered in the orbits of 2 of Saturn's moons (Tethys and Dione). Moreover, a temporary trojan has been discovered in Ceres's orbit and in the asteroid Vesta's orbit. The IAU Minor Planet Center maintains a list of all known trojans and their orbital elements, which is updated as new trojans are discovered.

Perhaps more interesting than these trojans is the discovery, announced by NASA in 2011, of a near-Earth asteroid designated as 2010 TK7. With a diameter of 1,000 feet, this trojan shares the same orbit as Earth! Earth is in no danger of colliding with 2010 TK7 because the 2 objects have nearly the same orbital period and are never closer to each other than 12.4 million miles.

In contrast to trojans, centaurs do not have stable orbits. Centaurs are thought to originate in the Kuiper Belt, but because of the strong gravitational forces of the giant planets in the Outer Planets region, centaurs periodically cross the orbits of 1 or more of those giants. Because they have unstable orbits, centaurs will eventually crash into the Sun or a planet, leave the Solar System entirely, or become a short-period comet.

2. Some astronomers consider the Scattered Disc region to be part of the Kuiper Belt.

3. In a 3-body problem, there are exactly 5 places, called Lagrange points, at which the gravitational and centripetal forces acting upon the smaller object are in balance, thus allowing the smaller object to remain in a stable position relative to the larger object whose orbit it shares. Trojans oscillate around the Lagrange points designated as L4 (60° ahead) and L5 (60° behind).

The first centaur was discovered in 1977 and named Chiron. The second one, Pholus, was not discovered until 1992. Chiron was originally classified as an asteroid, but subsequent analysis showed that it also exhibited the characteristics of a comet. Because objects such as Chiron and Pholus are "half-asteroid" and "half-comet," they are called centaurs after the half-man, half-horse creature in Greek mythology.

Although only a few hundred objects have been confirmed as centaurs since Chiron and Pholus were discovered, astronomers estimate that there may be tens of thousands of centaurs in the Solar System that are 1 km or larger in size. The largest known centaur, 10199 Chariklo, is estimated to have a diameter of 162 miles and lies in an orbit between Saturn and Uranus. Despite its relatively small size, 2 rings encircle Chariklo! There is also evidence that Phoebe may have been a centaur that wandered too close to Saturn and was captured by Saturn's gravity to become a Saturnian moon.

With so many exotic objects having been discovered, we now know that our immediate neighborhood, astronomically speaking, is indeed very different from the simple Solar System that the ancients knew. Our Solar System is a fascinating place with untold numbers of new discoveries awaiting us.

The next few sections will describe various objects and their relative brightnesses in the night sky. For example, when viewed from Earth, Venus has an average apparent visual magnitude of $-4.9$ while Jupiter has an average apparent visual magnitude of $-2.9$. This means that Venus will appear to be brighter than Jupiter. However, Jupiter is much larger and much farther away from Earth than Venus is. If both planets were viewed from the same distance, the much larger Jupiter would appear to be brighter than Venus (an apparent visual magnitude of $-9.4$ for Jupiter versus $-4.4$ for Venus). To account for differences in planetary sizes and their distance from Earth, astronomers typically give a planet's apparent visual magnitude as if measured from a standard distance of 1 AU. So, the data given in section 8.6 and the calculations in section 8.11 assume that an object is 1 AU from Earth. That is why the visual magnitudes given for the planets will seem to vary in this chapter and why one would state that Venus is brighter than Jupiter when in fact Venus only *seems* brighter because it is closer to the Earth.

## 8.1 The Search for Planets

With Pluto now relegated to being a dwarf planet, there are 8 confirmed planets in our Solar System. Proceeding outward from the Sun, they are Mercury, Venus, Earth, Mars, Jupiter, Saturn, Uranus, and Neptune. Pluto's orbit lies beyond that of Neptune, although Pluto is sometimes closer to Earth than Neptune is.

**Figure 8.1** NASA Montage of Planets
Images taken by various NASA spacecraft were combined to create this montage. From top to bottom are Mercury, Venus, Earth (the Moon is to the right of Earth), Mars, Jupiter, Saturn, Uranus, and Neptune. The planets are not shown to scale. (Image courtesy of NASA/JPL)

Whether other planets remain to be found within the Solar System has been a topic of considerable debate for many years. Searching for a new planet is difficult, even with sophisticated telescopes, powerful electronic sensors, and extensive mathematical tools at our disposal. Finding a needle in a haystack, to borrow a tired phrase, is almost certainly an easier job.

Mathematics and physics are arguably the most powerful tools that astronomers have at their disposal to explore the universe and search for new planets. To explain perturbations observed in Mercury's orbit, the French mathematician Urbain Le Verrier (1811–1877) hypothesized that an unknown planet would be found in an orbit between the Sun and Mercury. Le Verrier named the hypothetical planet Vulcan, after the Roman god of fire, because of its close proximity to the Sun. Since Le Verrier had earlier used mathematical analysis alone to successfully predict the existence of Neptune, astronomers took his predictions quite seriously and began to search in earnest for Vulcan.

In March 1859, the French physician and astronomer Edmond Modeste Lescarbault was studying the Sun when he noticed what he first thought was a sunspot. After observing it for some time, Lescarbault realized that whatever he was seeing was moving too rapidly across the Sun's surface to be a sunspot. He continued observing the object for over an hour and made calculations from which he concluded that he was observing the transit of a previously unknown planet. Lescarbault announced to the world that he had found the planet whose existence Le Verrier had predicted.

All subsequent efforts to find the object Lescarbault observed, including attempts made in this century by astronomers and NASA's STEREO spacecraft, have failed to find anything that could be the hypothesized planet Vulcan. Moreover, by applying Einstein's *Theory of General Relativity*, astrophysicists are able to explain the perturbations in Mercury's orbit without having to assume the existence of some new planet. It is possible that what Lescarbault saw was an asteroid transiting the Sun. Even today, some still search for small asteroids orbiting the Sun in the region of space between the Sun and Mercury. All attempts to date to find any such objects, referred to as vulcanoids, have been unsuccessful.

Despite our considerable knowledge about the Solar System, the search for new planets within our Solar System continues. In January 2016, Michael Brown and Konstantin Batygin of the California Institute of Technology (Caltech) announced they had mathematical evidence for a new planet, dubbed Planet 9,[4] that they predict lies in an orbit 20 times farther away from the Sun

---

4. Incidentally, Brown was a driving force behind the effort to downgrade Pluto to a dwarf planet. Perhaps discovering Planet 9 will be his penance for demoting Pluto!

than Neptune. Furthermore, they predict that Planet 9 has an orbital period of 10,000–20,000 years and a mass 10 times greater than Earth's mass.

Planet 9 is presently just a theory. It has yet to be observed or otherwise confirmed, and so it is not known whether Planet 9 will someday be celebrated as the ninth planet in our Solar System or if it will eventually be viewed as merely a 21st-century version of the search for Vulcan. Ironically, just as Le Verrier proposed Vulcan to explain perturbations in Mercury's orbit, Brown and Batygin proposed Planet 9, in part, to explain perturbations in the orbits of Uranus, Neptune, and certain objects that lie beyond the Kuiper Belt.

The quest for other planets is not restricted to our Solar System. Planets outside our Solar System are called exoplanets. The first confirmed exoplanet was discovered in 1992 when radio astronomers Aleksander Wolszczan and Dale Frail discovered 2 exoplanets orbiting a pulsar (designated as PSR B1257+12) in the constellation of Virgo.[5] A third exoplanet, located 2,300 light years from the Sun, was discovered in 1994 orbiting that very same pulsar.

As of February 2016, nearly 2,000 exoplanets, almost all of which are in the Milky Way Galaxy, have been confirmed. Most were discovered through NASA's Kepler Space Telescope, which was launched in 2009 specifically to search for exoplanets. Future missions and spacecraft are already in the planning stages to continue the search and to discover which exoplanets are in their corresponding star's so-called habitable zone. The habitable zone is the region of space around a star in which orbiting objects large enough to be planets can hold an atmosphere and support liquid water on their surface.

NASA maintains an online Exoplanet Archive with the equatorial coordinates and other basic data about all confirmed exoplanets. One such exoplanet, Kepler-186f, is located in the Cygnus constellation, some 490 light years away from Earth. Announced in 2014, Kepler-186f is the first confirmed exoplanet that is about the same size as Earth and whose orbit lies within its sun's habitable zone.

Located 50 light years away, exoplanet 51 Pegasi b (now officially named Dimidium) was discovered in 1995 in the constellation of Pegasus. Dimidium orbits a star that is very much like our Sun. However, with surface temperatures of 1,000 °C (1832 °F), Dimidium is hardly habitable. It is half as massive as Jupiter and orbits its sun at approximately the same distance as Mercury is from our Sun. Dimidium orbits its sun in only 4 days while its orbit is such that 1 side of Dimidium always faces its sun while the other side is in perpetual darkness.

---

5. A pulsar is a rotating star that emits a beam of electromagnetic energy. The beam can be detected only when it is pointing in the same direction as Earth. A pulsar is similar to a lighthouse in that a ship at sea can see the lighthouse's beam of light only when the beam is pointing in the same direction as the ship. The first pulsar was discovered in 1967 by Antony Hewish and Jocelyn Bell.

The real exploration of, and search for, exoplanets and star systems similar to our Solar System is just beginning. Each new discovery raises a host of new questions while continuing to demonstrate how truly immense and amazing our universe is. For example, the largest currently known "solar system" is 100 million light years away. It has an exoplanet, designated as 2MASS J2126-8140, in orbit around its sun at an astounding distance of 642 billion miles (6,900 AUs)! Estimated to be 12 times more massive than Jupiter, this exoplanet requires nearly a million years to complete just 1 orbit around its sun.

## 8.2 The Inner Planets

Named inner planets because they orbit so close to the Sun, the 4 inner planets Mercury, Venus, Earth, and Mars are terrestrial planets. Terrestrial objects are Earthlike, which means that they are similar to Earth in terms of their composition, not that they are necessarily capable of supporting life. Terrestrial objects have a solid surface composed primarily of silicon-based or metal-bearing rocks. To be classified as terrestrial, an object must also have a molten metallic (typically iron) core and surface features similar to those found on Earth (craters, mountains, volcanoes, canyons, etc.). Earth is the only terrestrial planet with liquid oceans, although Mars may have had liquid oceans in its remote past. Terrestrial objects are not limited to planets. For example, the dwarf planet Ceres is a terrestrial object because it meets the required geological criteria. Moreover, terrestrial objects are not limited to our Solar

**Figure 8.2** Terrestrial Planets
These images taken by various NASA spacecraft show the relative sizes of the Solar System's 4 terrestrial planets. Proceeding outward from the Sun (left to right) are Mercury, Venus, Earth, and Mars. (Image courtesy of NASA/JPL/HST/JHUAPL)

**Figure 8.3** Mercury's Caloris Basin
Mercury's Caloris Basin, the bright area at the upper right, was photographed by NASA's MESSENGER spacecraft. Much like our Moon, Mercury's surface is pitted with craters caused by the impact of numerous meteoroids. (Image courtesy of NASA/Johns Hopkins University Applied Physics Laboratory/Arizona State University/Carnegie Institution of Washington)

System. Data gathered by the Kepler Space Telescope indicates that there may be 40 billion terrestrial exoplanets in the Milky Way alone!

### 8.2.1 Mercury

Mercury is the closest planet to the Sun. It is named after the Roman god Mercury, the speedy winged messenger of the gods who was the god of commerce, travel, and thievery. Mercury sometimes appears in the sky as a "morning star" and sometimes as an "evening star." The Greeks called it Apollo when it appeared in the morning and Hermes when it appeared in the evening, although they eventually realized that both names referred to the same object. The Greeks and Romans were not the first to discover Mercury; it was well known long before the rise of the Greek and Roman civilizations. The earliest known reference to Mercury comes from the ancient Sumerians, whose writings about the planet on cuneiform tablets date back to 3000 BC.

Mercury's orbit is so close to the Sun that it is often difficult to see even though it is bright enough to be visible to the naked eye. While the *dwarf planet* Pluto is smaller than Mercury, Mercury is the smallest known *planet* in our Solar System. With a diameter of 3,030 miles, Mercury is slightly larger than the Moon, and like the Moon it exhibits phases when viewed from Earth. Orbiting at an average distance of 36 million miles (0.39 AU) from the Sun, Mercury's orbital period is 88 Earth days (a Mercurian year) while it takes nearly 59 Earth days (a Mercurian day) to complete 1 rotation on its axis relative to the stars. No moons have been detected orbiting Mercury.

Because Mercury is so close to the Sun, daytime surface temperatures soar to 800 °F while nighttime temperatures drop to as low as −280 °F. Mercury's density is nearly the same as Earth's density, but its mass is so much smaller (0.06 times that of the Earth) that Mercury's gravitational field is not strong enough to hold much of an atmosphere. Technically called an exosphere, Mercury's atmosphere consists of only trace amounts of oxygen, hydrogen, helium, and other gases. Its atmospheric pressure is so low that it is essentially a vacuum.

The first space probe to reach Mercury was Mariner 10, which sent back images of the planet in 1975 during the course of 3 different flybys. NASA launched the MErcury Surface, Space ENvironment, GEochemistry and Ranging (MESSENGER) space probe in 2004 to do a more detailed study of the planet. From data that MESSENGER captured, it was discovered that Mercury's surface is much like the Moon with maria-like plains, craters, and impact basins all over its surface. Mercury's largest crater is Caloris Basin, which has a diameter of 960 miles, making it larger than the state of Texas. MESSENGER also detected water in the form of ice in craters at Mercury's North Pole. After orbiting Mercury some 4,000 times over the course of 4 years, MESSENGER used its last remaining fuel to intentionally leave orbit and crash onto the planet.

The next mission to Mercury will most likely be the joint ESA-JAXA BepiColombo mission, which was launched in 2018 and consists of 2 spacecraft: the Mercury Planetary Orbiter (MPO) and the Mercury Magnetospheric Orbiter (MMO). Its ambitious scientific mission includes studying Mercury's geological structure, exosphere, and magnetic field. In what is perhaps a poetic nod to Le Verrier, who tried so valiantly to explain perturbations in Mercury's orbit, the BepiColombo mission includes performing experiments to confirm Einstein's theory of relativity, which astronomers believe is the true explanation for those perturbations.

### 8.2.2 Venus

Venus is named after the Roman goddess of love and beauty. The planet sometimes appears in the evening sky and sometimes in the morning sky, as Mercury does. Venus is usually the planet that is meant when a reference is made to the morning or evening star. The ancient Greeks gave Venus the name Phosphorus when it appeared as a morning star and Hesperus when it appeared

**Figure 8.4** Venus
A dense cloud cover perpetually obscures the surface of Venus. This picture is a composite of radar images taken by NASA's Magellan spacecraft to show the Venusian surface without the planet's dense cloud cover. (Image courtesy of NASA/JPL)

as an evening star. However, thanks to the Greek mathematician Pythagoras, the Greeks came to realize that both names referred to the same object.

Except for the Moon, Venus is the brightest natural object in the nighttime sky. Venus has a visual magnitude of $-4.9$, making it significantly brighter than Mercury, whose visual magnitude is $-2.6$. Both planets are brighter than Sirius, the brightest star in the sky with a visual magnitude of $-1.5$. By comparison, the Pole Star (Polaris), with a visual magnitude of 2.0, is much dimmer than Sirius or either planet.

Venus was well known in the ancient world. The oldest known reference comes from the Sumerians whose priests composed hymns to honor the goddess Inanna, which was their name for Venus. In the Western Hemisphere, the Aztecs offered human sacrifices to placate Venus because they believed Venus was the harbinger of disasters. Venus was the most important planet to the Mayans, who considered it to be a companion of the Sun and associated with war. Mayan astronomers made exacting observations from which they created a highly accurate calendar for predicting when Venus would appear in the morning sky. Mayan leaders tried to arrange battles to coincide with favorable movements of the planet. When victorious, they offered captured warriors as human sacrifices to Venus.

Venus is unique in the Solar System for at least 2 reasons. First, Venus rotates on its axis from east to west. All other planets in the Solar System, with the possible exception of Uranus, rotate in the opposite direction. If Earth rotated from east to west, the Sun would rise in the west and set in the east. Second, Venus is unique because its day is longer than its year! Venus rotates on its axis with respect to the stars once every 243 Earth days whereas it takes 225 Earth days to complete 1 orbit around the Sun.

In terms of size, mass, composition, and proximity to the Sun, Venus is so similar to Earth that it is sometimes called Earth's sister planet. Venus orbits the Sun at an average distance of 67 million miles (0.72 AU), has a diameter of 7,500 miles (Earth's diameter is 7,900 miles), and is 0.82 times as massive as Earth. There are no Venusian moons, but a trojan presently shares the same orbit as Venus.

Although Venus and Earth may share enough similarities to be considered sister planets, they are starkly different. Venus is the hottest planet in the Solar System with surface temperatures measured as high as 900 °F, which is hot enough to melt lead (whose melting point is 622 °F). Dense clouds perpetually cover the entire planet, causing a runaway greenhouse effect as heat from the Sun's rays is retained in a heavy atmosphere that is more than 95 percent carbon dioxide. Whereas Mercury has almost no atmospheric pressure, atmospheric pressure on the surface of Venus is 92 times greater than the atmospheric pressure on Earth's surface.

The ancients used the Latin name Lucifer to aptly refer to Venus. Space probes sent to Venus reveal a hellish world in which sulfuric acid rains down from billowy clouds of sulfur. Lightning bursts periodically light up the sky when storms rage across the planet. Over 1,000 volcanoes, some of which may still be active, dot the landscape with lava flows that extend for hundreds of miles. In ages past, the flowing lava carved out immense canals on the surface, one of which is more than 3,000 miles in length.

Space probes reveal that Venus has 2 large highland areas. The Aphrodite Terra extends for 6,000 miles at Venus's equator, making it about the size of South America. The Ishtar Terra located near Venus's north pole is approximately the size of Australia. Maxwell Montes, the tallest mountain on Venus, is located in the Ishtar Terra region and reaches a height of 6.8 miles, making it over a mile higher than Earth's Mt. Everest.

More than 40 space probes have been sent so far to explore this hostile alien world. Some of those probes were sent into the Venusian atmosphere, and some even landed on its surface. Because of the extreme surface temperatures, high atmospheric pressure, corrosive sulfuric acid from Venusian rainstorms, and generally harsh conditions on Venus, none of the probes sent toward the surface survived for more than a few hours.

The first space probe to reach Venus was Mariner 2, which reached the planet in December 1962 and sent back data showing the planet's extreme temperatures. Mariner 2 was a historic milestone because it was the first time that a space probe from Earth visited another planet. Between 1965 and 1978, the Soviets successfully landed 10 Venera space probes on Venus. Although the Soviets' Luna 9 successfully landed on the Moon in 1966, when Venera 7 landed on Venus in December 1970 it was the first space probe to ever land on the surface of another planet. The Soviets achieved another first in October 1975 when Venera 9 was the first space probe to send back photographs from the surface of another planet. Venera 13 followed in March 1982 and sent back the first ever color images of the Venusian surface.

In the early 1990s, NASA placed the Magellan and Galileo spacecraft in orbit around Venus to map its surface with radar and infrared sensors. Both ESA and JAXA sent spacecraft to Venus in the early 2000s to explore the Venusian surface and atmosphere. The joint ESA-JAXA BepiColombo mission destined for Mercury will make 2 flybys of Venus before proceeding on to its primary mission. Additionally, NASA's Solar Probe Plus will perform 7 flybys on its way to study the Sun. The Russians plan to return to Venus as well with the Venera-D program, which will continue the program that enjoyed so many successes in the mid-20th century. The first Venera-D space probe is expected to land on Venus around the year 2025.

### 8.2.3 Mars

Perhaps no planet has stirred as much controversy and imagination as has Mars, the so-called red planet whose distinctive color is due to the presence of iron oxide on its surface. Named for the Roman god of war, Mars has been scrutinized for signs of life ever since the 1870s when the Italian astronomer Giovanni Schiaparelli reported seeing *canali* (translated as "channels" or "canals") on the planet's surface. Science fiction writers capitalized on this and wrote fanciful stories about "little green men from Mars" intent on invading the Earth. The classic example is H. G. Wells's *War of the Worlds*, which was broadcast as a radio drama on October 30, 1938, and briefly succeeded in making some listeners believe that Earth really was being invaded by hostile Martians and their nefarious war machines. Mariner 9 reached Mars in 1971 but found no signs of Martian canals, built by "little green men" or otherwise. Likewise, the Viking space probes that landed in 1976 found no evidence of Martian canals or life, intelligent or otherwise. The *canali* Schiaparelli and others saw may have been the result of periodic dust storms on the surface of Mars.

With a diameter of 4,200 miles, Mars is the second smallest planet in our Solar System. It orbits the Sun at an average distance of 141.6 million miles (1.52 AU) and completes 1 orbit around the Sun in 687 Earth days. Thus, a Martian year is a little less than twice the length of an Earth year. A Martian day with respect to the stars is 24.7 Earth hours, which makes a Martian day a little more than a half hour longer than a day on Earth.

Both Mariner 9 and the Viking space probes sent back important data about Mars. With a mass 0.11 times that of Earth, the atmospheric pressure on Mars is 1/100 that of the Earth. This means that if humans ever walk on its surface, they will require pressurized space suits to survive. Moreover, they will need to bring their own oxygen because the Martian atmosphere is chiefly carbon dioxide with some argon and nitrogen, but only trace amounts of oxygen.

Surface temperatures on Mars are much colder than they are on Earth. This is partially due to Mars being farther away from the Sun, but primarily because Mars has such a thin atmosphere that it does not retain much of the Sun's heat energy. Temperatures near the Martian poles may be as low as $-195\,°F$. Daytime temperatures near the equator reach a comfortable $70\,°F$, but nighttime temperatures drop to $-100\,°F$.

Mars, like Earth, has seasons. Martian polar caps grow and shrink in much the same way that Earth's polar regions grow and shrink with the changing

**Figure 8.5** Mars
This picture taken by NASA's Viking 1 Orbiter shows 3 Martian volcanoes on the left. The topmost is Ascraeus Mons, which is 10 miles high; the base is 300 miles across. Below Ascraeus Mons are Pavonis Mons and Arsia Mons, the latter of which is barely visible at the left edge of the picture. The largest volcano, Olympus Mons, is located slightly above Ascraeus Mons and is just beyond the horizon in this view. The great rift Valles Marineris in the middle of the picture stretches for 2,500 miles across the planet's surface. At the left end of Valles Marineris is Noctis Labyrinthus, a mazelike area filled with steep valleys and canyons. (Image courtesy of NASA/USGS)

seasons. This, along with its dust storms, may account for the apparent changes on the surface of Mars.

Two Martian moons were discovered in 1877 and named Phobos and Deimos. With a diameter of 14 miles, Phobos is 7 times more massive than Deimos, whose diameter is a mere 8 miles. Phobos orbits at a distance of only 3,700 miles above the Martian surface. By comparison, the Moon orbits the Earth at an average distance of 384,400 miles. Phobos is so close to Mars that it completes an orbit in 7.7 hours, which means that Phobos orbits Mars faster than Mars rotates on its axis. Consequently, anyone on the surface of Mars would see Phobos rise and set twice a day. Phobos is the only moon in the Solar System that revolves about its parent planet faster than the planet itself rotates on its own axis.

As of February 2016, 13 man-made satellites are orbiting Mars. Of these, NASA's 2001 Mars Odyssey, ESA's Mars Express, NASA's Mars Reconnaissance Orbiter (MRO), ISRO's Mars Orbiter Mission, and NASA's Mars Atmosphere and Volatile EvolutioN Mission (MAVEN) are actively sending back important scientific data. NASA also has 3 rovers on Mars (Curiosity, Spirit, and Opportunity) that are providing an unprecedented amount of data about the planet.

Much to the surprise of astronomers, the first space probes to reach Mars revealed that its surface is pitted with craters that are partially filled in with windblown sand. The craters demonstrate that like our Moon and Mercury, Mars had a violent past in which it was repeatedly struck by meteoroids. The largest visible impact crater is Hellas Planitia, whose rim rises up from the surface by more than a mile. The crater's floor is over 4 miles lower than the Martian surface while radar data from the MRO space probe suggest that glaciers may exist beneath that floor.

The Vastitas Borealis basin in Mars's northern hemisphere is a lowland area that lies 2–3 miles below the surface. It covers about 40 percent of the planet and may be the result of an ancient collision with an immense meteoroid. Located to the east of the Argyre Planitia impact basin is the Galle crater. First photographed by the Viking 1 Orbiter, Galle is known as the "Happy Face Crater" because a curved mountain range in the crater's interior gives the illusion that there is a giant "smiley face" on the surface of Mars.

The most infamous optical illusion discovered on Mars is undoubtedly the "Face on Mars," which is in the Cydonia region at 40.8° N latitude, 9.6° W longitude, about halfway between the Arandas and Bamberg craters. Somewhat resembling a human face, it was discovered in images captured by Viking 1 and was quickly seized upon by some as proof of the existence of life on Mars. However, better images taken by NASA's Mars Global Surveyor (MGS) in September 1997 revealed that the formation is actually a mesa. What appeared to some to be a huge monument and a surrounding complex built by an alien race was simply a product of fortuitous lighting and human imagination.

The Tharsis region is a large mountainous area centered near Mars's equator in its western hemisphere. This region is 1,120 miles across and contains 3 enormous volcanoes first discovered by Mariner 9. More recently, NASA's Mars Odyssey discovered a group of 7 caves near Arsia Mons, the southernmost of those 3 volcanoes. The smallest of the 3 volcanoes is Pavonis Mons,

**Figure 8.6** Face on Mars
The "Face on Mars" is located in the Cydonia region. The black dots in this image are errors in the original data from the Viking 1 Orbiter. More recent imaging confirms that this formation is nothing more than a mesa. (Image courtesy of NASA/JPL)

which is located between the other 2. Ascraeus Mons is the tallest of the 3 with a height of 10–11 miles and a diameter of 300 miles. By comparison, Earth's largest volcano is the Mauna Loa volcano in Hawaii. Mauna Loa has a maximum diameter of 75 miles and rises about 2.6 miles above sea level. The extinct volcano Olympus Mons is the tallest volcano on Mars and the tallest volcano in the Solar System. Located northwest of the Tharsis region, it is 370 miles, wide and reaches a height of 13–17 miles, making it about 3 times higher than Mount Everest.

Seemingly everything on Mars is larger than its counterparts on Earth. The same is true of Valles Marieneris, the Martian equivalent of Earth's Grand Canyon. Valles Marieneris is a canyon system located just below the Martian equator that stretches east to west for 2,500 miles. It is 120 miles wide in some places and over 6 miles deep at its deepest point. Earth's Grand Canyon is a mere 227 miles long with a maximum width of 18 miles and a maximum depth of 1 mile.

Several space probes sent to Mars have confirmed the presence of ice at the poles and even liquid water at other places on its surface. With the confirmed

# Our Solar System

presence of water and the discovery of natural shelters, such as the caves near Arsia Mons, Mars is likely to be the first planet that humankind will colonize. Several proposals for a manned mission to Mars have been made, but the first such visit may not be until 2025. In the meantime, ESA plans to send their own rover in 2020, while NASA plans to send an astrobiology rover to Mars in 2020. The United Arab Emirates has also announced their plans to launch a Mars probe in 2020 to study the Martian atmosphere.

With active probes orbiting Mars and multiple robotic rovers on its surface, our understanding of the planet continues to grow by leaps and bounds. Much of the imagery and data are available online for us armchair explorers to enjoy and explore firsthand. Detailed maps can be viewed online at Google Mars. NASA's Mars Trek and Experience Curiosity online tools and ESA's Mars Express website allow one to "fly" around Mars to explore the planet in 3D. These types of tools and ready access to the treasure trove of images garnered from Martian space probes provide amazing views of the Martian surface that earlier generations with even their best instruments could only dream of.

## 8.3 The Outer Planets

The outer planets (Jupiter, Saturn, Uranus, and Neptune) are those Solar System planets whose orbits lie in the region of space between the Asteroid and Kuiper Belts. They are sometimes called the Jovian planets because they are so dominated by the sheer size and mass of Jupiter. The outer planets are also called the Gas Giants because of their composition, although that terminology is falling out of favor. Some astronomers categorize only Jupiter and

**Figure 8.7** Outer Planets
This grouping of images shows the sizes of the outer planets relative to Earth. Earth is on the left, followed by Jupiter, Saturn, Uranus, and Neptune. Although not an outer planet, Pluto is the small dot on the extreme right. (Original image courtesy of NASA/Lunar and Planetary Institute, edited by the author)

Saturn as Gas Giants while categorizing Uranus and Neptune as Ice Giants. "Giant" is indeed appropriate for all 4 planets, all of which are at least 10 times more massive than Earth and have a diameter at least 3.8 times greater than that of Earth. By any measure, relative to Earth the outer planets are giants indeed.

In terms of their composition, the outer planets are substantially different from the inner planets. The inner planets all have a solid surface primarily composed of rocky materials. While it is true that about two-thirds of the Earth's surface is covered by water, that water exists on top of an underlying rocky surface. Space probes can, and have, landed on the surface of all the inner planets. By contrast, the outer planets do not have a solid, well-defined surface on which a space probe could land. Instead, the outer planets are composed primarily of hydrogen, helium, and water in various physical states (gas, liquid, ice). The outer planets may have some rocky materials, but such materials are largely found only within the planets' cores. Trying to land on the "surface" of Jupiter or Saturn would be like trying to "land" in Earth's atmosphere, while landing on Uranus or Neptune would be like landing on a block of ice that has nothing solid underneath it.

### 8.3.1 Jupiter

Jupiter is by far the largest planet in our Solar System. With a diameter exceeding 86,800 miles, it is 11 times larger in diameter than Earth and has 318 times more mass. Jupiter is so enormous that it is 2.5 times more massive than all the other planets in the Solar System *combined*. It is therefore appropriate that Jupiter is named after the Roman ruler of the gods. Jupiter is the same as the Greek god Zeus, who, in addition to being the king of the gods, was also the god of the sky and thunder.

With an apparent visual magnitude of $-2.9$, Jupiter is the third brightest natural object in the nighttime sky. Only the Moon and Venus, and sometimes Mars (depending on how close it is to Earth), appear brighter. Jupiter is easily visible to the naked eye and was well known to the ancient world. The earliest written references to Jupiter come from the Babylonians who recorded their astronomical observations in the 7th or 8th century BC. The ancient Babylonians knew the planet as Marduk, who in their theology was both the king of the gods and the patron god of the city Babylon.

Jupiter orbits the Sun at an average distance of 483.9 million miles (5.20 AUs), completing a single orbit with respect to the stars in 4,333 Earth days (11.86 Earth years). Jupiter rotates on its axis in 9.9 hours, making it the planet with the shortest day in our Solar System. Jupiter's atmosphere is 90 percent hydrogen with some helium, methane, and ammonia. The temperature at

# Our Solar System

**Figure 8.8** Jupiter and Its Moons
Jupiter with its 4 planet-size moons, called Galilean satellites, were photographed by NASA's Voyager 1 and assembled into this collage. Although not shown to scale, they are in their correct relative positions. Io (center left) is nearest Jupiter, Europa is in the center, Ganymede is bottom center, and Callisto is on the lower right. Four much smaller satellites circle Jupiter inside Io's orbit while Jupiter's other satellites lie millions of miles away. (Image courtesy of NASA/JPL)

the top of the clouds in its atmosphere is estimated to be as low as $-234\,°F$ while temperatures at the planet's core may be $43,000\,°F$. Jupiter's core is thus about 4 times hotter than the surface of the Sun.

As one of the gas giants, Jupiter has no clearly definable surface. Even if it were possible to land on Jupiter, it would hardly be a hospitable place. In December 1995, the Galileo orbiter parachuted a titanium space probe about 100 miles into Jupiter's atmosphere. The probe transmitted data for nearly an hour before it was destroyed by an atmospheric pressure 23 times greater than the pressure on Earth's surface and an atmospheric temperature exceeding $300\,°F$. The Galileo orbiter was itself sent into Jupiter's atmosphere in September 2003. Before vaporizing in Jupiter's hostile atmosphere, the orbiter recorded temperatures in excess of $570\,°F$ and wind speeds in excess of 400 miles per hour.

**Figure 8.9** Jupiter's Great Red Spot
Voyager 1 took this image of Jupiter's famous Great Red Spot in December 1998 through a series of color filters to highlight subtle features. The Great Red Spot is a raging, structurally complex storm in Jupiter's upper atmosphere that is 3.5 times the size of Earth. (Image courtesy of NASA/JPL)

The clouds in Jupiter's atmosphere form alternating dark and light beltlike zones that encircle the planet. These zones are bands of high- and low-pressure areas that run parallel to Jupiter's equator because of the planet's rapid rotation. Because Jupiter rotates so rapidly, objects at its equator whirl around at nearly 28,000 miles per hour. By comparison, objects at Earth's equator move at a mere 1,000 miles per hour as Earth rotates.

Jupiter's most prominent feature is the Great Red Spot. Located south of Jupiter's equator, the Great Red Spot is a massive anticyclonic (a cyclone that rotates in a counterclockwise direction) weather system in Jupiter's upper atmosphere. With wind speeds reaching 350 miles per hour, the phenomenon has been observed from Earth for over 300 years. However, it is noticeably shrinking in size at about a rate of 580 miles per year. In 1831, Samuel Heinrich Schwabe measured the Great Red Spot to be 25,000 miles long. The Pioneer and Voyager space probes of the 1970s showed that it had shrunk to about 15,000 miles long. Recent images from the Hubble Space Telescope show that it is now just over 10,000 miles long. Even so, the Earth would easily fit into the Great Red Spot with plenty of room to spare.

What caused the Great Red Spot, why it is red, and why it is shrinking are unsolved mysteries. If it continues shrinking at its present rate, the Great Red Spot will be gone in about 20 years. However, some computer simulations suggest that the Great Red Spot weather system is relatively stable and will eventually stop shrinking. Even if it does disappear, recent images from the Hubble Space Telescope show 2 smaller red spots adjacent to the Great Red Spot. Perhaps red spots, however and for whatever reason they form, are a permanent feature of Jupiter's atmosphere.

With its 67 moons, Jupiter has the most moons of any object in the Solar System. The 4 largest Jovian moons, as large as some planets, are called Galilean satellites in honor of Galileo, who saw them in 1610 through his telescope. Simon Mayr, a German astronomer, discovered them at about the same time as Galileo and gave them the names by which we know them today. The names of the Galilean satellites are, in order of their orbits around Jupiter, Io, Europa, Ganymede, and Callisto. Ganymede has a diameter of 3,300 miles and therefore has a larger diameter than that of the planet Mercury. Ganymede is the largest Jovian moon and the largest moon in the Solar System.

As of February 2016, 8 spacecraft have reached Jupiter: Pioneer 10 and 11, Voyager 1 and 2, Galileo, Cassini, Ulysses, and New Horizons. Of these, the only one to orbit the planet was the Galileo orbiter, which orbited Jupiter for 7 years before it was sent plunging into Jupiter's atmosphere. The remaining spacecraft captured their images and data in flybys as they made their way to other parts of the Solar System. These spacecraft have revealed many surprises, such as Voyager 1's discovery that Jupiter has rings. Subsequent investigation has revealed that Jupiter has 4 distinct sets of rings: a 7,500-mile-wide halo ring that is nearest to the planet, a brighter but much narrower 4,000-mile-wide main ring, and 2 very faint rings called the Amalthea and Thebe gossamer rings. These rings are made up mostly of tiny dust particles and are too faint to be seen from an Earthbound observatory.

Data returned by these spacecraft has greatly increased our knowledge of Jupiter. Onboard cameras have allowed astronomers to witness events that cannot be observed from Earth. For example, Voyager photographed lightning flashes in the nighttime Jovian sky. The New Horizons flyby in 2007 also captured lightning strikes near Jupiter's poles that, for some unknown reason, were occurring at a rate of about one strike per second. Perhaps the most phenomenal event occurred in July 1994 when the Galileo orbiter captured

**Figure 8.10** Jupiter's South Pole
The Cassini space probe took this remarkable picture of Jupiter's South Pole in December 2000. It shows the complex atmosphere characteristic of this distant planet. (Original image courtesy of NASA/JPL/SSI, enhanced by the author)

the moment that a fragment from the Shoemaker-Levy 9 comet collided with Jupiter.

Spacecraft sent to Jupiter also made flybys of several of Jupiter's moons. From those flybys, astronomers discovered that Ganymede has a magnetic field, making it the only moon in the Solar system known to have this characteristic. The presence of a magnetic field suggests that Ganymede's core is composed of iron. Ganymede also appears to have an underground ocean as well as exposed water-ice at its north pole.

Images captured of Callisto show that it is the most cratered object in the Solar System. It is so heavily cratered that any new impact creates a new crater that obliterates an older one. Callisto's most prominent feature is its 1100-mile-wide Valhalla impact basin.

In contrast to Callisto, Europa is so smooth that it has been compared to a billiard ball. Europa may have an ocean underneath its surface that is mainly salty water. Europa's icy crust is cracked, making it possible that water from Europa's underground ocean may have seeped through to the surface.

Io is unlike any other known moon in the Solar System. It has more than 400 active volcanoes that are constantly erupting and spewing out massive amounts of sulfur, sulfur dioxide, and ash. Some of the volcano plumes on Io are 125 miles high while some of Io's mountains are as high as Mount Everest. Io is the most volcanic object in the Solar System.

Three new missions are under way to continue our exploration of Jupiter, and those missions will undoubtedly reveal even more surprises. NASA's Juno spacecraft reached Jupiter in July 2016 and is now in a polar orbit around the planet. ESA plans to launch the Jupiter Icy Moon Explorer (JUICE) spacecraft in 2022 to study the Galilean moons. It will be a long journey, requiring 8 years before JUICE will be in orbit around Jupiter. NASA plans to launch the Europa Multiple-Flyby Mission in 2025 with the specific mission of making a detailed study of Europa. The spacecraft will include a robotic lander that will land on Europa to study the moon's surface and see if there really is an ocean of liquid water beneath the crust. The lander will also perform various experiments to determine if conditions are favorable for life on Europa.

### 8.3.2 Saturn

Bright enough to be visible to the naked eye, Saturn is the most distant planet known to the ancients. The first reference to Saturn comes from the Assyrians who called the planet the Star of Ninib after their god of war, hunting, and agriculture. In 700 BC, the ancient Assyrians described the Star of Ninib as a god surrounded by a ring of serpents. Why the Assyrians and other ancient civilizations came to believe that Saturn has a ring around it is an intriguing mystery. The rings of Saturn cannot be seen with the naked eye, yet the telescope was not invented until the 17th century, several hundred years after the Assyrians wrote about the Star of Ninib.

Soon after the telescope was invented, Galileo turned his attention to Saturn. Because of the limitations of his telescope, Galileo could not distinguish rings around Saturn and instead thought that he was seeing a triple planet. In 1655, the Dutch mathematician, physicist, and astronomer Christiaan Huygens (1629–1695), using a higher quality 50-power telescope that he designed himself, saw that a ring encircled Saturn. Huygens also discovered the Saturnian

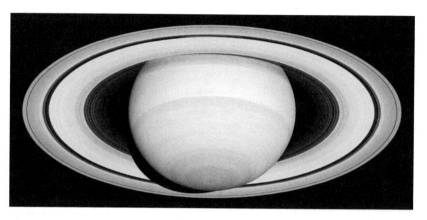

**Figure 8.11** Saturn's Rings
The rings of Saturn are quite evident in this image taken by the Hubble Space Telescope in March 2004. The wide, bright ring closest to the planet is the B ring while the narrower adjacent bright ring is the A ring. The dark gap between the A and B rings is the Cassini Division. (Image courtesy of NASA/ESA/E. Karkoschka [University of Arizona])

moon Titan and explained why Saturn's rings seem to appear and disappear. The reason, as Huygens correctly surmised, is that Saturn's rings are so thin they cannot be seen when viewed from their edge. They can only be seen when the rings are inclined with respect to an Earth-bound viewer. Twenty years later, the Italian astronomer and mathematician Giovanni Domenico Cassini (1625–1712) discovered a dark gap separating what Huygens thought was a single ring into 2 distinct rings. This gap is called the Cassini Division in his honor. Cassini also discovered 4 more moons orbiting Saturn.

Saturn is hardly the only ringed object in our Solar System, but its rings are the most impressive. Early data from Voyager 1 and Voyager 2 clearly demonstrated that the rings of Saturn are far more numerous and complex than astronomers imagined. Astronomers generally agree that there are 7–8 main ring groups encircling the planet, but those groups are themselves made up of many smaller rings and divisions, including the Cassini Division and several other divisions and gaps that have only recently been discovered. Determining how many rings Saturn has is difficult because it is a matter of how one decides where 1 ring ends and another begins.

Regardless of how many rings there are, they are comprised of billions of rocks and chunks of ice ranging in size from a grain of sand to several feet in diameter. Objects in the rings are thought to be remnants of a comet, asteroid, or an exploded Saturnian moon. The main rings, according to some estimates, are only 30 feet thick. However, the Cassini-Huygens space probe discovered that particles in some of the rings form bumps and ridges that may be as much as 2 miles high. The space probe also discovered spokes in the rings that form and disperse over a period of only a few hours. Perhaps because of the gravitational interaction of 2 small shepherd satellites, Saturn's F ring even appears to be braided! Astronomers continue to struggle to understand the incredibly complex structure of Saturn's rings.

Saturn is the second largest planet in our Solar System and lies about 887 million miles (9.54 AUs) from the Sun. It has a diameter of 72,400 miles and is 95 times more massive than Earth. Saturn orbits the Sun with respect to the stars in 29.45 Earth years. As with Jupiter, Saturn has a very short day compared to Earth's. Saturn rotates on its axis in only 10.7 hours. Like Jupiter, Saturn's atmosphere is composed mostly of hydrogen and helium with small amounts of methane and water ice. Saturn is the least dense planet in our Solar System with a density that is only 0.7 times that of water. This means that objects with the same density as Saturn would float in water.[6]

Atmospheric pressure on Saturn is 100 times greater than the atmospheric pressure on Earth's surface. Temperatures at the top of Saturn's clouds average −350 °F, while temperatures in the lower layers of Saturn's atmosphere may reach 135 °F. Saturn has a highly active weather system that spawns tremendous storms that may last for months or even years. Winds have been measured in some storms to reach 1,100 miles per hour. Monitored by the Cassini-Huygens space probe in 2004, the Dragon Storm located in Saturn's southern hemisphere generated lightning that was 1,000 times more powerful than lightning on Earth. The space probe also monitored a large band of white clouds in Saturn's northern hemisphere that astronomers named the Northern Electrostatic Disturbance. Such giant storms appear to be cyclical in nature, appearing about every 30 years. This suggests that Saturn's reoccurring storms may somehow be tied to the length of its year.

An unusual weather pattern near Saturn's north pole was first discovered by Voyager 1 and Voyager 2. This strange weather system has a hexagonal, honeycomb-like structure that is 15,000 miles across and extends downward 60 miles into the atmosphere. More than 20 years later, the Cassini-Huygens space probe captured the same weather pattern, suggesting that it may be a permanent feature of Saturn's atmosphere, or that it is periodic and by chance Cassini-Huygens reached Saturn in time to see the storm reappear.

---

6. Would Saturn actually float in water? Well, sort of. While it is true that Saturn is less dense than water, it is physically impossible for a body of water large enough to float Saturn to exist. For a detailed explanation, see Rhett Allain's "No. Saturn Wouldn't Float in Water" in the July 2013 issue of *Wired* magazine. The safest statement to make (which is true), is that objects immersed in water will float if they are less dense than water.

**Figure 8.12** Saturn's Hexagonal Storm
Numerous storms have been detected in Saturn's atmosphere. The Cassini spacecraft captured this strange hexagonal-shaped storm at Saturn's North Pole in April 2014 from a distance of 1.4 million miles. The storm is twice as wide as the Earth! (Image courtesy of NASA/JPL-Caltech/SSI)

Saturn has a periodic feature called the Great White Spot that has been observed approximately every 20–30 years since 1876. The Great White Spot, also called the Great White Oval, is a gigantic storm system analogous to Jupiter's Great Red Spot. It was last seen in 2010 when the Cassini-Huygens probe sent back data and high-resolution images of the phenomenon. Located at that time in Saturn's northern hemisphere, the Great White Spot was 25 miles above Saturn's clouds and covered a surface area of more than 600 million square miles. It had sustained winds of 300 miles per hour that lasted over 7 months before dissipating

The largest of Saturn's moons, and the first to be discovered, is Titan. Titan has a diameter of 3,200 miles making it a little smaller than Ganymede, the Solar System's largest moon, and 1.5 times larger than our Moon. Titan's atmosphere is 95 percent nitrogen with trace amounts of methane. Its atmosphere extends over 370 miles above its surface, as compared to Earth whose atmosphere extends less than 40 miles above the surface.

Liquid seas and rivers abound on Titan's surface, but they flow with liquid methane instead of water. Methane rivers are not the only hazards abounding on Titan. The Cassini-Huygens space probe discovered a cloud of hydrogen cyanide 185 miles up in Titan's atmosphere that is roughly the size of Egypt. Despite such dangers, Titan has been described by some as the *least* hostile place in the Outer Planets region of our Solar System!

As of 2015, 62 moons, including Titan, are known to orbit Saturn; 16 of those moons always present the same side toward Saturn. Titan may seem bizarre to those of us more accustomed to Earth's Moon, but Saturn has even stranger moons in its diverse collection of companions.

- Prometheus, Pandora, Pan, and several others are shepherd moons. This means that they interact gravitationally with objects in Saturn's rings to keep the rings in their orbit.
- Four of Saturn's moons are also trojans. Telesto and Calypso share the same orbit as Tethys while Helene and Polydeuces share the same orbit as Dione.
- Tethys has a huge rift called Ithaca Chasma that runs nearly three-quarters of the way around the moon.
- Janus and Epimetheus sometimes pass close to each other with the result that they swap orbits about every 4 years.
- Enceladus has more than 100 geysers near its southern pole. They are like ice volcanoes and they eject plumes of water when they erupt, forming a huge cloud of water vapor over the moon's south pole.
- Iapetus is like "Dr. Jekyll, Mr. Hyde" in that 1 of its sides is as bright as snow while the other side is as dark as coal. Besides having several craters, Iapetus has a large ridge on its dark side that almost perfectly follows its equator for nearly three-fourths of the way around the moon.
- Anthe and Methone have partial rings around them while Pallene has a complete ring encircling it. Rhea may also have a ring system around it.
- Phoebe is irregularly shaped and may be an object captured from the Kuiper Belt by Saturn's strong gravitational pull. Phoebe has a retrograde orbit, which means that it orbits Saturn in a direction opposite to that of Saturn's rotation.
- Mimas is about the size of Spain and has a large bullseye crater named Herschel whose diameter is about one-third that of the moon itself. The impact that created Herschel almost shattered Mimas.

Four space probes have been sent to explore Saturn and its moons. The first was Pioneer 11, which arrived in 1979 and was followed a short time later by Voyager 1 and Voyager 2. The last spacecraft sent to Saturn was the Cassini-Huygens spacecraft jointly sponsored by ESA, NASA, and the Italian space agency, Agenzia Spaziale Italiana (ASI). In 2005, the Huygens probe separated from the larger spacecraft and landed on Titan, making it the first, and presently the only, spacecraft to land on another planet's moon. The probe transmitted data from the surface for 90 minutes before going silent. Cassini remained on station near Saturn until it reached the end of its mission in September 2017. It was then sent into Saturn's atmosphere, where it was destroyed.

From the early Pioneer and Voyager flybys to the present day Cassini-Huygens spacecraft, our knowledge of Saturn and its neighboring moons has greatly expanded. For all our efforts, however, our understanding of this part of the Solar System is still very primitive. Saturn remains as mysterious as ever, still beckoning us like a beautiful siren.

### 8.3.3 Uranus

Sir William Herschel is credited with discovering Uranus on March 13, 1781. His discovery marked the first time in history that a planet was discovered with a telescope. Herschel was conducting a sky survey, which eventually became the basis for the widely used New General Catalog (NGC), when he thought he had discovered a comet. In fact, he first announced his discovery as being a comet. But as he studied the object in subsequent weeks, Herschel came to the startling realization that instead of a new comet he had found a previously unknown planet. Discovering Uranus assured Herschel of a respected place in the history of astronomy. A star, a crater on the Moon, a crater on Mimas, an impact basin on Mars, an asteroid, and a space telescope are all named in Herschel's honor.

Despite the acclaim, Herschel was not the first to view Uranus, which is just barely visible to the naked eye. The Greek astronomer Hipparchus may have viewed Uranus, and Ptolemy may have recorded it as a star in *The Almagest*, an ancient catalog of the visible stars in each constellation. The British Astronomer Royal John Flamsteed definitely included Uranus as a star, which he designated as 34 Tauri, on a star chart produced in 1690. Also, 4 different astronomers viewed Uranus on multiple occasions between 1690 and 1771. Even so, Herschel was the first to recognize that Uranus is a planet.

As the one who discovered Uranus, Herschel wanted to name it Georgian Sidus (George's Star) in honor of King George III. Instead, the German astronomer Johann E. Bode suggested naming the new planet Uranus after the Greek god of the sky. Uranus is the only planet in our Solar System named after

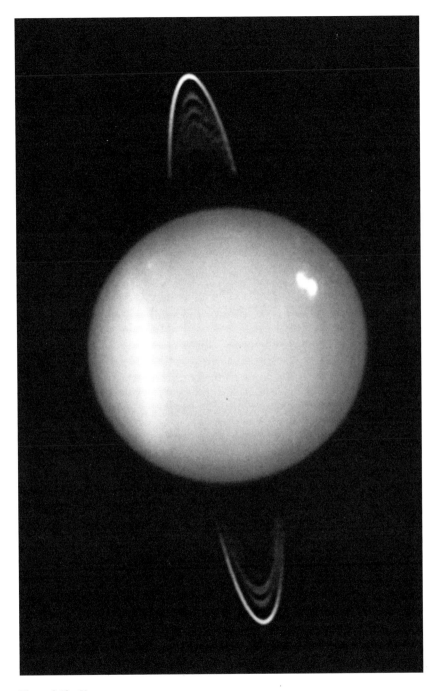

**Figure 8.13** Uranus
This image of Uranus taken in 2005 through the Hubble Space Telescope shows a storm in the northern atmosphere and a faint set of rings. Note that Uranus is "lying down" in its orbit around the Sun. (Image courtesy of NASA/ESA/M. Showalter [SETI Institute])

a Greek god instead of a Roman god. Had the convention of using Roman gods to name the planets been followed, Uranus might well have been called Caelus since that is the Roman counterpart of the Greek god Uranus.

Uranus is one of the ice giants, so called because it is composed mostly of icy materials such as water, ammonia, and methane. It is 14.5 times more massive than Earth, and it has a diameter of 31,500 miles. Uranus lies 1.8 billion miles (19.19 AUs) from the Sun and takes 84 Earth years to make 1 trip around the Sun. It rotates on its axis in just 17.2 hours. Uranus has seasons, but seasons on Uranus last for 20 years!

The atmosphere on Uranus consists mostly of methane, hydrogen, and helium. The methane in Uranus's atmosphere absorbs the red light rays from the Sun, thereby giving the planet its characteristic blue-green color. The tops of the methane clouds that cover Uranus have an average temperature of about $-350$ °F. The lower layers of those clouds may be composed of water vapor.

An interesting feature of Uranus is that it is "lying down" in its orbit. That is, its axis of rotation lies almost entirely in its orbital plane. This means that the planet's poles point almost directly toward the Sun rather than being more nearly perpendicular to the Sun, as is the case for all the other planets in the Solar System. Astronomers do not presently agree on which of Uranus's poles is its north pole and which is its south pole. For that reason, they do not agree on whether the planet rotates east to west or west to east. In any case, because of the orientation of Uranus's poles with respect to the Sun, some areas of the planet do not face the Sun for over 40 years. A night on Uranus is very long indeed!

In 1977, astronomers at the Australian Perth Observatory discovered that Uranus has rings. Voyager 2 came within 50,600 miles of Uranus in a 1986 flyby and confirmed the existence of rings around the planet as well as discovering 10 moons. Voyager 2 is the only spacecraft to date that has visited Uranus; consequently, most of what we know about the planet is limited to what can be discovered from Earth. Fortunately, astronomers periodically use the Hubble Space Telescope and various Earthbound tools to study Uranus. Because of such efforts, we know that Uranus has 27 moons and 13 faint rings. Unfortunately, no new missions to send spacecraft to Uranus have been announced, so our knowledge of this ice giant may well remain limited for the foreseeable future.

### 8.3.4 Neptune

Before the discovery of Neptune, the only hope for discovering a new planet was by chance. No one pointed their telescope to a spot in the sky and expected to find a new planet there. However, 2 astronomer/mathematicians, John C.

# Our Solar System

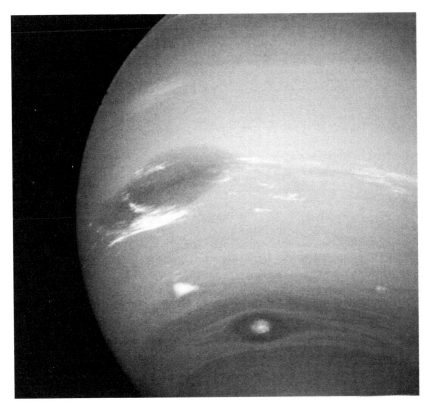

**Figure 8.14** Neptune
Voyager 2 took this image of Neptune in 1989. It shows 3 large anticyclonic storms: the Great Dark Spot is on the left near Neptune's equator, the Scooter storm is the white area just below the Great Dark Spot, and the Small Dark Spot is at the bottom. (Image courtesy of NASA/JPL)

Adams of England and Urbain Le Verrier of France, independently and almost concurrently calculated the position of where Neptune should be. The German astronomer Johann G. Galle sighted the planet where it was predicted to be on September 23, 1846. To date, Neptune is the only planet whose discovery was predicted by mathematics before it was actually observed.

There is, however, evidence that perhaps Galileo should be given credit as Neptune's discoverer. Neptune and Jupiter were at about the same place in the sky in December 1612 when Galileo was studying the moons of Jupiter. He included what he may have thought to be a star in a sketch of Jupiter and its moons, but he noted that the object had moved when he again turned his telescope to Jupiter in January 1613. There is also evidence that at least 2 other astronomers saw the planet prior to Galle's sighting. Even so, Galle was the first to announce that he had observed a new planet, and he had done so by turning his telescope to the very location that Adams and Le Verrier had predicted.

Following the tradition of naming planets after Roman gods, Neptune was named after the Roman god of the seas. That is perhaps appropriate because the methane in Neptune's atmosphere gives the planet its characteristic blue color, which is reminiscent of the color of Earth's oceans. The temperature in Neptune's atmosphere has been measured as low as $-360\,°F$, while storms observed in Neptune's atmosphere have been measured with speeds in excess of 1,500 miles per hour. It is truly a giant planet with a mass more than 17 times that of Earth and a diameter of 30,600 miles. Neptune lies 2.8 billion miles from the Sun (30.07 AUs) and has an orbital period of 165 Earth years. Neptune rotates on its axis every 16.1 hours.

To date, the only spacecraft to visit Neptune has been Voyager 2, which discovered 3 large anticyclonic storms in Neptune's atmosphere. Neptune has the most violent storms ever discovered on any planet in our Solar System. The largest storm Voyager 2 discovered, the Great Dark Spot, extended 8,000 miles in the east-west direction and 4,100 miles in the north-south direction. Voyager 2 also discovered a storm that was named Scooter because of how quickly it moves across the planet's surface. Slightly larger than Scooter but substantially smaller than the Great Dark Spot, the Small Dark Spot is another storm in Neptune's southern hemisphere that Voyager 2 imaged during its 1989 flyby. Because of its appearance, the Small Dark Spot is sometimes known as the Wizard's Eye. Some have compared Neptune's Great Dark Spot to Jupiter's Great Red Spot. While storms on Jupiter appear to have long lives, storms are relatively short lived on Neptune. The Hubble Space Telescope was trained on Neptune in 1994, but by then all 3 storms Voyager 2 discovered had disappeared.

In 1982, astronomer Edward Guinan announced the tentative discovery of a ring encircling Neptune. However, Voyager 2 data showed that Guinan had actually discovered Neptune's moon Larissa. Then in 1984, astronomers at the Chilean La Silla Observatory and the Chilean Cerro Tololo Interamerican Observatory announced that they had observed rings around Neptune. Voyager 2 confirmed that the 1984 observations were indeed of rings around Neptune, thus bringing the number of confirmed ringed planets in our Solar System from 1 (Saturn) to 4 (Saturn, Jupiter, Uranus, and Neptune).

Unlike the other ringed planets in our Solar System, the objects encircling Neptune are unique in that they do not form a complete ring, but instead form 1 or more incomplete arcs. Three faint arcs comprise the Adams ring, named in honor of John C. Adams who had predicted Neptune's position. Two other rings with incomplete arcs are named after Le Verrier, who had also predicted Neptune's position, and Galle, who is credited with discovering Neptune.

As of January 2016, 14 moons have been discovered orbiting Neptune. The largest is Triton, which is larger than our Moon and may be an object from the Kuiper Belt captured by Neptune's gravity. Triton, like Saturn's moon Phoebe, has a retrograde orbit. Triton's orbit is rapidly decaying. In 1966, Caltech student Thomas McCord predicted that Triton will collide with Neptune in 100 million years because it will be within Neptune's Roche Limit. The Roche Limit is the distance at which an orbiting body will disintegrate due to the overpowering tidal forces of the body around which it orbits. Modern astronomers agree with McCord's conclusion, although they predict that Triton's orbit will not decay to the Roche Limit for another 3.6 billion years. When that happens, the result may be another asteroid belt in the outer reaches of our Solar System, or the creation of a ringed planet whose rings rival those of Saturn.

## 8.4  The Dwarf Planets

The discovery of Eris created quite a stir within the astronomy community. When it was discovered, astronomers thought that Eris was 27 percent more massive than Pluto and had a diameter a little larger than Pluto's diameter. However, subsequent study has shown that Eris is very close to being the same size as Pluto. Even so, the obvious question that arose was whether Eris should be designated as the 10th planet in our Solar System; in fact, the IAU met in 2006 to resolve that very question. In so doing, they formally defined a planet and created the new category of dwarf planets. Under the definitions formulated by the IAU, Eris is classified as a dwarf planet. At present, the IAU officially recognizes 5 Solar System objects as dwarf planets: Pluto, Ceres, Haumea, Makemake, and Eris. A dozen or more other objects are being evaluated to determine if they, too, meet the qualifications for a dwarf planet.

NASA's New Horizons space probe reached the Kuiper Belt in 2015 while NASA's Dawn space probe reached the Asteroid Belt in the same year. A major mission for both probes is to study the known dwarf planets in their proximity and to look for other dwarf planets. Some estimate that 200 more dwarf planets may be found within the Kuiper Belt when the New Horizons probe completes its mission. When the search is extended to regions of space outside the Kuiper Belt, there may be tens of thousands more dwarf planets patiently waiting to be discovered. Consequently, the data about the dwarf planets shown in table 8.1, as well as the number of officially recognized dwarf planets, is very likely to change substantially as these space probes carry out their missions. As the

**Figure 8.15** Pluto's "Heart"
This image of Pluto was taken by NASA's New Horizons spacecraft in July 2015. A prominent feature of Pluto's surface is the heart-shaped area seen at the bottom of the image. (Image courtesy of NASA/JHUAPL/SwRI)

table shows, all known dwarf planets (and indeed all currently suspected dwarf planets) are much smaller than Earth's Moon. The two largest, Pluto and Eris, are only about two-thirds the size of our Moon.

We now briefly turn our attention to the 5 official dwarf planets, starting with Pluto. We begin with Pluto because of its historical importance and in deference to those of us who cannot resist still fondly thinking of it as a full-fledged planet.

### 8.4.1 Pluto

Relatively little is known about Pluto. Even with the largest Earthbound telescopes, Pluto appears only as a point of light. Fortunately, NASA's New Horizons space probe has arrived in the vicinity and is currently studying Pluto and other objects in the Kuiper Belt. New information about Pluto will surely be revealed in the coming years.

Discovered by the American astronomer Clyde Tombaugh in 1930 at the Lowell Observatory, Pluto was named after the Greek god of the underworld. Tombaugh did not suggest the name Pluto; it was suggested by an 11-year-old girl, Venetia Burney, from Oxford, England. The name was considered

# Our Solar System

**Table 8.1** Our Solar System's Dwarf Planets
This table provides basic information about the 5 officially recognized dwarf planets.

| Object | Distance from Sun (AUs) | Diameter (Miles) | Orbital Period (Years) |
|---|---|---|---|
| Ceres | 2.77 | 590 | 4.6 |
| Pluto | 39.48 | 1,464 | 247.9 |
| Haumea | 43.13 | 1,218 | 283.3 |
| Makemake | 45.79 | 890 | 309.9 |
| Eris | 68.01 | 1,445 | 560.9 |

especially appropriate because not only did it continue the tradition of naming celestial objects after mythological characters, but its first 2 letters (PL) would also honor Percival Lowell, who founded the Lowell Observatory. Lowell had also predicted Pluto's existence in 1915 because of perturbations he had observed in the orbits of Uranus and Neptune.

Pluto lies 3.7 billion miles (39.48 AUs) away from the Sun. It takes Pluto 247.9 years to complete 1 orbit around the Sun while rotating on its axis in 6.4 Earth days. Like Uranus, Pluto's axis of rotation is nearly in the plane of its orbit around the Sun. Its diameter is 1,430 miles, making it much smaller than the smallest planet (Mercury) and smaller even than our Moon. Its mass is only 0.002 times that of the Earth. Because of Pluto's relatively small size, some astronomers suggested that it might once have been one of Neptune's moons. It is true that Pluto's orbit overlaps Neptune's orbit, and therefore sometimes it is closer to the Sun and the Earth than Neptune is. However, Pluto's orbit is highly inclined relative to Neptune's orbit and their orbits do not intersect. Thus, it seems unlikely that Pluto was ever one of Neptune's moons.

The surface of Pluto is a frozen mixture of nitrogen, methane, carbon monoxide, and water with surface temperatures that reach as low as −400 °F. As Pluto's orbit takes it closer to the Sun, the surface temperature rises and thaws some of the surface ice to give Pluto a measurably larger atmosphere. The atmosphere refreezes as Pluto gets farther away from the Sun.

NASA's New Horizons space probe has revealed an amazing variety of geological features on Pluto's surface. Although those features have been given appropriately dignified scientific names such as Tombaugh Regio and Sputnik Planum, Pluto's surface features have also been assigned more descriptive names such as the Heart, Whale, and Brass Knuckles. The Sputnik Planum region, which is the so-called Heart, is a frozen plain of nitrogen and carbon monoxide ices. In other places, there are frozen nitrogen mountains that rise as much as 11,000 feet above the surface.

**Figure 8.16** Pluto's "Snakeskin" Terrain
This image taken in 2015 by the New Horizons spacecraft shows Tartarus Dorsa, Pluto's "snakeskin" terrain. (Image courtesy of NASA/JHUAPL/SwRI)

The New Horizons space probe also imaged an area near Sputnik Planum that has been named Tartarus Dorsa. The 330-mile-wide terrain in this area resembles a snakeskin, and it may actually be tightly packed elevations of frozen methane that rise some 1,650 feet above the surrounding area that give the terrain a scaly appearance. In other areas on Pluto's surface, there appear to be hills of frozen water that float on a surface of frozen nitrogen. According to NASA scientists, "These water-ice hills are floating in a sea of frozen nitrogen and move over time like icebergs in Earth's Arctic Ocean."

Pluto has 5 known moons, the first of which was discovered in 1978 by James Christy and Robert Harrington. Named Charon by its discoverers, it is the largest of Pluto's moons with a diameter of 730 miles. As we mentioned in section 6.1, Pluto and Charon form a binary system. They are locked in a synchronous orbit so that they always face each other with the same side. The remaining 4 moons orbiting Pluto are named Hydra, Nix, Kerberos, and Styx in keeping with the theme of naming objects related to Pluto after the mythological Hadean world of the Greeks.

### 8.4.2 Ceres

Astronomers in the latter part of the 18th century suspected that there must be a planet between Mars and Jupiter. The gap between the 2 planets just seemed too big for there not to be at least 1 planet located there. After extensive

**Figure 8.17** Ceres' Occator Crater
Taken in September 2015 at an altitude of 915 miles, this image from NASA's Dawn spacecraft shows the bright spots in Ceres's Occator crater. The Occator crater is 60 miles wide and 2 miles deep. (Image courtesy of NASA/JPL-Caltech/UCLA/MPS/DLR/IDA)

searching by several astronomers, in 1801 the Italian astronomer, mathematician, and Catholic priest Giuseppe Piazzi discovered a "planet" that he named Ceres. Astronomers later reclassified Ceres as an asteroid because it was too dissimilar to the other known planets to be considered a planet. Ceres is now categorized as both an asteroid and a dwarf planet.

Piazzi named his discovery after the Roman goddess of the harvest. Ceres is the largest known asteroid, the closest dwarf planet to Earth, and the smallest of the currently recognized dwarf planets. With a mass 0.012 times the mass of our Moon, Ceres constitutes about 25–33 percent of the entire mass of the Asteroid Belt. Ceres does not have any known moons, but there are multiple temporary trojans that share its orbit.

Orbiting the Sun at an average distance of 257.6 million miles (2.77 AUs), Ceres has an orbital period of 4.6 Earth years and rotates on its axis in 9.1 hours. Unlike other asteroids, Ceres is nearly spherical in shape and is bright enough that those with exceptionally good eyes may be able to see it with their naked eyes on very dark nights.

In 2014 ESA's Herschel Space Observatory detected water vapor being ejected from 2 different regions on the surface of Ceres. NASA's Dawn space probe is currently in the area and has imaged 2 "bright spots" inside the Occator crater in Ceres's northern hemisphere. Despite these findings, there does not appear to be volcanic activity on Ceres. The escaping water vapor may be due to temperature changes or underground pressure. The bright spots are presently thought to be water ice or salt deposits, but astronomers in the La Silla Observatory in Chile have noted that they randomly change their appearance on a daily basis. Ceres's bright spots are presently an intriguing mystery that astronomers are actively studying to understand.

### 8.4.3 Haumea

Haumea was discovered in the Kuiper Belt in 2004 or 2005, the date depending on who should receive credit for its discovery. It was named after the Hawaiian goddess of childbirth and has 2 known moons: Hi'iaka, named after the goddess of the island of Hawaii, and Nakama, named after the Hawaiian water spirit.

Haumea lies 4 billion miles (43.13 AUs) from the Sun and requires 283.3 Earth years to complete 1 orbit. With a mass a third of that of Pluto and a rotational period of 3.9 hours, Haumea is the fastest known spinning object in our Solar System. It is unique among the dwarf planets because of its football-like shape, with a minimum width of 620 miles and an end-to-end length of 950–1,200 miles. Because its surface features are difficult to detect, astronomers were forced to deduce Haumea's shape by analyzing data from sensors. Additionally, a red spot has been detected on its surface.

Little else is known about Haumea because it is so far away and so small. With a visual magnitude of 17.3, under optimum viewing conditions it can be viewed through a telescope with a minimum aperture of 25–35 inches. Even then, Haumea will appear as a mere speck of light. No spacecraft have yet visited Haumea, although NASA's Dawn space probe is currently surveying objects in the Kuiper Belt. Perhaps Dawn will soon provide more detailed information about this odd object and provide images of its surface.

### 8.4.4 Makemake

Makemake was discovered in 2005 by a team of astronomers at the Palomar Observatory. The team decided, somewhat in keeping with how Haumea was named, to name the newly discovered dwarf planet after the Rapa Nui god of fertility. The Rapa Nui are the native inhabitants of Easter Island.

Makemake lies in the Kuiper Belt at an average distance from the Sun of 4.3 billion miles (45.79 AUs). It requires 309.9 Earth years to complete 1 orbit around the Sun, and it rotates on its axis in an estimated 7.8 hours. Makemake has no known moons, which makes its mass difficult to calculate. Its diameter is estimated to be two-thirds the size of Pluto, making it a respectable-sized dwarf planet. With a visual magnitude of 17.0, it is brighter and easier to view with an Earthbound telescope than Haumea. Still, it is unlikely that much more will be known about Makemake until the Dawn space probe or some other space probe is sent to gather data.

### 8.4.5 Eris

By some estimates, Eris is the largest dwarf planet, although other estimates name Pluto as the largest. Eris is 28 percent more massive than Pluto, but it has a slightly smaller diameter. Because it is so close in size to Pluto, it is often described as Pluto's twin. In any case, Eris is large enough that if all the objects in the Asteroid Belt were coalesced into 1 object, it would easily fit inside Eris.

Eris was discovered in 2005 in the Scattered Disc region of space by a team of astronomers led by Michael Brown. It was originally named Xena after a popular television character, but its name was later changed to Eris, the Greek goddess of discord and strife. It orbits the Sun at a distance of 6.3 billion miles (68.01 AUs) and has an orbital period of 560.9 Earth years. Eris's orbital plane is the most inclined of any dwarf planet with a tilt of 47°. It has a rotational period of 25.9 hours and 1 known moon named Dysnomia.

Exploration of Eris is just beginning, so little is known about this remote dwarf planet. The surface of Eris is more reflective than snow, which suggests that its surface may be covered in ice.

## 8.5 Belts, Discs, and Clouds

Astronomers have historically concentrated on studying the Sun, Moon, and planets more than other Solar System objects, primarily because those objects are larger, have more predictable orbits, and are more easily within the range of Earthbound instruments. There are, however, other fascinating objects throughout our Solar System that are revealing their secrets to increasingly sophisticated Earthbound and spaceborne instruments. With such tools we are poised to study even the most remote objects in the Solar System.

### 8.5.1 The Asteroid Belt

When Ceres was discovered, astronomers soon found that what appeared to be a new planet was actually one of several large objects orbiting between Mars and Jupiter. Between 1801 and 1807, 3 more objects were found: Pallas, Vesta, and Juno. All were discovered in the region of space that we know as the Asteroid Belt.

Unlike planets and dwarf planets, there is no generally accepted definition of what properties an object must possess to be classified as an asteroid. Also known as planetoids or minor planets, asteroids are small, rocky objects that are typically highly irregular in shape, although some, such as Vesta, are nearly spherical. Asteroids are pitted and cratered just as the planets are. Vesta, for example, has an impact crater 250 miles wide, which is more than 80 percent of the diameter of Vesta itself. The asteroid/centaur Chariklo has 2 rings, while more than 150 asteroids are accompanied by a companion moon. Some asteroids even have 2 moons orbiting them.

Most, but not all, asteroids lie between Mars and Jupiter. Over 200 asteroids with a diameter greater than 60 miles have been discovered in the Asteroid Belt. Moreover, it is estimated that there may be 2 million more asteroids with diameters larger than 1 km. Because of their small size—with the exception of the brightest asteroid, Vesta, and on occasion some near-Earth asteroids—asteroids cannot be seen with the naked eye. Table 8.2 lists the 10 largest asteroids, their size, and their average distance from the Sun. The asteroids

**Table 8.2** Largest Asteroids
This table is representative of the thousands of cataloged objects in the Asteroid Belt. These are the 10 largest asteroids, as determined by their diameter, along with their average distance from the Sun.

| Asteroid | Diameter (miles) | Distance (AUs) |
|---|---|---|
| Ceres | 480 | 2.77 |
| Vesta | 326 | 2.36 |
| Pallas | 318 | 2.77 |
| Hygiea | 268 | 3.14 |
| Interamnia | 203 | 3.06 |
| Europa | 196 | 3.10 |
| Davida | 180 | 3.17 |
| Sylvia | 178 | 3.49 |
| Cybele | 170 | 3.44 |
| Eunomia | 167 | 2.64 |

listed in the table are representative of the thousands of cataloged objects in the Asteroid Belt.

Beginning with the German physician and astronomer Wilhelm Olbers, who discovered both Pallas and Vesta, many have suggested that the Asteroid Belt is the remnant of an exploded planet. Such a planet could have exploded because of a collision with another massive object or due to gravitational pressures generated by Jupiter. However, since the total mass of all the asteroids is less than the mass of the Moon, such a preexploded planet would have been relatively small. A more recent theory is that the Asteroid Belt is the result of thousands of planetesimals that subsequently collided and fragmented.

The sheer number of asteroids makes it impossible to assign each the name of a mythological god, person, mountain, or some otherwise interesting object. Instead, astronomers use a provisional naming scheme to name an asteroid until it is given a formal name. The scheme consists of a number followed by 2 letters followed by an optional number. The first number is the year in which the asteroid was discovered while the first letter indicates the month in which the asteroid was discovered. Instead of using A for January, B for February, and so on, astronomers use the letter A for January 1–15, B for January 16–31, C for February 1–15, D for February 16–29, and so forth. The second letter in the naming convention is used in case more than 1 asteroid is discovered within the same time period with A representing the first one discovered, B the second, and so on. For example, asteroid 1984 BC was the third asteroid (letter C) discovered between January 16 and January 31 (letter B) in 1984. The optional number is given in case more than 26 asteroids are discovered within a time period covered by a single letter.

The great number of asteroids and their sizes might lead one to think that the Asteroid Belt poses a serious threat to space travel. Attempts to visualize the Asteroid Belt often depict a zone of densely packed boulders that barely leave enough room for a hapless spaceship to pass through. This view is wrong. Isaac Asimov estimated that the asteroids, except for occasional clusters, might have an average distance of 10 million miles between them, which makes the probability of accidentally encountering an asteroid, much less colliding with one, small. Voyager passed through the Asteroid Belt on its way to Jupiter and Saturn without mishap, which is further evidence that the Asteroid Belt poses little real navigational danger.

Astronomers classify asteroids based on their orbits and spectral reflectivity. About 95 percent of all known asteroids lie totally within the Asteroid Belt and are known as Main Belt asteroids. Near-Earth asteroids have orbits that carry them outside the Asteroid Belt and are subdivided into 3 major subgroups.

Asteroids in the Amor subgroup's asteroids have orbits between Earth and Mars, the Apollo subgroup have orbits that cross Earth's orbit, and those in the Aten subgroup have orbits that cause them to spend most of their time inside Earth's orbit.

As of November 2014, 11,600 near-Earth asteroids have been classified as Earth-crossers, meaning that their orbits cross Earth's orbit at 1 or more points. Of these, approximately 1,000 are 1 km or larger in diameter. Astronomers are interested in tracking these objects because if they were to collide with Earth, the results could be catastrophic. For example, on February 15, 2013, an asteroid estimated to be only 65 feet in diameter entered the Earth's atmosphere over Chelyabinsk, a city in the northeast part of Russia. The asteroid created a shock wave that left a trail of damage for 55 miles and injured 1,200 people, mostly from flying glass as the shock wave blew windows out of buildings.

The asteroid that struck near Chelyabinsk was a member of the Apollo subgroup of asteroids. Earth's trojan asteroid, 2010 TK7, also belongs to the Apollo subgroup, but it is in no danger of striking Earth. More ominously, as of August 2015, more than 1,000 Apollo asteroids have been found that are large enough and close enough to Earth to be potentially dangerous. In addition, over 100 Aten asteroids have been identified as potentially dangerous to Earth. The IAU Minor Planets Center has the responsibility of identifying and tracking all potentially hazardous asteroids.

Besides classifying asteroids by their orbital trajectory, astronomers also classify asteroids by spectral reflectivity, which offers clues to an asteroid's composition. There are presently 14 classes of asteroids based on spectral reflectivity. It is estimated that over 75 percent of asteroids are C-type asteroids. They are gray in color and their reflectivity suggests that they are largely composed of carbonaceous materials. S-type asteroids are reddish or greenish in color and are stony, silicate rocks. Some 17 percent of all known asteroids are S-type. M-type asteroids are metallic, F-type asteroids have a flat reflectivity, R-type asteroids are reddish in color, and V-type asteroids appear to be basaltic in nature, which suggests that they are a result of volcanic activity.

In a remarkable demonstration of space navigation and engineering, NASA's Near Earth Asteroid Rendezvous (NEAR) Shoemaker space probe orbited the asteroid 433 Eros and then landed on it in February 2001. In 2006, JAXA's Hayabusa spacecraft performed an even more amazing feat; it landed on the near-Earth asteroid 25143 Itokawa, collected samples, and then returned to Earth. Images, data, and samples collected from these space probes have done much to increase our knowledge of asteroids.

In perhaps the most ambitious undertaking since the manned Apollo Moon missions, in 2013 NASA announced the early planning stages for a mission to capture a near-Earth asteroid and move it into an orbit around the Moon. Once in a lunar orbit, it could be visited by astronauts and studied first hand. The audacity of such a mission is truly the stuff of which science fiction writers dream.

### 8.5.2 The Kuiper Belt

The Kuiper Belt does not have well-defined boundaries. Astronomers generally agree that it begins beyond Neptune's orbit and extends 50 AUs from the Sun. Like the Asteroid Belt, the Kuiper Belt is a massive debris field containing thousands of objects following their own orbits around the Sun. However, the Kuiper Belt differs from the Asteroid Belt in at least 2 significant ways. First, the Kuiper Belt is 20 times larger than the Asteroid Belt and may contain 200 times more mass than the Asteroid Belt. Second, most asteroids are composed of rocks and metals whereas the denizens of the Kuiper Belt are largely frozen objects composed of methane, water, and ammonia.

Objects within the Kuiper Belt are sometimes referred to as trans-Neptunian objects (TNOs), or more often as Kuiper Belt Objects (KBOs). Pluto is the largest known KBO, although it is estimated that there may be 200 or more other dwarf planets in this region of space. Little hard data is available about KBOs because they are so far away. NASA's New Horizons spacecraft is currently in the Kuiper Belt collecting important data, but it will be several years before the data is fully understood. At present, astronomers generally believe that some of the moons orbiting the Outer Planets, such as Saturn's Phoebe and Neptune's Triton, originated in the Kuiper Belt.

The total amount of mass estimated to be in the Kuiper Belt is small in comparison to the planets and moons that populate the rest of the Solar System. The total mass is estimated to be as much as a tenth to as little as a hundredth of the Earth's mass. Because they have such little mass, KBOs are dominated by and heavily impacted by Neptune's gravitational pull. The cumulative effects of Neptune's gravity over eons of time have created gaps in the Kuiper Belt. Even so, Neptune's gravity alone is insufficient to explain perturbations observed in some KBOs. Attempting to explain such perturbations is part of what prompted Caltech astronomers Brown and Batygin to postulate the existence of Planet 9.

### 8.5.3 The Scattered Disc Region

The Scattered Disc region of space, which some include as part of the Kuiper Belt, begins 30–35 AUs from the Sun and extends up to 100 AUs away from the Sun. Objects within this region of space, called Scattered Disc Objects (SDOs), are icy objects much like those in the Kuiper Belt. However, SDOs typically have an orbital eccentricity that may be as high as 0.8 and orbital planes inclined by as much as 40° with respect to the ecliptic plane. Astronomers believe that the strong gravitational influence of Neptune causes SDOs to have erratic orbits. Although perihelion for most SDOs lies within the Kuiper Belt, perihelion for some may be as much as 150 AUs from the Sun.

Because of the erratic nature of a typical SDO orbits, the Scattered Disc region is thought to be the source of short-period comets. A short-period comet is one that may take up to 200 years to complete 1 orbit around the Sun. Halley's Comet, whose orbital period is 75–76 years, is one such short-period comet thought to originate in the Scattered Disc region of space.

Sedna is an example of an SDO. Discovered in 2004, Sedna is three-fourths the size of Pluto. Its highly elliptical orbit causes its distance from the Sun to range between 8 billion (86 AUs) and 84 billion miles (937 AUs). This makes Sedna the most distant and coldest known object within the confines of our Solar System. Because of its vast distance away from the Sun, Sedna takes 10,500–11,400 years to complete just 1 orbit around the Sun. With an estimated diameter of 1,100 miles, Sedna is quite likely to be a dwarf planet.

The individual sometimes credited as having discovered the most comets is American astronomer Carolyn Shoemaker. She was the codiscoverer of the Shoemaker-Levy 9 comet that was observed crashing into Jupiter in 1994. Besides discovering or codiscovering 32 comets, Shoemaker is also credited with discovering 800 asteroids and 377 minor planets.

### 8.5.4 The Oort Cloud

The Oort Cloud is a theoretical region of space that begins at about 50,000 AUs (0.8 light years) from the Sun and extends to 200,000 AUs (3.2 light years) from the Sun.[7] Instead of being flat like the Asteroid and Kuiper Belts, the Oort Cloud is theorized to be spherical in shape with the Sun at the center of the sphere. No one has yet seen the Oort Cloud or found any objects within that region of space. However, that fact does not keep astronomers from theorizing what may be in the Oort Cloud! It is not surprising that because of the vast distances involved, Oort Cloud objects are expected to be icy objects varying in size from a few feet to perhaps a few hundred miles across. The Oort Cloud may contain trillions of objects that are at least 0.5 miles across, and perhaps billions of objects 10 or more miles across.

Why do astronomers believe there is an Oort Cloud? In 1932, the Estonian astronomer Ernst Öpik suggested that long-period comets might originate in a cloud located at the outermost boundary of our Solar System. Öpik's idea did

---

7. Some astronomers limit the farthest extent of the Oort Cloud to 100,000 AUs.

not catch on until the Dutch astronomer Jan Oort independently suggested the same idea in 1950. Hence, this theoretical region of space is named in Oort's honor.

Long-period comets, such as Comet Hale-Bopp, take thousands of years to complete an orbit around the Sun. Hale-Bopp's orbital period is 2,500 years and it is up to 25 miles wide. When it made its way toward Earth in the late 1990s, it was considerably brighter than Halley's Comet and was easily visible to the naked eye for about 18 months.

Another example of a long-period comet that may have originated in the Oort Cloud is Comet Hyakutake. It was discovered in January 1996 and passed near Earth in March of that same year. When it was first spotted, astronomers calculated Hyakutake's orbital period as 17,000 years. However, as it passed through the Solar System, the gravitational effects of the objects it came near to altered its orbital period. Astronomers calculate Hyakutake's orbital period to now be in the range of 70,000 years!

## 8.6 Locating the Planets

Calculating the location of a planet is a lengthy undertaking.[8] As with locating the Sun and Moon, the main idea is to calculate a planet's true anomaly from which it can be determined how far the planet has moved along its orbit. The process is more complicated than for the Sun or Moon because we must know whether the planet under consideration is an inferior or superior planet,[9] and we must determine the location of the Earth.

Determining where a planet will appear in the sky begins by computing the planet's position with respect to the Sun. We essentially did this with Earth as an example planet in section 6.2, although we used a geocentric model to make the computations easier. Unfortunately, following the techniques in section 6.2 alone gives a planet's heliocentric ecliptic coordinates (i.e., with respect to the Sun) when what we really want are a planet's geocentric ecliptic coordinates (i.e., with respect to Earth). The need to convert from heliocentric to geocentric coordinates is why it is necessary to compute Earth's location.

---

8. For the remainder of this chapter, we will treat Pluto as a planet because determining the location, angular size, etc. of an object depends only on its orbital elements, not an arbitrary classification (planet, dwarf planet, etc.) it falls under.

9. The adjectives inferior and superior refer to where an orbit falls with respect to the Earth. Orbits for the inferior planets (Mercury and Venus) are between Earth and the Sun. Orbits for the superior planets (Mars, Jupiter, Saturn, Uranus, Neptune, and Pluto) are farther away from the Sun than Earth's orbit. Inferior and superior must not be confused with Inner and Outer Planets, which refer to regions of space.

When determining a planet's position, it can be confusing whether a calculation is being done for Earth or a planet. To avoid confusion, we will use the subscript $e$ when referring to Earth (e.g., $M_e$ for Earth's mean anomaly) and the subscript $p$ when referring to a planet (e.g., $M_p$ for a planet's mean anomaly). The overall steps required are:

1. Compute the planet's mean anomaly $M_p$ with respect to the Sun.
2. Compute the planet's true anomaly $\upsilon_p$.
3. Compute the planet's heliocentric ecliptic longitude $L_p$ and latitude $\Lambda_p$.
4. Compute Earth's position in the ecliptic plane to get its mean anomaly $M_e$.
5. Compute Earth's true anomaly $\upsilon_e$.
6. Compute Earth's heliocentric ecliptic coordinates $L_e$ and $\Lambda_e$.
7. Given the heliocentric ecliptic location for Earth and the planet, project the planet's location onto the ecliptic plane to obtain its geocentric ecliptic coordinates.

The process for projecting a planet's heliocentric coordinates onto the ecliptic plane depends on whether the planet is an inferior or superior planet. Projections for both types of planets will be demonstrated in this section.

Several equations are required for this lengthy process. A planet's mean anomaly measured from the standard epoch is given by

$$M_p = \frac{360° D_e}{365.242191 T_p} + \varepsilon_p - \varpi_p, \tag{8.6.1}$$

where $T_p$ is the planet's orbital period in tropical years, $D_e$ is the number of days since the standard epoch, $\varepsilon_p$ is the planet's ecliptic longitude at the standard epoch, and $\varpi_p$ is the ecliptic longitude of the planet at the moment of perihelion. The values for $T_p$, $\varepsilon_p$, and $\varpi_p$ are provided in tables 8.3 (Inferior Planets) and 8.4 (Superior Planets).[10]

We can solve the equation of the center or Kepler's equation to obtain the true anomaly from the mean anomaly. We will approximate the equation of the center for planets by

$$E_p \approx \frac{360°}{\pi} e_p \sin M_p, \tag{8.6.2}$$

---

[10]. The data in the 2 tables has been rounded to no more than 6 decimal digits to fit on the page. Although only 6 decimal digits may be shown, the data is provided with more digits of accuracy in the orbital elements data file for this book, and the full accuracy of the data is used in the example calculations for this chapter.

**Table 8.3** Data for the Inferior Planets

This snapshot of orbital data, taken from *The Astronomical Almanac 2000* and various NASA sources, is for the standard epoch J2000. Solar and lunar data are included, but refer to chapter 6 (for the Sun) and chapter 7 (for the Moon) to calculate their angular diameter and distance.

| Data Item | Sun, Moon, Inferior Planets, and Earth | | | | |
|---|---|---|---|---|---|
| | Sun | Moon | Mercury | Venus | Earth |
| $T_p$, orbital period in tropical years | | | 0.240847 | 0.615197 | 1.000017 |
| $m_p$, mass relative to Earth | 333,000 | 0.0123 | 0.055274 | 0.814998 | 1.0 |
| $r_p$, radius in km | 695,700 | 1738.1 | 2439.7 | 6051.8 | 6378.14 |
| Length of day relative to Earth | 25.449 | 27.322 | 58.6462 | 243.018 | 1.0 |
| $e_p$, orbital eccentricity | 0.016708 | 0.0549 | 0.205636 | 0.0067767 | 0.0167112 |
| $a_p$, length of the orbital semi-major axis in AUs | 1.0 | 0.00257 | 0.3870993 | 0.723336 | 1.000003 |
| $\theta_p$, angular diameter at 1 AU | $1919''$ | | $6.74''$ | $16.92''$ | |
| $V_p$, visual magnitude at 1 AU | $-26.74$ | $-12.74$ | $-0.42$ | $-4.40$ | |
| $\mu_p$, standard gravitational parameter in km$^3$/s$^2$ | $1.32712E11$ | 4,900 | 22,032 | 324,860 | 398,600 |
| $\iota_p$, inclination of the orbital plane w.r.t. the ecliptic | $0.00005°$ | $5.145397°$ | $7.004979°$ | $3.394676°$ | $-0.000015°$ |
| $\varepsilon_p$, ecliptic longitude at the epoch | $280.466069°$ | $218.316433°$ | $252.250324°$ | $181.979100°$ | $100.464572°$ |
| $\varpi_p$, ecliptic longitude at perihelion | $282.938346°$ | $83.353451°$ | $77.457796°$ | $131.602467°$ | $102.937682°$ |
| $\Omega_p$, ecliptic longitude of the ascending node at the epoch | | $125.044522°$ | $48.330766°$ | $76.679843°$ | |

**Table 8.4** Data for the Superior Planets

This snapshot of orbital data for the Superior Planets and Pluto at the standard epoch J2000 was taken from *The Astronomical Almanac 2000* and various NASA sources.

| Data Item | Superior Planets and Pluto | | | | | |
|---|---|---|---|---|---|---|
| | Mars | Jupiter | Saturn | Uranus | Neptune | Pluto |
| $T_p$, orbital period in tropical years | 1.880848 | 11.862615 | 29.447498 | 84.016846 | 164.79132 | 247.92065 |
| $m_p$, mass relative to Earth | 0.107447 | 317.828133 | 95.160904 | 14.535757 | 17.147813 | 0.002219 |
| $r_p$, radius in km | 3389.5 | 69,911 | 58,232 | 25,362 | 24,622 | 1,151 |
| Length of day relative to Earth | 1.025957 | 0.41354 | 0.44401 | 0.71833 | 0.67125 | 6.3872 |
| $e_p$, orbital eccentricity | 0.093394 | 0.048393 | 0.053862 | 0.0472574 | 0.008590 | 0.248827 |
| $a_p$, length of the orbital semi-major axis in AUs | 1.523710 | 5.202887 | 9.536676 | 19.189165 | 30.069923 | 39.482117 |
| $\theta_p$, angular diameter at 1 AU | 9.36″ | 196.74″ | 165.60″ | 65.80″ | 62.20″ | 8.20″ |
| $V_p$, visual magnitude at 1 AU | −1.52 | −9.40 | −8.88 | −7.19 | −6.87 | −1.00 |
| $\mu_p$, standard gravitational parameter in km$^3$/s$^2$ | 42,828 | 126,687,000 | 37,931,000 | 5,794,000 | 6,835,100 | 870 |
| $\iota_p$, inclination of the orbital plane with respect to the ecliptic | 1.849691° | 1.3043975° | 2.485992° | 0.772638° | 1.770043° | 17.140012° |
| $\varepsilon_p$, ecliptic longitude at the epoch | −4.553432° | 34.396441° | 49.954244° | 313.232810° | −55.120030° | 238.929038° |
| $\varpi_p$, ecliptic longitude at perihelion | −23.943630° | 14.728480° | 92.598878° | 170.954276° | 44.964762° | 224.068916° |
| $\Omega_p$, ecliptic longitude of the ascending node at the epoch | 49.559539° | 100.473909° | 113.662424° | 74.016925° | 131.784226° | 110.303937° |

where $e_p$ is the planet's orbital eccentricity. The true anomaly is then approximated by

$$\upsilon_p \approx M_p + E_p. \tag{8.6.3}$$

A planet's heliocentric ecliptic longitude is given by

$$L_p = \upsilon_p + \varpi_p \tag{8.6.4}$$

while the heliocentric ecliptic latitude is given by

$$\Lambda_p = \sin^{-1}\left[\sin(L_p - \Omega_p)\sin\iota_p\right], \tag{8.6.5}$$

where $\Omega_p$ is the planet's ecliptic longitude of the ascending node and $\iota_p$ is the planet's orbital inclination.

Except for equation 8.6.5, the preceding equations should look familiar. They are the same as those in section 6.2 for computing the Sun's mean anomaly $M_\odot$, equation of the center $E_c$, true anomaly $\upsilon_\odot$, and ecliptic longitude $\lambda_\odot$. Of course, the orbital elements for the planet in question are used here instead of those for the Sun or Earth.

Once a planet's heliocentric coordinates are computed, we repeat the same computations for Earth. We could save ourselves 1 step by assuming $\Lambda_e$ is 0°, which would be true if the inclination of Earth's orbit with respect to the ecliptic were exactly 0°. This is easily seen by substituting $\iota_e = 0°$ into equation 8.6.5. However, table 8.3 shows that Earth's orbit is actually inclined with respect to the ecliptic by a small amount. So, we will compute Earth's heliocentric ecliptic latitude as a matter of principle, even though it will not improve our answer by very much.

When the coordinates for the planet and Earth have been calculated, we next need to know both the distance from the Sun to Earth and the distance from the Sun to the planet in question. This is because we will use spherical geometry to project the planet's heliocentric coordinates into a geocentric coordinate system, and we need to know the radius of the spheres involved to do so. The distance from the Sun in AUs, called the radius vector $R_p$, is given by

$$R_p = \frac{a_p\left(1 - e_p^2\right)}{1 + e_p \cos \upsilon_p}, \tag{8.6.6}$$

where $a_p$ is the length (measured in AUs) of the object's orbital semi-major axis, $e_p$ is the orbital eccentricity, and $\upsilon_p$ is the true anomaly.

The remaining detail is to project the heliocentric coordinates into the ecliptic plane to produce geocentric ecliptic coordinates. The equation

$$L'_p = \Omega_p + \tan^{-1}\left[\frac{\sin(L_p - \Omega_p)\cos\iota_p}{\cos(L_p - \Omega_p)}\right] \qquad (8.6.7)$$

calculates an adjustment to the planet's ecliptic longitude, so a quadrant adjustment may need to be applied to resolve the ambiguity of the arctangent function. As a longitude, $L'_p$ must be in the range $[0°, 360°]$.

If the planet is an inferior planet, compute the planet's geocentric ecliptic longitude $\lambda_p$ via the equation

$$\lambda_p = 180° + L_e + \tan^{-1}\left[\frac{R_p \cos\Lambda_p \sin(L_e - L'_p)}{R_e - R_p \cos\Lambda_p \cos(L_e - L'_p)}\right]. \qquad (8.6.8)$$

If the planet is a superior planet, use the following equation instead:

$$\lambda_p = L'_p + \tan^{-1}\left[\frac{R_e \sin(L'_p - L_e)}{R_p \cos\Lambda_p - R_e \cos(L_e - L'_p)}\right]. \qquad (8.6.9)$$

A quadrant adjustment may need to be made when the arctangent function is applied in either of these 2 equations to ensure that the longitude is in the proper range. Moreover, $\lambda_p$ may need to be adjusted after applying the other terms in the equations to ensure that it is still in the range $[0°, 360°]$.

The equation for determining a planet's geocentric ecliptic latitude, whether an inferior or a superior planet, is

$$\beta_p = \tan^{-1}\left[\frac{R_p \cos\Lambda_p \tan\Lambda_p \sin(\lambda_p - L'_p)}{R_e \sin(L'_p - L_e)}\right]. \qquad (8.6.10)$$

No quadrant adjustment is needed in this last equation since the arctangent returns a value that is already in the proper range. Computing a quadrant adjustment in this case will give an incorrect ecliptic latitude!

This lengthy process will be demonstrated for both an inferior and a superior planet. For the purposes of both examples, assume that an observer is in the Eastern Standard Time zone at 78° W longitude, 38° N latitude. Assume that it is January 3, 2016 at $22^h$ LCT, and the observer is not on daylight saving time.

### 8.6.1 Inferior Planet Example

For the stated observer, determine the location of Venus.

# Our Solar System

1. Use the equations from chapter 3 to convert the LCT to UT, GST, and LST times, and adjust the date if necessary.
   (Ans: $UT = 3.0^h$, $GST = 9.878053^h$, $LST = 4.678053^h$, Date = 1/4/2016.)

2. Compute the Julian day number $JD_e$ for the standard epoch.
   (Ans: $JD_e = 2,451,545.0$ for J2000.)

3. Compute the Julian day number $JD$ for the desired date. Be sure to use the Greenwich date and UT from step 1, *not* the LCT time and date, and be sure to include the fractional part of the day.
   (Ans: $JD = 2,457,391.625$.)

4. Compute $D_e$, the total number of elapsed days since the standard epoch, by subtracting $JD_e$ from $JD$.
   (Ans: $D_e = 5846.625$ days.)

Steps 5–8 calculate the planet's mean anomaly $M_p$ and true anomaly $\upsilon_p$. For this example, we will use equations 8.6.2 and 8.6.3 to approximate the equation of the center to obtain the true anomaly.

5. Use equation 8.6.1 to compute the mean anomaly for Venus. Since Venus is an inferior planet, use table 8.3 to get the tropical period $T_p$, ecliptic longitude at the epoch $\varepsilon_p$, and the ecliptic longitude at perihelion $\varpi_p$.
   (Ans: $T_p = 0.615197$, $\varepsilon_p = 181.979100°$, $\varpi_p = 131.602467°$, $M_p = 9417.633039°$.)

6. If necessary, adjust $M_p$ so that it falls in the range $[0°, 360°]$.
   (Ans: $M_p = 57.633039.°$)

7. Use equation 8.6.2 to solve the equation of the center for Venus.
   (Ans: $E_p = 0.655907°$.)

8. Apply equation 8.6.3 to compute the true anomaly from the equation of the center.
   (Ans: $\upsilon_p = 58.288946°$.)

Steps 9–13 calculate the planet's heliocentric ecliptic coordinates $(L_p, \Lambda_p)$ and radius vector length $R_p$.

9. Use equation 8.6.4 to compute the planet's heliocentric longitude $L_p$.
   (Ans: $L_p = 189.891413°$.)

10. If necessary, adjust $L_p$ so that it falls in the range $[0°, 360°]$.
    (Ans: no adjustment is necessary.)

11. Use equation 8.6.5 to compute the planet's heliocentric latitude $\Lambda_p$.
    (Ans: $\Omega_p = 76.679843°$, $\iota_p = 3.394676°$, $\Lambda_p = 3.119613.°$)

12. If necessary, adjust $\Lambda_p$ so that it falls in the range $[0°, 360°]$.
    (Ans: no adjustment is necessary.)
13. Use equation 8.6.6 to compute the planet's radius vector length $R_p$.
    (Ans: $a_p = 0.723336$ AUs, $e_p = 0.006777$, $R_p = 0.720735$ AUs.)

Steps 14–22 are identical to steps 5–13 except that the calculations are done for Earth.

14. Compute Earth's mean anomaly. Use table 8.3 to get the necessary data for Earth.
    (Ans: $T_e = 1.000017$, $\varepsilon_e = 100.464572°$, $\varpi_e = 102.937682°$, $M_e = 5760.137095°$.)
15. If necessary, adjust $M_e$ so that it falls in the range $[0°, 360°]$.
    (Ans: $M_e = 0.137095°$.)
16. Use equation 8.6.2 to solve the equation of the center for Earth.
    (Ans: $E_e = 0.004582°$.)
17. Apply equation 8.6.3 to compute Earth's true anomaly from the equation of the center.
    (Ans: $\upsilon_e = 0.141677°$.)
18. Compute Earth's heliocentric longitude $L_e$.
    (Ans: $L_e = 103.079359°$.)
19. If necessary, adjust $L_e$ so that it falls in the range $[0°, 360°]$.
    (Ans: no adjustment is necessary.)
20. Compute Earth's heliocentric latitude $\Lambda_e$.
    (Ans: $\Omega_e = 0.0°$, $\iota_e = -0.000015°$, $\Lambda_e = -0.000015°$.)
21. If necessary, adjust $\Lambda_e$ so that it falls in the range $[0°, 360°]$.
    (Ans: $\Lambda_e = 359.999985°$.)
22. Compute Earth's radius vector length $R_e$.
    (Ans: $a_e = 1.000003$ AUs, $e_e = 0.016711$, $R_e = 0.983291$ AUs.)

Now that we have the heliocentric positions and radius vector lengths for Earth and the planet, we need to project the planet's location onto the ecliptic plane with respect to Earth. Steps 23–33 perform that projection.

23. Use equation 8.6.7 to calculate an adjustment to the planet's ecliptic longitude. Because a quadrant adjustment may be necessary, we will first form just the argument to the arctangent function with $y$ being the numerator and $x$ being the denominator.
    (Ans: $y = 0.917443$, $x = -0.394128$.)

24. Let $T = \tan^{-1}\left(\frac{y}{x}\right)$.
    (Ans: $T = -66.751959°$.)

25. Apply a quadrant adjustment factor to $T$ to remove the ambiguity of the arctangent function. This is necessary because $T$ will be used to calculate a longitude, which must be in the range $[0°, 360°]$.
    (Ans: $Adjustment = 180°$ because $y$ is $+$ and $x$ is $-$, $T = 113.248041°$.)

26. Finish computing $L'_p$ via equation 8.6.7.
    (Ans: $L'_p = 189.927883°$.)

27. Adjust $L'_p$ if necessary to ensure it is in the range $[0°, 360°]$.
    (Ans: no adjustment is necessary.)

    Steps 28–32 are for an inferior planet only.

28. This is an inferior planet, so use equation 8.6.8 to compute the planet's geocentric ecliptic longitude. Since an arctangent is again involved, we will first compute the numerator $y$ and the denominator $x$ for the argument to the arctangent function.
    (Ans: $y = -0.718579$, $x = 0.943727$.)

29. Let $T = \tan^{-1}\left(\frac{y}{x}\right)$.
    (Ans: $T = -37.286603°$.)

30. Apply a quadrant adjustment factor to $T$ to remove the ambiguity of the arctangent function. This is necessary because $T$ will be used to calculate a longitude, which must be in the range $[0°, 360°]$.
    (Ans: $Adjustment = 360°$ because $y$ is $-$ and $x$ is $+$, $T = 322.713397°$.)

31. Finish computing $\lambda_p$ using equation 8.6.8.
    (Ans: $\lambda_p = 605.792756°$.)

32. Adjust $\lambda_p$ if necessary to ensure it is in the range $[0°, 360°]$.
    (Ans: $\lambda_p = 245.792756°$.)

33. Use equation 8.6.10 to compute the planet's geocentric ecliptic latitude. No quadrant adjustment is needed here because the arctangent returns a value in the proper range for a latitude. Computing a quadrant adjustment will give an incorrect answer!
    (Ans: $\beta_p = 1.893914°$.)

    The remaining steps convert the ecliptic coordinates to their corresponding equatorial and horizon coordinates.

34. Convert $\lambda_p$ and $\beta_p$ to their corresponding equatorial coordinates.
    (Ans: $\alpha_p = 16.283091^\text{h}$, $\delta_p = -19.407214°$.)

35. Finally, convert the equatorial coordinates to the appropriate horizon coordinates.

(Ans: $h_p = -70°42'$, $A_p = 17°08'$.)

Instead of using the equation of the center to compute the true anomalies in steps 7–8 (for the planet) and steps 16–17 (for Earth), we could have solved Kepler's equation. Had we used the simple iterative scheme to do so, the results would be

$v_p = 58.291919°$, $v_e = 0.141774°$.

Using these values and carrying through the calculations, steps 32 and 33 would yield

$\lambda_p = 245.793823°$, $\beta_p = 1.893824°$,

which then leads to horizon coordinates of

$h_p = -70°42'$, $A_p = 17°08'$

for the stated observer.

The values obtained for the Earth's and planet's true anomalies and ecliptic coordinates by using the equation of the center versus Kepler's equation are fairly close when rounded to 1 or 2 decimal places. Since our purpose in this example was only to get horizon coordinates for Venus within a few arcminutes of precision, the additional effort required to solve Kepler's equation was not worth it. However, the values obtained by solving Kepler's equation versus the equation of the center are different enough that using the equation of the center may not be sufficiently accurate for many purposes.

How accurate is this algorithm? NASA's online Horizon program gives the equatorial coordinates for Venus as

$\alpha_p = 15.952147^h$, $\delta_p = -18.316056°$

for an observer at 0° latitude, 0° longitude and $0^h$ UT on January 1, 2000. Using the algorithm presented here and solving Kepler's equation, the results obtained are

$\alpha_p = 15.952334^h$, $\delta_p = -18.317877°$,

for a difference of $0.7^s$ in right ascension and $6.6''$ in declination.

Determining the position of an inferior planet is a long process indeed! We now turn our attention to locating a superior planet. The process for doing so is not any shorter.

## Our Solar System

### 8.6.2 Superior Planet Example

For the stated observer, determine the position of Saturn.

1. Use the equations from chapter 3 to convert the LCT to UT, GST, and LST times, and adjust the date if necessary.
   (Ans: $UT = 3.0^h$, $GST = 9.878053^h$, $LST = 4.678053^h$, Date = 1/4/2016.)

2. Compute the Julian day number $JD_e$ for the standard epoch.
   (Ans: $JD_e = 2,451,545.0$ for J2000.)

3. Compute the Julian day number $JD$ for the desired date. Be sure to use the Greenwich date and UT from step 1, *not* the LCT time and date, and be sure to include the fractional part of the day.
   (Ans: $JD = 2,457,391.625$.)

4. Compute $D_e$, the total number of elapsed days since the standard epoch, by subtracting $JD_e$ from $JD$.
   (Ans: $D_e = 5846.625$ days.)

Steps 5–8 calculate the planet's mean anomaly $M_p$ and true anomaly $v_p$. For this example, we will use equations 8.6.2 and 8.6.3 to approximate the equation of the center to obtain the true anomaly.

5. Use equation 8.6.1 to compute the mean anomaly for Saturn. Since Saturn is a superior planet, use table 8.4 to get the tropical period $T_p$, ecliptic longitude at the epoch $\varepsilon_p$, and the ecliptic longitude at perihelion $\varpi_p$.
   (Ans: $T_p = 29.447498$, $\varepsilon_p = 49.954244°$, $\varpi_p = 92.598878°$, $M_p = 153.049767°$.)

6. If necessary, adjust $M_p$ so that it falls in the range $[0°, 360°]$.
   (Ans: no adjustment is necessary.)

7. Use equation 8.6.2 to solve the equation of the center for Saturn.
   (Ans: $E_p = 2.797300°$.)

8. Apply equation 8.6.3 to compute the true anomaly from the equation of the center.
   (Ans: $v_p = 155.847067°$.)

Steps 9–13 calculate the planet's heliocentric ecliptic coordinates $(L_p, \Lambda_p)$ and radius vector length $R_p$.

9. Use equation 8.6.4 to compute the planet's heliocentric longitude $L_p$.
   (Ans: $L_p = 248.445945°$.)

10. If necessary, adjust $L_p$ so that it falls in the range $[0°, 360°]$.
    (Ans: no adjustment is necessary.)

11. Use equation 8.6.5 to compute the planet's heliocentric latitude $\Lambda_p$.
    (Ans: $\Omega_p = 113.662424°$, $\iota_p = 2.485992°$, $\Lambda_p = 1.764216°$.)
12. If necessary, adjust $\Lambda_p$ so that it falls in the range [0°, 360°].
    (Ans: no adjustment is necessary.)
13. Use equation 8.6.6 to compute the planet's radius vector length $R_p$.
    (Ans: $a_p = 9.536676$ AUs, $e_p = 0.053862$, $R_p = 10.000499$ AUs.)

Steps 14–22 are identical to steps 5–13 except that the calculations are done for Earth.

14. Compute Earth's mean anomaly. Use table 8.3 to get the necessary data for Earth.
    (Ans: $T_e = 1.000017$, $\varepsilon_e = 100.464572°$, $\varpi_e = 102.937682°$, $M_e = 5760.137095°$.)
15. If necessary, adjust $M_e$ so that it falls in the range [0°, 360°].
    (Ans: $M_e = 0.137095°$.)
16. Use equation 8.6.2 to solve the equation of the center for Earth.
    (Ans: $E_e = 0.004582°$.)
17. Apply equation 8.6.3 to compute Earth's true anomaly from the equation of the center.
    (Ans: $\upsilon_e = 0.141677°$.)
18. Compute Earth's heliocentric longitude $L_e$.
    (Ans: $L_e = 103.079359°$.)
19. If necessary, adjust $L_e$ so that it falls in the range [0°, 360°].
    (Ans: no adjustment is necessary.)
20. Compute Earth's heliocentric latitude $\Lambda_e$.
    (Ans: $\Omega_e = 0.0°$, $\iota_e = -0.000015°$, $\Lambda_e = -0.000015°$.)
21. If necessary, adjust $\Lambda_e$ so that it falls in the range [0°, 360°].
    (Ans: $\Lambda_e = 359.999985°$.)
22. Compute Earth's radius vector length $R_e$.
    (Ans: $a_e = 1.000003$ AUs, $e_e = 0.016711$, $R_e = 0.983291$ AUs.)

Now that we have the heliocentric positions and radius vector lengths for Earth and the planet, we need to project the planet's location onto the ecliptic plane with respect to Earth. Steps 23–33 perform that projection.

23. Use equation 8.6.7 to calculate an adjustment to the planet's ecliptic longitude. Because a quadrant adjustment may be necessary, we will first form

just the argument to the arctangent function with $y$ being the numerator and $x$ being the denominator.
(Ans: $y = 0.709105$, $x = -0.704430$.)

24. Let $T = \tan^{-1}\left(\frac{y}{x}\right)$.
(Ans: $T = -45.189506°$.)

25. Apply a quadrant adjustment factor to $T$ to remove the ambiguity of the arctangent function.
(Ans: *Adjustment* $= 180°$ because $y$ is $+$ and $x$ is $-$, $T = 134.810494°$.)

26. Finish computing $L'_p$ via equation 8.6.7.
(Ans: $L'_p = 248.472919°$.)

27. Adjust $L'_p$ if necessary to ensure it is in the range $[0°, 360°]$.
(Ans: no adjustment is necessary.)

Steps 28–32 are for a superior planet only.

28. This is a superior planet, so use equation 8.6.9 to compute the planet's geocentric ecliptic longitude. Since an arctangent is again involved, we will first compute the numerator $y$ and the denominator $x$ for the argument to the arctangent function.
(Ans: $y = 0.558447$, $x = 10.805079$.)

29. Let $T = \tan^{-1}\left(\frac{y}{x}\right)$.
(Ans: $T = 2.958627°$.)

30. Apply a quadrant adjustment factor to $T$ to remove the ambiguity of the arctangent function.
(Ans: no adjustment is necessary.)

31. Finish computing $\lambda_p$ using equation 8.6.9.
(Ans: $\lambda_p = 251.431546°$.)

32. Adjust $\lambda_p$ if necessary to ensure it is in the range $[0°, 360°]$.
(Ans: no adjustment is necessary.)

33. Use equation 8.6.10 to compute the planet's geocentric ecliptic latitude. No quadrant adjustment is needed here.
(Ans: $\beta_p = 1.629973°$.)

The remaining steps simply convert the ecliptic coordinates to their corresponding equatorial and horizon coordinates.

34. Convert $\lambda_p$ and $\beta_p$ to their corresponding equatorial coordinates.
(Ans: $\alpha_p = 16.675230^h$, $\delta_p = -20.537509°$.)

35. Finally, convert the equatorial coordinates to the appropriate horizon coordinates.

(Ans: $h_p = -72°32'$, $A_p = 0°08'$.)

Had we used the simple iterative method to solve Kepler's equation in steps 7–8 and 16–17 instead of approximating the equation of the center, the results for Saturn would be

$v_p = 155.687438°$, $v_e = 0.141774°$, $\lambda_p = 251.284449°$, $\beta_p = 1.634685°$.

These lead to horizon coordinates $h_p = -72°31'$, $A_p = 0°37'$ for the stated observer. The additional accuracy obtained by solving Kepler's equation appears warranted in this example, particularly with regards to the result obtained for the azimuth.

NASA's online Horizon program gives the equatorial coordinates for Jupiter as

$\alpha_p = 4.870576^h$, $\delta_p = 22.131222°$

for an observer at 0° latitude, 0° longitude and $0^h$ UT on May 5, 2001. Using the algorithm presented here and solving Kepler's equation, the results obtained are

$\alpha_p = 4.873366^h$, $\delta_p = 22.137182°$,

for a difference of $10.0^s$ in right ascension and $21.5''$ in declination. Better accuracy can be obtained by using a standard epoch closer to the desired date (e.g., 2001.0), but this accuracy is more than sufficient for the purposes of this book.

The astute reader has probably wondered why the gravitational attraction of various planets is not considered in the preceding steps. Various orbital perturbations must be considered to calculate a planet's location with a high degree of accuracy, especially for the giant planets Jupiter and Saturn, which can account for a discrepancy of as much as 1°. If higher accuracy is desired, orbital corrections would need to be added for each of the planets in much the same manner that orbital corrections were considered during the calculations for the Moon's position.

## 8.7 Planet Rise and Set Times

The examples worked out in the previous section show why it is useful to know when a planet will rise and set. After all that work, neither Venus nor

Saturn is visible to our observer because in both cases, the planets are below the observer's horizon.

Calculating sunrise, sunset, moonrise, and moonset required an interpolation scheme since the equatorial coordinates of the Sun and Moon change so rapidly. Fortunately, this is not the case with the planets. Calculating the position of a planet only once during the day is sufficient to determine its rise and set times to a reasonable degree of accuracy.

To calculate a planet's rising and setting times, we will use the same method that we used in section 5.2 to calculate the rising and setting times for a star. To illustrate, let us again refer to the observer from the previous section. When will Saturn appear above the observer's horizon? The required steps are as follows:

1. Calculate the geocentric ecliptic coordinates of the desired planet at $0^h$ UT on the date of interest.
    (Ans: $\lambda_p = 251.311100°$, $\beta_p = 1.629447°$.)
2. Convert the geocentric ecliptic coordinates to equatorial coordinates.
    (Ans: $\alpha_p = 16.666734^h$, $\delta_p = -20.521694°$.)
3. Using the equatorial coordinates from the previous step and treating the planet as if it were a star, use the algorithm from section 5.2 to compute LST rising and setting times.
    (Ans: $LST_r = 11.800372^h$, $LST_s = 21.5333096^h$.)
4. Convert the LST times to LCT times.
    (Ans: $LCT_r = 5.168407^h$, $LCT_s = 14.874561^h$.)
5. Convert the LCT times to HMS format.
    (Ans: $LCT_r = 5^h 10^m$, $LCT_s = 14^h 52^m$.)

## 8.8 Planetary Distance and Angular Diameter

Calculating the distance from Earth to a planet falls out quite readily from the work done to calculate a planet's position. The equation required is

$$Dist = \sqrt{R_e^2 + R_p^2 - 2 R_e R_p \cos(L_p - L_e)}, \tag{8.8.1}$$

where $R_e$ and $R_p$ are the radius vector lengths of the Earth and planet, respectively, and $L_e$ and $L_p$ are the *heliocentric* ecliptic longitudes of the Earth and planet, respectively.

To illustrate, calculate the distance from Earth to Saturn for the observer from our previous example.

1. Using the steps from section 8.6, calculate the radius vector length for Earth and the planet, and the heliocentric ecliptic longitude for Earth and the planet at $0^h$ UT.

   (Ans: $R_p = 10.000361$ AUs, $R_e = 0.983294$ AUs, $L_p = 248.411905°$, $L_e = 101.933467°$.)

2. Apply equation 8.8.1 to calculate the distance in AUs from Earth to the planet.

   (Ans: $Dist = 10.833730$ AUs.)

3. If desired, convert the distance to kilometers and miles.

   (Ans: $Dist = 1.6207E09$ km, or $1.0071E09$ miles.)

Calculating a planet's angular diameter requires the distance just calculated. The equation required is

$$\theta = \frac{\theta_p}{Dist}, \tag{8.8.2}$$

where $\theta_p$ is the angular diameter when the planet is 1 AU from Earth (see tables 8.3 and 8.4) and *Dist* is the distance from the planet to Earth measured in AUs. Since $\theta_p$ is given in arcseconds, the result from equation 8.8.2 will also be in arcseconds.

Again using Saturn and the observer from the previous section, we have

$$\theta = \frac{\theta_p}{Dist} = \frac{165.60''}{10.833730 \text{ AUs}} = 15.29''.$$

Recall that sections 6.5 and 7.5 presented a method for calculating distance and angular diameter directly from an object's orbital elements. Why have we not used those equations here? Those equations are, of course, still valid, but they are stated in a form relative to the object being orbited, which is not Earth in the case of the planets. Thus, if we directly apply the equations in section 6.5 to a planet, we will obtain a planet's distance from the Sun and its angular diameter when viewed from the Sun whereas our interest here is a planet's distance and angular diameter with respect to Earth.

Theoretically, it is possible to adjust a planet's orbital elements to be with respect to Earth rather than the Sun so that the equations presented in section 6.5 will give us what we want. Unfortunately, because the planets and Earth are constantly in motion, a planet's orbital elements with respect to Earth would have to be constantly recomputed to account for that motion. Moreover, to account for their motion, we have to calculate the positions of Earth and the planet in question to get the data we need to reference everything with respect to Earth. Since we have to compute the position of Earth and the planet in the

# Our Solar System

first place, we might as well take advantage of that fact and use the method presented here for determining distance and angular diameter as a byproduct of calculating a planet's position.

## 8.9 Perihelion and Aphelion

As the planets move in their respective orbits, the distance between them and the Sun varies. In this section we calculate when a planet will pass through perihelion and aphelion. These are the points at which a planet is closest to (perihelion) or farthest away (aphelion) from the Sun. The method presented here is based on the technique Jean Meeus presents in *Astronomical Algorithms*, 2nd edition. First, $K$ is calculated by the equation

$$K = k_0(Y - k_1), \tag{8.9.1}$$

where $k_0$ and $k_1$ are chosen from table 8.5 for the planet of interest, and $Y$ is the date of interest obtained from

$$Y = Year + \frac{Days\ into\ year}{365.25}. \tag{8.9.2}$$

The value obtained for $K$ in equation 8.9.1 must be rounded to the nearest integer for computing perihelion or to the nearest number ending in 0.5 for aphelion.

Given $K$, the next step is to use it to compute the Julian day number for perihelion ($K$ is an integer) or aphelion ($K$ ends in 0.5). The Julian day number is given by

$$JD = j_0 + j_1 K + j_2 K^2, \tag{8.9.3}$$

**Table 8.5** Perihelion and Aphelion
These coefficients are used to calculate a planet's passage through perihelion and aphelion.

| Planet  | $k_0$   | $k_1$   | $j_0$         | $j_1$        | $j_2$      |
|---------|---------|---------|---------------|--------------|------------|
| Mercury | 4.15201 | 2000.12 | 2,451,590.257 | 87.96934963  | 0.0        |
| Venus   | 1.62549 | 2000.53 | 2,451,738.233 | 224.7008188  | −3.27E-8   |
| Earth   | 0.99997 | 2000.01 | 2,451,547.507 | 365.2596358  | 1.56E-8    |
| Mars    | 0.53166 | 2001.78 | 2,452,195.026 | 686.9957857  | −1.187E-7  |
| Jupiter | 0.08430 | 2011.20 | 2,455,636.936 | 4332.897065  | 1.367E-4   |
| Saturn  | 0.03393 | 2003.52 | 2,452,830.120 | 10,764.21676 | 8.27E-4    |
| Uranus  | 0.01190 | 2051.10 | 2,470,213.500 | 30,694.8767  | −5.41E-3   |
| Neptune | 0.00607 | 2047.50 | 2,468,895.100 | 60,190.3300  | 3.429E-2   |

where $j_0$, $j_1$, and $j_2$ are obtained from table 8.5 for the planet of interest. Once we have the Julian day number, it is easy to convert it to the corresponding calendar date and UT via the methods presented in section 3.6. The Julian day number obtained from equation 8.9.3 will be the date closest to the date used in equations 8.9.1 and 8.9.2, but it may well be significantly later or earlier than the date used in those 2 equations.

For example, compute perihelion and aphelion for Mars closest to October 30, 1938, the date of the infamous *War of the Worlds* broadcast. The steps are:

1. Compute $Days$, the number of days into the year (see section 3.7).
   (Ans: $Days = 303$.)

2. Use equation 8.9.2 to compute $Y$.
   (Ans: $Y = 1938.829569$.)

3. Use equation 8.9.1 to compute $K$ with the appropriate values for Mars taken from table 8.5.
   (Ans: $k_0 = 0.53166$, $k_1 = 2001.78$, $K = -33.468226$.)

4. Let $K_{per}$ be the integer value closest to $K$, and let $K_{aph}$ be the fraction ending in 0.5 that is closest to $K$.
   (Ans: $K_{per} = -33$, $K_{aph} = -33.5$.)

5. Use equation 8.9.3, the appropriate values from table 8.5, and $K_{per}$ to compute the Julian day number corresponding to perihelion.
   (Ans: $j_0 = 2,452,195.026$, $j_1 = 686.9957857$, $j_2 = -1.187\text{E-}7$, $JD_{per} = 2,429,524.16494$.)

6. Use equation 8.9.3, the appropriate values from table 8.5, and $K_{aph}$ to compute the Julian day number corresponding to aphelion.
   (Ans: $JD_{aph} = 2,429,180.66705$.)

7. Convert the Julian day numbers from the previous step to calendar dates.
   (Ans: $Date_{per} = 09/17.664943/1939$, $Date_{aph} = 10/9.167046/1938$.)

8. Convert the fractional days from the previous step to an integer day and UT.
   (Ans: $Day_{per} = 17$ at $15.958623^h$ UT, $Day_{aph} = 9$ at $4.009100^h$ UT.)

9. Convert the UT results to HMS format.
   (Ans: Perihelion on 9/17/1939 at $15^h58^m$ UT, Aphelion on 10/9/1938 at $4^h01^m$ UT.)

Thus, Mars passed through perihelion on September 17, 1939, at $15^h58^m$ UT and through aphelion on October 9, 1938, at $4^h01^m$ UT.

Meeus indicates that his equations are not highly accurate, with errors that may range from a few hours for Mars to a month or more for Saturn and the more distant planets. Hence, computing UT in step 7 just shown implies a

## Our Solar System

greater accuracy than is warranted; therefore, one may wish to truncate the day in step 6 to an integer and stop.

Using noninteger values of $K$ in equation 8.9.3 or that do not end in 0.5 will produce meaningless results. Furthermore, greater error is incurred for years significantly distant from 2000. In practice, one would typically consult *The Astronomical Almanac* or similar resource for highly accurate perihelion and aphelion times.

If one knows the times at which a planet will pass through perihelion and aphelion, it is natural to ask is how far away will the planet be from Earth and from the Sun at those times. Determining a planet's distance from Earth when the planet is at perihelion or aphelion is straightforward, although the calculations are quite lengthy:

- Use the equations in this section to determine the dates and times at which a planet will pass through perihelion and aphelion.

- Once the dates and times are known for perihelion and aphelion, use section 8.6 to determine the position of the planet and Earth at those times.

- Finally, use section 8.8 and the positions from the prior step to determine the distance from Earth to the planet.

Be careful to note that these steps determine how far away a planet is from Earth when the planet is farthest from (aphelion) or closest to (perihelion) the Sun. These steps do *not* determine when a planet is farthest or closest to the Earth, which is an entirely different and more difficult problem that we will not address.[11]

A planet's distance from the Sun at perihelion and aphelion can be determined quite easily and directly from just the characteristics of the planet's orbit. If $e_p$ is a planet's orbital eccentricity and $a_p$ is the length of its orbital semi-major axis, then the planet's distance from the Sun at perihelion and aphelion are

$$Dist_{per} = a_p(1 - e_p) \qquad (8.9.4)$$

$$Dist_{aph} = a_p(1 + e_p). \qquad (8.9.5)$$

Since $a_p$ is given in AUs, the result of both of these equations will also be in AUs.

Let us determine how far away Jupiter is from the Sun at perihelion and aphelion. The required steps are:

---

11. It is improper to say apogee and perigee here because the planets orbit the Sun, not Earth. Hence, apogee and perigee are meaningless in this context.

1. Use equation 8.9.4 and the appropriate $e_p$ and $a_p$ values (from table 8.4 for this example) to compute the distance at perihelion.
   (Ans: $e_p = 0.0483927$, $a_p = 5.202887$ AUs, $Dist_{per} = 4.951105$ AUs.)
2. Use equation 8.9.5 to compute the distance at aphelion.
   (Ans: $Dist_{aph} = 5.45669$ AUs.)
3. Convert the distances to km or miles if desired.
   (Ans: $Dist_{per} = 7.46067E08$ km, or $4.6023E08$ miles, $Dist_{aph} = 8.1601E08$ km, or $5.0704E08$ miles.)

Although we will not consider the problem of determining when a planet is closest to or farthest from Earth, equations 8.9.4 and 8.9.5 *can* be applied to determine how far away the Moon is from Earth at perigee and apogee. In fact, those equations can be used to determine the apside distances for any object that orbits another.

To determine lunar perigee and apogee distances, the required steps are:

1. Obtain $e_p$ and $a_p$ from table 7.1.
   (Ans: $e_p = 0.0549$, $a_p = 384{,}400$ km.)
2. Use equation 8.9.4 to compute the Moon's distance at perigee. Since $a_p$ is given in km, the answer will be in km.
   (Ans: $Dist_{per} = 363{,}296$ km, or $225{,}742$ miles.)
3. Use equation 8.9.5 to compute the Moon's distance at apogee. The answer is in km.
   (Ans: $Dist_{aph} = 405{,}504$ km, or $251{,}968$ miles.)

## 8.10 Planet Phases

Planets go through phases just as the Moon does. Instead of computing phases and age as we did for the Moon, in this section we will deal only with percent illumination because phase, age, and percent illumination are really the same thing. The percent illumination for a planet is

$$K_\% = 100 \left[ \frac{(R_p + Dist)^2 - R_e^2}{4 R_p Dist} \right], \qquad (8.10.1)$$

where $R_p$ and $R_e$ are the radius vector lengths computed in section 8.6, and $Dist$ is the distance from Earth to the planet as computed in section 8.8.

To illustrate, consider Saturn on January 3, 2016. The steps required to compute how much of Saturn is illuminated as seen from Earth are:

# Our Solar System

1. Compute the radius vector length for the planet and for Earth for the given date at $0.0^h$ UT (see section 8.6).
   (Ans: $R_p = 10.00361$ AUs, $R_e = 0.983294$ AUs.)
2. Compute the distance from Earth to the planet (section 8.8).
   (Ans: $Dist = 10.833730$ AUs.)
3. Apply equation 8.10.1 to calculate percent illumination.
   (Ans: $K_\% = 99.9\%$.)

## 8.11  Planetary Magnitude

How bright an object appears to be is obviously related to how far away it is from the viewer. Thus, as the Earth and planets move in their respective orbits, how bright a planet appears to be depends upon how far away it is from Earth. We saw how to calculate how far away a planet is from Earth in section 8.8. We can use that information along with the following equations to approximate a planet's visual magnitude:

$$PA = \frac{1 + \cos(\lambda_p - L_p)}{2} \tag{8.11.1}$$

$$mV = V_p + 5\log_{10}\left(\frac{R_p Dist}{\sqrt{PA}}\right), \tag{8.11.2}$$

where $\lambda_p$ is the planet's geocentric ecliptic longitude, $L_p$ is the planet's heliocentric ecliptic longitude, $R_p$ is the planet's radius vector length, $Dist$ is the distance from Earth to the planet, and $V_p$ is the visual magnitude of the planet when it is 1 AU from the Earth (see tables 8.3 and 8.4). Note that $PA$ is the planet's phase angle while $\lambda_p - L_p$ is its elongation. Both concepts were introduced in section 7.6.

As an example, what was Saturn's visual magnitude on January 3, 2016?

1. Compute the planet's position for the given date at $0.0^h$ UT to obtain $L_p$, $\lambda_p$, and $R_p$.
   (Ans: $L_p = 248.411905°$, $\lambda_p = 251.311100°$, $R_p = 10.000361$ AUs.)
2. Compute the distance from Earth to the planet.
   (Ans: $Dist = 10.833730$ AUs.)
3. Apply equation 8.11.1 to calculate the planet's phase angle.
   (Ans: $PA = 0.999360°$.)
4. Obtain $V_p$ from the appropriate table (table 8.4 for this example) and apply equation 8.11.2 to calculate the planet's visual magnitude.
   (Ans: $V_p = -8.88$, $mV = 1.29$.)

It is easy to compare the relative brightness of 2 objects from their respective visual magnitudes. If $m_1$ and $m_2$ are the magnitudes of 2 objects, then the brightness of the first object ($m_1$) with respect to the second ($m_2$) is

$$\Delta_b = 10^{0.4(m_2 - m_1)}. \tag{8.11.3}$$

For example, how much brighter does the Sun appear to be than a Full Moon? The visual magnitude of the Sun is $-26.7$ while the Moon's visual magnitude is $-12.7$. Using these values for $m_1$ and $m_2$, we have

$$\Delta_b = 10^{0.4(m_2 - m_1)} = 10^{0.4(-12.7 - (-26.7))}$$

$$= 10^{0.4(14)} = 10^{5.6} \approx 398{,}107.$$

Thus, the Sun appears to be nearly 400,000 times brighter than a Full Moon.

On January 3, 2016, how much brighter was Saturn than the Andromeda Galaxy whose visual magnitude is about 3.4? In this case, we have

$$\Delta_b = 10^{0.4(m_2 - m_1)} = 10^{0.4(3.4 - 1.29)}$$

$$= 10^{0.4(2.11)} = 10^{0.844} \approx 6.98,$$

so Saturn was about 7 times brighter in the sky at that time than the Andromeda Galaxy.

Be careful to note that an object's *apparent* magnitude (or visual magnitude) is not the same as its *absolute* magnitude. The apparent magnitude of an object is simply a measure of how bright it appears in the sky. An object's absolute magnitude $M$ is how bright the object would appear to be if it were at a standard distance of 10 parsecs (approximately 32.64 light years) away from Earth.

Astronomers use absolute magnitude because it measures the intrinsic brightness of an object and therefore does not depend on the object's distance from Earth, as apparent magnitude does. In other words, apparent magnitude indicates only how bright an object appears to be while absolute magnitude indicates how bright the object really is. For example, the apparent magnitude of the Sun is $-26.7$, while its absolute magnitude is $+4.7$. This means that if the Sun were 10 parsecs away from Earth, its visual magnitude would appear to be $+4.7$. The Pole Star (Polaris) has an apparent magnitude of $+1.97$ but an absolute magnitude of $-3.64$. So, although Polaris appears much dimmer in the nighttime sky than the Sun ($mV$ of $-26.7$ compared to $+1.97$), Polaris is actually a much brighter object than the Sun ($M$ of $+4.7$ compared to $-3.64$). Polaris only *appears* dimmer because it is so much farther away. Applying equation 8.11.3, in terms of apparent magnitude the Sun appears to be

290 billion times brighter than Polaris, while in terms of absolute magnitude, Polaris is in reality about 2,000 times brighter than the Sun.

Apparent and absolute magnitude are related by the equation

$$M = mV + 5 - 5\log_{10}(d), \tag{8.11.4}$$

where $d$ is the distance in parsecs away from Earth. We will not consider absolute magnitude any further.

## 8.12 Miscellaneous Calculations

It is perhaps fair to say that the Space Age began with the launch of Sputnik 1 in 1957. The progress made since then to explore the cosmos has been truly amazing. Although manned flights into space are still relatively infrequent, launching manned flights and space probes are so routine that a rocket launch rarely makes the evening news. With so many countries actively engaged in space exploration, and with serious discussions being held about manned flights to the Moon and Mars, space exploration may be even more dramatic in the very near future. Unfortunately, most of us will not have the opportunity to participate in these exciting adventures. Even so, as a poor man's substitute we can at least look at some of the interesting factors that have to be considered by those who are fortunate enough to be engaged in a space program.

We will begin this section by determining what an astronaut or spacecraft would weigh on some distant celestial object and then calculate how long it takes light to reach us from that object to understand the delays inherent in radio transmissions. Just for fun, we will also calculate the length of a planet's year and how fast a planet is moving. Finally, one of the most important calculations we will make is to determine how fast a spacecraft must travel to escape a planet's gravitational pull.

### 8.12.1 Weight on a Celestial Object

Newton's *Law of Universal Gravitation* provides a way to determine what an object will weigh on a celestial body. Newton's law states that objects are attracted to each other with a force directly proportional to the product of their masses and inversely proportional to the square of the distance between them. Mathematically, this is expressed as

$$F = G\frac{m_1 m_2}{R^2}, \tag{8.12.1}$$

where $G$ is the gravitational[12] constant, $m_1$ and $m_2$ are the masses of the 2 objects, $R$ is the distance between them, and $F$ is the resulting force of attraction.

To illustrate this important equation, let $m$ be an astronaut's mass, let $m_e$ be Earth's mass, and let $m_p$ be the mass of a celestial object such as a planet. Let $r_e$ and $r_p$ be the radius of Earth and the celestial object, respectively. Also, let $W_e$ and $W_p$ be the weight of the astronaut on the Earth and celestial object, respectively. Substituting these values into equation 8.12.1, we have

$$W_e = G \frac{mm_e}{r_e^2} \qquad (8.12.2)$$

and

$$W_p = G \frac{mm_p}{r_p^2}. \qquad (8.12.3)$$

Now if we divide $W_p$ by $W_e$, that will produce the ratio of the astronaut's weight on the celestial object to the astronaut's weight on Earth. That is,

$$\frac{W_p}{W_e} = \left( G \frac{mm_p}{r_p^2} \right) \div \left( G \frac{mm_e}{r_e^2} \right) = \left( \frac{Gmm_p}{r_p^2} \right) \left( \frac{r_e^2}{Gmm_e} \right) = \frac{m_p r_e^2}{r_p^2 m_e}. \qquad (8.12.4)$$

Notice that this ratio depends only upon the size and mass of Earth and the celestial object. The gravitational constant and astronaut's mass do not appear at all in the final equation. This should not be surprising because it says that the percentage by which 1 astronaut's weight is affected by being on some celestial object is the same as the percentage by which another astronaut's weight is affected, regardless of how much each individual astronaut weighs. Once the weight-factor ratio is known for a celestial object, one only has to multiply that ratio by an object's weight on Earth to obtain how much the object weighs on that celestial object.

For example, how much did Neil Armstrong's spacesuit weigh on the Moon? (The weight of the multilayer space suit he wore was a hefty 180 pounds on Earth!) To answer this question, we apply equation 8.12.4 to find the ratio of weight on the Moon to weight on Earth. From table 8.3, we find that $m_e = 1.0$, $m_p = 0.0123$, $r_e = 6378.14$ km, and $r_p = 1738.1$ km. (Actually, table 8.3 gives masses relative to Earth, but this makes no difference in solving the problem because we are only interested in ratios.) Applying equation 8.12.4 to these

---

12. Although scientists have accurately determined the value of the gravitational constant $G$, we do not need to concern ourselves with its value.

## Our Solar System

values, we obtain

$$\frac{W_p}{W_e} = \frac{(0.0123)(6378.14^2)}{(1738.1^2)(1.0)} \approx 0.166.$$

This ratio can be rewritten as $W_p = 0.166 W_e$, which says that weight on the Moon is 16.6 percent of what it is on Earth. Since Armstrong's spacesuit weighed 180 pounds on Earth, its weight on the Moon was $W_p = (0.166)(180\,\text{lbs}) \approx 30\,\text{lbs}$.

Assuming it were possible to land on the surface of the Sun, how much would Armstrong's spacesuit weigh on the Sun? We proceed as before to apply equation 8.12.4. From table 8.3, we find that $m_e = 1.0$, $m_p = 333{,}000$, $r_e = 6378.14$ km, and $r_p = 695{,}700$ km, which gives us

$$\frac{W_p}{W_e} = \frac{(333{,}000)(6378.14^2)}{(695{,}700^2)(1.0)} \approx 27.989.$$

This can be rewritten as $W_p = 27.989 W_e$. So, the spacesuit would weigh an astonishing $W_p = (27.989)(180\,\text{lbs}) \approx 50{,}368\,\text{lbs}$ on the surface of the Sun! Clearly, dealing with the Sun's extreme temperatures is only one of the issues involved in attempting to explore the Sun from close proximity.

More realistic than landing on the Sun would be a manned mission to Mars, which is likely to take place within the next 20 years or so. How much would Neil Armstrong's spacesuit weigh on Mars? From table 8.3, we find that $m_e = 1.0$, $m_p = 0.107447$, $r_e = 6378.14$ km, and $r_p = 3389.5$ km. Working through the math as in the previous examples, we arrive at $W_p = 0.380 W_e$, so Armstrong's spacesuit would weigh about 68 lbs on Mars.

### 8.12.2 Radio Transmission Delays

During the Voyager missions, there was a noticeable communications delay between the time a command was sent from Earth until the Voyager space probe received and acted upon that command. This is because radio waves, although traveling at the speed of light, travel at a finite speed. In typical electromagnetic communications on Earth we do not notice a delay because the distances involved are so relatively short. However, for a distant object such as a planet or a galaxy, the delay is very noticeable. Let us find out how much of a delay there is.

Suppose an object travels uniformly at $s$ miles per second for $t$ seconds. The distance traveled (in miles) is

$$Dist = st. \tag{8.12.5}$$

We want how long it takes to travel some distance for a given speed, so solving for $t$ gives

$$t = \frac{Dist}{s}. \tag{8.12.6}$$

To make future calculations easier, let us use this last equation to find out how long it takes light to travel 1 AU. Radio waves propagate at the speed of light, so the results will be applicable to radio waves too. The speed of light is 186,400 miles per second while 1 AU is $9.29 \times 10^7$ miles. Substituting these values into equation 8.12.6 gives

$$t = \frac{9.29 \times 10^7 \text{ miles}}{186,400 \text{ miles/sec}} \approx 498.39 \text{ seconds.}$$

Converting seconds to hours, we find that light travels 1 AU in about 0.1384 hours.

Given this information, finding out how long it takes light and radio waves to reach Earth from another planet is simple: multiply the distance from Earth to the planet (expressed in AUs) by 0.1384, as indicated in the equation

$$t = 0.1384 Dist \tag{8.12.7}$$

where the resulting time is in hours. Alternatively, we can use

$$t = 498.39 Dist \tag{8.12.8}$$

to obtain a result in seconds. Because of how we derived both equations, $Dist$ must be expressed in AUs.

Suppose a space probe reached Saturn on January 3, 2016. How long will it take to send a radio message from Earth to the space probe? We already calculated the distance to Saturn in section 8.8 to be 10.833730 AUs. Applying equation 8.12.7, the result is $1.499388^h$ ($1^h 30^m$). Any message sent from Earth to the probe would take $1^h 30^m$ to get there, while a reply from the probe would also take $1^h 30^m$ to reach Earth. Hence, if a command were sent to the probe that required a reply, there would be a net communications delay of $3^h 00^m$.

As another example, consider sending a radio message from New York City to London. The distance between the 2 cities is about 3,500 miles, or $3.767^{-5}$ AUs. This time we apply equation 8.12.8 and find that it will take 0.02 seconds for the transmission to occur. Of course, we could also have applied equation 8.12.6, which gives

$$t = \frac{Dist}{s} = \frac{3,500 \text{ miles}}{186,400 \text{ miles/sec}} \approx 0.02 \text{ seconds.}$$

Our Solar System  257

Regardless of which equation is used, such a small delay is hardly noticeable, except to very precise instruments, which is the reason we typically do not notice delays in Earthbound communications. For the relatively close Moon, however, the 1-way communications delay is about 2.1 seconds, which is probably noticeable to most humans and is certainly noticeable to most instruments.

### 8.12.3 Length of a Planetary Year

As presented in section 3.3, a year is the time it takes for Earth to revolve around the Sun. We also pointed out in that section that the length of a year depends on how one chooses a reference point for defining when a year starts and ends (a tropical year uses the vernal equinox as a reference, a sidereal year uses a star, etc.). For the discussion here, we will only be concerned about a Julian year (exactly 365.25 days) and a tropical year.

A planetary year, like an Earth year, is defined as the time it takes a planet to make 1 complete orbit around the Sun. That is, the length of a planet's year is simply another way of referring to a planet's orbital period. Tables 8.3 and 8.4 give the orbital period for each of the planets in terms of Earth's tropical years. Since a tropical year is 365.242191 mean solar Earth days, the length of a planet's year in Earth days can be calculated by multiplying its orbital period $T_p$ by 365.242191. Stated as an equation, we have

$$Year_p = 365.242191 T_p. \qquad (8.12.9)$$

For example, Saturn's orbital period is 29.447498 tropical years. Applying equation 8.12.9, this means Saturn's year is 10,755.468689 Earth days. Dividing this number by 365.25, Saturn's year is thus equal to 29.446868 Julian years, or 29 years, 163 days, $5^h 15^m$. In actual practice, given the limited accuracy of the data in tables 8.3 and 8.4, one would not express the result in terms of hours and minutes because Saturn's orbital period and the factor for converting to tropical Earth years are not sufficiently precise to warrant such an implied level of accuracy.

How do astronomers determine a planet's orbital period in the first place? The answer is to apply Kepler's laws. Specifically, Kepler's third law relates a planet's orbital period to its distance from the Sun and is stated mathematically as

$$P^2 \propto A^3, \qquad (8.12.10)$$

where $P$ is a planet's period, $A$ is the planet's average distance to the Sun (its semi-major axis), and $\propto$ is the symbol mathematicians use for "is proportional to." In the case of planets and other objects orbiting the Sun, Kepler's third law can be stated as an equality. That is,

$$T_p = \sqrt{a_p^3}, \qquad (8.12.11)$$

where $T_p$ is a planet's orbital period in tropical years and $a_p$ is the length of the planet's semi-major axis in AUs.

For example, getting the length of Saturn's semi-major axis $a_p$ from table 8.4 and applying equation 8.12.11, we obtain

$$T_p = \sqrt{9.536676^3} \approx 29.450694 \text{ years.}$$

This result differs from table 8.4 by less than 1.2 days. The difference is due to the precision and accuracy with which a planet's semi-major axis is measured, how precisely an AU is measured, and how precisely the conversion factor for tropical years is stated. If one applies equation 8.12.11 to Mercury and compares the result to table 8.3, the difference is 2 minutes. For Neptune, the difference is 37 days.

Equation 8.12.11 can be used to determine the length of a planet's orbital semi-major axis when its orbital period is known. Rewriting equation 8.12.11 to solve for the semi-major axis, we have

$$a_p = \sqrt[3]{T_p^2}. \qquad (8.12.12)$$

This equation is very useful because it provides a way to calculate the length of a planet's orbital semi-major axis from observational data: that is, by observing a planet long enough to determine its orbital period.

Using equation 8.12.12 and Pluto's orbital period from table 8.4, we have

$$a_p = \sqrt[3]{247.92065^2} \approx 39.464669 \text{ AUs.}$$

This result differs from the semi-major axis shown for Pluto in table 8.4 for the same reasons that the earlier calculation of Saturn's orbital period from its semi-major axis differs from what the table shows for Saturn.

### 8.12.4 Orbital Velocity

In the previous subsection, we used Kepler's laws to determine a few characteristics of a planet's orbit. The discovery of Kepler's laws was a major milestone in astronomy and physics, and they remain indispensable tools for understanding the motion of celestial objects. We will now use Kepler's second law to determine how fast a planet is moving.

Kepler's second law states that as a planet orbits the Sun, it sweeps out equal areas of space in equal amounts of time. This means that a planet's orbital

# Our Solar System

velocity constantly changes as it moves along its elliptical orbit, getting faster as it approaches perihelion and slowing down as it approaches aphelion. The only way a planet's orbital velocity would be constant is if the planet's orbit is a perfect circle. We will address circular orbits in chapter 9 when we deal with satellites.

A planet's orbital velocity in km/s at perihelion is given by

$$V_{per} = \sqrt{\frac{\mu(1+e_p)}{a_p(1-e_p)(1.5 \times 10^8)}}, \quad (8.12.13)$$

where $\mu$ is the standard gravitational parameter for the object being orbited (the Sun in this case), $e_p$ is the planet's orbital eccentricity, and $a_p$ is the length of the planet's semi-major axis in AUs. The denominator in this equation is the planet's distance from the Sun at perihelion (see section 8.9), except that multiplying by $1.5 \times 10^8$ gives the distance in kilometers rather than AUs.

The equation for determining a planet's orbital velocity at aphelion is very similar:

$$V_{aph} = \sqrt{\frac{\mu(1-e_p)}{a_p(1+e_p)(1.5 \times 10^8)}}. \quad (8.12.14)$$

In this case the denominator in the equation is the planet's distance from the Sun at aphelion (see section 8.9), and we have also expressed the distance in kilometers.

The standard gravitational parameter $\mu$ that appears in these 2 equations requires some explanation. Because the value $Gm$ is so frequently encountered in astrophysics (due to Newton's law of universal gravitation), it is called the standard gravitational parameter and denoted by the symbol $\mu$. We encountered this value earlier in subsection 8.12.1 when we calculated an object's weight on some planet. Recall that $G$ is the gravitational constant while $m$ is the mass of the celestial object in which we are interested. The standard gravitational parameter is provided for the Sun, Moon, and each of the planets in tables 8.3 and 8.4, and it is expressed in km cubed per second squared ($km^3/s^2$).

As an example, let us calculate the orbital velocity for Mercury at perihelion and aphelion. The required steps are:

1. Apply equation 8.12.13 to determine the planet's orbital velocity at perihelion.
(Ans: $\mu_\odot = 1.32712 \times 10^{11}$, $e_p = 0.205636$, $a_p = 0.387099$ AUs, $V_{per} = 58.897$ km/s, which is 36.597 miles/s or 131,749 mph.)

2. Apply equation 8.12.14 to calculate the planet's orbital velocity at aphelion. (Ans: $V_{aph} = 38.806$ km/s, which is 24.113 miles/s or 86,807 mph.)

It is important to be sure that the proper value for $\mu$ is used to calculate orbital velocity. The standard gravitational parameter for the object being orbited (e.g., Sun) *must* be used and *not* the standard gravitational parameter for the object that is doing the orbiting.

If we use the correct $\mu$ and know the other orbital parameters required, we can determine the orbital velocity of any object that orbits another when the orbiting object is at its closest point or its most distant point. To illustrate, let us determine the Moon's orbital velocity at perigee and apogee. The steps are:

1. Apply equation 8.12.13 to determine the Moon's orbital velocity at perigee. (Ans: $\mu_p = 398{,}600$ (Earth!), $e_p = 0.0549$, $a_p = 0.00257$ AUs, $V_{perigee} = 1.074$ km/s, which is 0.667 miles/s or 2,401 mph.)
2. Apply equation 8.12.14 to calculate the Moon's orbital velocity at apogee. (Ans: $V_{apogee} = 0.962$ km/s, which is 0.598 miles/s or 2,153 mph.)

We can also compute a planet's average orbital velocity. Deriving a suitable equation is straightforward. Assuming the planet orbits the Sun in a perfect circle, the total distance it travels around the Sun is simply the circle's circumference, which is $2\pi r$ where $r$ is the radius of the planet's assumed circular orbit. If $t$ is the time it takes for the planet to travel around the circle (i.e., the orbital period), then the planet's speed is

$$V = \frac{2\pi r}{t}. \tag{8.12.15}$$

Since planets orbit the Sun in elliptical orbits instead of circular orbits, we will replace the radius $r$ in this equation with an "average radius." There are several possible ways to define an average radius, but the definition we will use is the length of the orbit's semi-minor axis, which can be easily computed as

$$b_p = a_p\sqrt{1 - e_p^2}. \tag{8.12.16}$$

This is a reasonable definition because the semi-minor axis is the average length of the distance from the occupied focus to the ellipse boundary. As for $t$, we will replace it with $T_p$ because that is the planet's orbital period for its true elliptical orbit.

## Our Solar System

Combining these concepts, equation 8.12.15 becomes

$$V_{avg} = \frac{2\pi \left(a_p\sqrt{1-e_p^2}\right)(1.5 \times 10^8)}{T_p(365.242191)(24)(3600)}.$$

Multiplying by $1.5 \times 10^8$ in the numerator converts the distance from AUs to kilometers. Multiplying by 365.242191 in the denominator converts tropical years to days, multiplying by 24 converts days to hours, and multiplying by 3600 converts hours to seconds. This allows us to obtain a result expressed in km/s. Carrying out the multiplications to simplify the equation gives

$$V_{avg} = 29.865958 \left(\frac{a_p\sqrt{1-e_p^2}}{T_p}\right). \tag{8.12.17}$$

Applying this equation to Mercury as an example, we find that its average orbital speed is 46.976 km/s, which is 29.190 miles/s or 105,084 mph. Combining this result with our earlier calculations for Mercury, we see that

$V_{per} = 58.897 \text{ km/s},$

$V_{avg} = 46.976 \text{ km/s},$

and

$V_{aph} = 38.806 \text{ km/s}.$

As we should expect, Mercury is indeed going faster as it approaches perihelion but is slowing down as it approaches aphelion. By comparison, for the Earth we have

$V_{per} = 30.246 \text{ km/s},$

$V_{avg} = 29.861 \text{ km/s},$

and

$V_{aph} = 29.252 \text{ km/s}.$

Suppose that instead of average velocity, we want to know how fast a planet is moving at an arbitrary point in its orbit. The required equation is

$$V_{date} = \sqrt{\left(\frac{\mu}{1.5 \times 10^8}\right)\left(\frac{2}{R_p} - \frac{1}{a_p}\right)}, \tag{8.12.18}$$

where $R_p$ is the distance from the planet to the Sun in AUs at the desired date and time.

To illustrate, let us determine the velocity for Venus at the date and time from the example in subsection 8.6.1 (January 3, 2016, at $22^h$ LCT). The steps are:

1. Calculate the planet's radius vector length.
   (Ans: $R_p = 0.720605$ AUs.)
2. Apply equation 8.12.18 to calculate the orbital velocity for the given date and time.
   (Ans: $\mu_\odot = 1.32712 \times 10^{11}$, $a_p = 0.723336$ AUs, $V_{date} = 35.11$ km/s, which is 21.82 miles/s or 78,552 mph.)

### 8.12.5  Escape Velocity

Our final calculation is to determine a planet's escape velocity, which is the speed necessary for 1 object to escape the gravitational field of another. From Newton's law of universal gravitation, it is clear that the force acting upon an object is greater for a more massive planet than for a smaller one. Hence, we would expect that the velocity required to leave a more massive planet is greater than it is for a less massive one.

The equation for the escape velocity in km/s is

$$V_{escape} = \sqrt{\frac{2\mu_p}{r_p}}, \qquad (8.12.19)$$

where $\mu_p$ is the planet's standard gravitational parameter and $r_p$ is its radius in km.

As an example, calculate the velocity required for a rocket to escape Earth. From table 8.3, for Earth we have $\mu_p = 398{,}600$ and $r_p = 6{,}378.14$ km. Putting these values into equation 8.12.19 yields an escape velocity of 11.18 km/second, which is 40,248 km/h or 25,009 mph. For another example, calculate the Moon's escape velocity. From table 8.3, for the Moon we have $\mu_p = 4{,}900$ and $r_p = 1{,}738$ km. Using these values yields an escape velocity of 2.37 km/second, which is 8,532 km/h or 5,302 mph. As expected from Newton's law of universal gravitation, the Moon's escape velocity is indeed less than Earth's escape velocity because the Moon is less massive.

Surprisingly, escape velocity depends only upon the mass of the object being escaped, which is captured in the equation by the standard gravitational parameter $\mu_p$. The escape velocity does not depend upon the mass of the object that is trying to escape. So, from a velocity point of view, it doesn't matter whether an astronaut is trying to escape the Moon via a jet pack or a huge rocket. The escape velocity in both cases is still 5,302 mph. However, the energy (i.e., the amount of fuel required) necessary to reach escape velocity is significantly different and very much dependent upon the total mass trying to reach escape velocity.

## 8.13 Program Notes

The program for this chapter provides the software necessary to locate inferior and superior planets (and the dwarf planet Pluto!), determine when the planets will rise and set, calculate their phases and visual magnitude, and make the sundry calculations described in the previous section. The program builds on the software from chapter 5 to plot star locations as well as the position of the Sun, Moon, and planets. This will allow you to produce your own star charts showing the location of the planets and major stars for your location.

Although the standard epoch J2000 is used for the examples in this chapter, this chapter's program reads the required orbital elements from a data file so that any standard epoch can be supported. See the README.TXT file included with the source code for more details about this data file. Also, this chapter's program lets you choose between using the equation of the center or Kepler's equation to calculate the position of the planets.

## 8.14 Exercises

For these problems, solve the equation of the center when required rather than Kepler's equation.

1. The date is January 22, 2015, at $22^h$ LCT. Assume an observer is not on daylight saving time but is within the Eastern Standard Time zone. If his location is 78.3° W longitude, 37.8° N latitude, calculate the equatorial and horizon coordinates for Mercury and Jupiter.
   (Ans: For Mercury: $\alpha = 21.249399^h$, $\delta = -14.171598°$, $h = -40°03'$, $A = 284°19'$. For Jupiter: $\alpha = 9.477288^h$, $\delta = 15.875033°$, $h = 38°26'$, $A = 100°00'$.)

2. For the previous problem, calculate the rising and setting times for Mercury and Jupiter.
   (Ans: For Mercury: $LCT_r = 8^h09^m$, $LCT_s = 18^h35^m$.
   For Jupiter: $LCT_r = 18^h44^m$, $LCT_s = 8^h27^m$.)

3. For problem number 1, calculate the distance from Earth and the angular diameter for Mercury and Jupiter.
   (Ans: For Mercury: $Dist = 0.785535$ AU, $\theta = 08.6''$.
   For Jupiter: $Dist = 4.381208$ AUs, $\theta = 44.9''$.)

4. Calculate the times of perihelion and aphelion closest to January 22, 2015, for Mercury and Jupiter. Also calculate their perihelion and aphelion distances.

(Ans: For Mercury: perihelion occurred on 1/21/2015 at $20^h34^m$, aphelion occurred on 3/6/2015 at $20^h12^m$, $Dist_{per} = 0.307498$ AUs, $Dist_{aph} = 0.466701$ AUs. For Jupiter: perihelion occurred on 3/16/2011 at $10^h28^m$, aphelion occurred on 2/18/2017 at $21^h14^m$, $Dist_{per} = 4.951105$ AUs, $Dist_{aph} = 5.454669$ AUs.)

5. For problem number 1, calculate the percent illumination for Mercury and Jupiter.

(Ans: For Mercury: $K_\% = 23.4\%$. For Jupiter: $K_\% = 99.9\%$)

6. For problem number 1, calculate the visual magnitude for Mercury and Jupiter.

(Ans: For Mercury: $mV = -1.93$. For Jupiter: $mV = -2.56$.)

7. For problem number 1, calculate the weight factor, time for light to reach Earth, orbital period, length of semi-major axis, orbital velocities, and escape velocity for Mercury and Jupiter.

(Ans: For Mercury: $W_p = 0.38W_e$, $t = 7^m$, $T_p = 0$ years 87 days $23^h$, $T_p = 0.240843$ tropical years (when calculated from the semi-major axis), $a_p = 0.387104$ AUs (when calculated from the orbital period), $V_{per} = 58.90$ km/s, $V_{avg} = 46.98$ km/s, $V_{aph} = 38.8$ km/s, $V_{date} = 58.89$ km/s, $V_{escape} = 4.25$ km/s. For Jupiter: $W_p = 2.65W_e$, $t = 36^m$, $T_p = 11$ years 314 days $23^h$, $T_p = 11.867701$ tropical years (when calculated from the semi-major axis), $a_p = 5.201400$ AUs (when calculated from the orbital period), $V_{per} = 13.69$ km/s, $V_{avg} = 13.08$ km/s, $V_{aph} = 12.42$ km/s, $V_{date} = 12.74$ km/s, $V_{escape} = 60.20$ km/s.)

# 9 Satellites

It is not known who first thought of launching artificial satellites to orbit the Earth. Sir Isaac Newton alluded to artificial satellites in his landmark work *Principia Mathematica*, published in 1687. The German physicist Hermann Oberth wrote his doctoral dissertation on rocket travel in 1922, but his ideas were initially dismissed as fantasy. He was later awarded a doctoral degree in physics based on the very same ideas that were originally dismissed as impractical. It may be fair to credit Oberth as being the first to develop realistic concepts—grounded in solid mathematics and physics—for using rockets to reach space.

Wernher von Braun, the chief engineer behind the Saturn V rockets that sent Apollo astronauts to the Moon, was greatly influenced by Oberth's ideas. Oberth and von Braun briefly worked together in 1929 to test Oberth's first liquid-fueled rocket engine, and then later at Germany's Peenemünde research facility during World War II. It is almost certain that von Braun and other scientists at Peenemünde considered using rockets to launch instruments and people into space.

The scientist credited as being the first to publish a workable concept for space-based communications is Arthur C. Clarke. Near the end of World War II in a *Wireless World* article entitled "Extra-Terrestrial Relays," he proposed using German V2 rockets to launch communications satellites into orbit. Few took Clarke's concept seriously at the time, but 20 years later, the Intelsat I Early Bird satellite became the world's first commercial satellite to employ Clarke's ideas. The pioneering Intelsat I provided live TV coverage of the Gemini 6 splashdown and the Apollo 11 lunar mission. Although still in orbit today, Intelsat I is no longer in service.

Regardless of who first conceived of them, satellites are relatively new in history, with the Russians having launched Sputnik 1, the world's first satellite, in October 1957. In the short decades since that historic event, satellites

266                                                                      Chapter 9

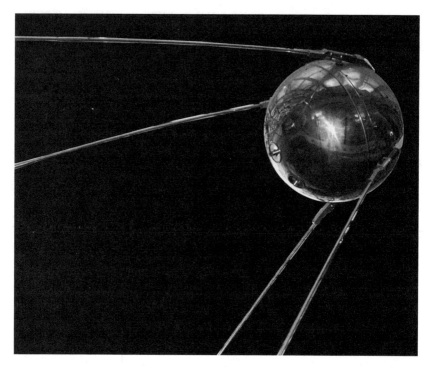

**Figure 9.1**  Sputnik 1
Launched in October 1957, Sputnik 1 was the world's first artificial satellite. This replica of Sputnik 1 is in the National Air and Space Museum in Washington, DC. (Image courtesy of NASA)

have become indispensable tools that in one way or another impact all our lives. It is difficult to imagine living in the modern world without them. Tens of thousands of active satellites encircle our globe; they perform a variety of vital tasks including providing precise worldwide navigation, capturing data and imagery for weather forecasting, allowing nearly instantaneous communications to every corner of the globe, and supporting scientists and astronomers as they explore Earth, the Solar System, and deep space.

   We can only briefly look at the fascinating world of satellites, but it is important to note from the outset that Kepler's and Newton's laws still apply. Satellites obey the same laws of physics that all orbiting bodies do. However, there are at least 3 important differences between satellites and other celestial objects that directly affect how we must approach their study.

   First, satellites are relatively close to the Earth. Because shorter distances are involved with them than with the planets and stars, greater accuracy and precision are required when measuring or computing distances and angles. Whereas

# Satellites

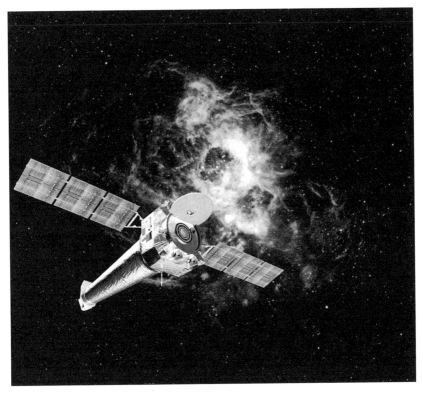

**Figure 9.2** Chandra X-ray Observatory
This artist's rendition is the Chandra X-ray Observatory, which was launched in 1999 by the space shuttle *Columbia* (STS-93) crew. It can detect X-ray sources 100 times fainter than any previous telescope making it, at present, the most sensitive X-ray telescope ever built. This important space-based telescope allows astronomers to analyze distant supernovas, stars, and galaxies, and search for black holes unhindered by Earth's atmosphere. (Image courtesy of NASA)

a distance of a few thousand miles is usually insignificant when dealing with distant celestial objects, a measurement that is off by only a few dozen feet can be very important when dealing with satellites. Moreover, we will be concerned about whether a stated distance is with respect to the *center* of Earth or Earth's *surface*. The difference between the center of the Earth and sea level is a matter of nearly 4,000 miles! Unless otherwise indicated, we will use the center of the Earth as our reference point when stating distances to satellites.

It is also important to account for an observer's distance from the center of the Earth. This was unnecessary in the preceding chapters because the distance an observer is from the center of the Earth is so small relative to the distance to a star or planet as to be negligible except for the most exacting measurements. An observer's distance from the Earth's center is *not* negligible for satellites. For greatest accuracy one should also use geodetic coordinates to specify an observer's location instead of the terrestrial latitude/longitude coordinates we used in prior chapters. This is because geodetic coordinates are based on an ellipsoid model of the Earth (e.g., World Geodetic System 84 [WGS84]) whereas the terrestrial latitude/longitude system presented in chapter 4 is based on a spherical model. Although we will consider an observer's height above sea level, we will continue to assume a spherical Earth to simplify calculations, albeit at the price of some loss in accuracy.

Second, because satellites are so close to Earth, they appear to Earth-bound observers to move much faster than the Moon, planets, or other celestial objects. This reality means that while the principles of physics for tracking satellites are the same as for any other orbiting object, practical techniques for tracking satellites differ from tracking celestial objects. For stars and celestial objects within the Solar System, we essentially record their position at some point in time (a standard epoch) and then use an object's orbital elements to calculate where it will be at some future point in time. Once equatorial coordinates are calculated for a star or planet, the object can be treated as being in a fixed location for a night of viewing, and we have to be concerned about only the Earth's rotation to track the object.

This is not true for satellites! Although we will still capture a satellite's position at an instant in time and use its orbital elements to determine how far it has moved, attempting to state a standard epoch for all satellites is meaningless. The epoch to which a satellite is referenced *and* its orbital elements must be updated weekly or even daily, depending on a satellite's particular orbit, in order to maintain a reasonable degree of accuracy. It is almost a certainty that the epoch used to reference one satellite's position will be different from every other satellite.

Third, unlike orbiting objects found in nature, satellites can, and must, periodically maneuver. Satellites usually have thrusters to periodically adjust their orbit so that they stay within their assigned orbit. This is called orbital station keeping, and it is necessary because satellite orbits decay over time because of the gravitational forces of the Sun and Moon, the effects of space weather, and drag from Earth's atmosphere. Moreover, a satellite's mission may require it to maneuver to change its orientation, such as to change where an onboard camera is pointing. A satellite may also need to adjust its orbit to avoid colliding with another satellite or debris from an expended rocket that has not yet fallen back to Earth. Colliding with even a small object orbiting the Earth would be disastrous because the speed of objects orbiting Earth, regardless of their size, is thousands of miles per hour.

**Satellites**

We must now digress briefly to cover some preliminaries about vectors, ellipses, coordinate systems, orbital elements, and types of orbits. Once armed with the information from the preceding chapters as supplemented here and noting the aforementioned differences between satellites and natural celestial objects, we are well prepared for the topics ahead. By the end of this chapter, we will have seen how to apply the principles used to locate the Moon and planets to locate a satellite and calculate some of its flight dynamics.

## 9.1  Vectors

In this chapter, it will often be convenient to express coordinates as a vector. A vector is a 1-dimensional array used to encapsulate information about something.[1] For instance, assume an object's Cartesian coordinates are $(x, y, z)$. Then that point can be represented as a column vector by

$$\vec{R} = \begin{bmatrix} x \\ y \\ z \end{bmatrix}$$

or as a row vector by

$$\vec{R} = \begin{bmatrix} x & y & z \end{bmatrix}.$$

A vector of Cartesian coordinates actually provides both direction (i.e., in what direction to point from the coordinate system's origin to the object) and magnitude (i.e., the object's distance from the origin). To calculate a vector's magnitude (also called its length or norm), we simply sum up the squares of the vector's elements and then take the square root of the sum. $\vec{R}$'s magnitude, whether expressed as a row or a column vector, is

$$\overline{R} = \sqrt{x^2 + y^2 + z^2}. \tag{9.1.1}$$

We will sometimes need to rotate a vector about the $x$ and $z$ axes. To do so, let us define 2 families of functions. The $f$ family of functions are

$$f_x(\theta, x, y, z) = x \tag{9.1.2}$$

$$f_y(\theta, x, y, z) = y\cos\theta + z\sin\theta \tag{9.1.3}$$

$$f_z(\theta, x, y, z) = z\cos\theta - y\sin\theta. \tag{9.1.4}$$

---

1. An arrow is placed above a variable's name to indicate that it represents a vector (e.g., $\vec{R}$) while a bar is placed over its name to indicate the vector's length (e.g., $\overline{R}$). We will use $\overline{R}$ and $R_{len}$ interchangeably to represent a vector's length because $R_{len}$ is easier to see than $\overline{R}$ when a vector length is in the numerator of a fraction.

These functions rotate a vector in a Cartesian coordinate system by $\theta$ degrees about the $x$-axis. The set of functions in the $g$ family are

$$g_x(\theta, x, y, z) = x \cos\theta + y \sin\theta \qquad (9.1.5)$$

$$g_y(\theta, x, y, z) = y \cos\theta - x \sin\theta \qquad (9.1.6)$$

$$g_z(\theta, x, y, z) = z. \qquad (9.1.7)$$

These functions rotate a vector by $\theta$ degrees about the $z$-axis. Rotations such as those performed by our $f$ and $g$ families of functions are usually described in terms of linear algebra, but we will use these 2 families of functions instead to avoid introducing matrix operations. The mathematical results are exactly the same.

The $f$ and $g$ families of functions take as input the $x$, $y$, and $z$ elements of a vector in a Cartesian coordinate system. They produce as output what the vector elements are after the rotation has been performed. Rotations do not change the length of a vector, so we can calculate a vector's length before or after a rotation is performed.

When applying these 2 families of functions, take care to use the same values of $x$, $y$, and $z$ for *each* function in the family. That is, the sequence

$x = 1000$,

$y = -5000$,

$z = 2000$,

$x = g_x(60°, x, y, z)$,

$y = g_y(60°, x, y, z)$,

$z = g_z(60°, x, y, z)$

is incorrect because the new $x$ value produced by applying function $g_x$ is used for functions $g_y$ and $g_z$ rather than the original $x$ value. A further error is that the modified $y$ value produced by applying $g_y$ is used as input to $g_z$. Applying the sequence of steps *exactly* as shown will produce the *erroneous* result (rounded to 2 decimal places)

$\vec{R}' = [-3830.13 \ 816.99 \ 2000.00]$

when the correct answer is

$\vec{R}' = [-3830.13 \ -3366.03 \ 2000.00]$.

# Satellites

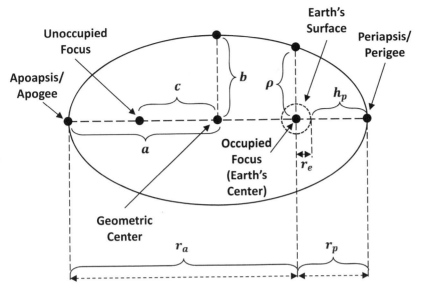

**Figure 9.3** Ellipse for Satellites
This figure shows the components of an ellipse in the context of a satellite orbiting Earth.

## 9.2 Ellipses Revisited

As is true for natural celestial objects, satellites can travel in circular, elliptical, or parabolic orbits. Therefore, all the properties of circles, ellipses, and orbits discussed in chapter 4 still apply. Mathematicians have derived a number of useful equations for ellipses, some of which we will enumerate in this section. Do not be overwhelmed by the number of equations in this section! We will have occasion to use only a few of them, but the list is provided as a ready reference that you can use in designing your own programs. Keep in mind that all the equations in this section are true for any ellipse. They are not unique to satellites, but they apply equally well to any object that orbits another.

Figure 9.3 shows the components of an ellipse that are of interest with respect to satellites while figure 9.4 shows how a satellite's true anomaly is defined. Both figures should be familiar because they appeared earlier in chapter 4, although the nomenclature has been adjusted here to be specific to satellites orbiting Earth. For example, the occupied focus in this case is the Earth's center and is so labeled. Additionally, this chapter references apogee and perigee rather than aphelion and perihelion because we are dealing specifically with the Earth.

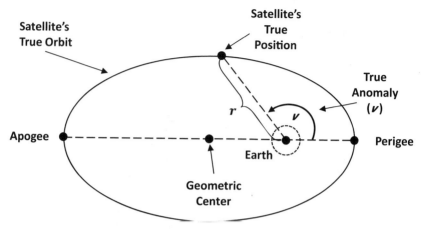

**Figure 9.4** Satellite True Anomaly
This shows a satellite's true anomaly and radius to the satellite.

In the context of satellites, the distances shown in figures 9.3 and 9.4 are:

1. Semi-major axis ($a$)
2. Semi-minor axis ($b$)
3. Linear eccentricity ($c$)
4. Semi-latus rectum ($\rho$)
5. Perigee height ($h_p$)
6. Apogee radius ($r_a$)
7. Perigee radius ($r_p$)
8. Radius to the satellite ($r$)
9. Radius of the Earth ($r_e$)

For the purposes of this chapter, we will assume that Earth is a perfect sphere whose radius $r_e$ is 6378.135 km.

Carefully note that $r$ in figure 9.4 is the distance from the *center* of the Earth to the *center of mass* of the satellite. Knowing the precise location of a satellite's center of mass is usually unimportant, but it is important to realize that the radius to the satellite is measured with respect to the *center* of the Earth, not the *surface* of the Earth! Because of the relatively short distances involved when dealing with satellites orbiting Earth, we cannot assume that Earth is a geometric point, but we must deal with it as a sphere (or even better, as an ellipsoid). Therefore, we must constantly remember to distinguish between Earth's center and its surface, as shown in the 2 figures.

# Satellites

The distance a satellite is above an observer when the satellite is at the observer's zenith is given by

$$r_{dist} = r - r_e - r_{sea}, \qquad (9.2.1)$$

where $r_{sea}$ is an observer's distance above sea level. (This equation only holds for a truly spherical Earth *and* when the satellite is at the observer's zenith.) When $r_{sea} = 0$, $r_{dist}$ is typically called the satellite's altitude.

Suppose a satellite is 25,000 km above the center of the Earth when it is precisely overhead the Palomar Observatory in California. How far is the satellite above the observatory? Palomar Observatory is 1.702 km above sea level, so applying equation 9.2.1, we have

$$r_{dist} = r - r_e - r_{sea} = (25{,}000 - 6378.135 - 1.702)\text{ km} = 18{,}620.163\text{ km},$$

which equates to 11,570.033 miles.

A very important attribute of an ellipse that is not shown in figures 9.3 and 9.4 is an ellipse's eccentricity. Recall from section 4.4 that eccentricity is the ratio of the distance between a focus and the geometric center to the length of the semi-major axis. In terms of the nomenclature of figure 9.3, this means

$$e = \frac{c}{a}. \qquad (9.2.2)$$

Although equation 9.2.2 is the definition of eccentricity, there are several ways to determine eccentricity based upon what other attributes of the ellipse are known. Some additional equations for calculating an ellipse's eccentricity are:

$$e = \sqrt{1 - \left(\frac{b}{a}\right)^2} \qquad (9.2.3)$$

$$e = \sqrt{1 - \left(\frac{p}{a}\right)^2} \qquad (9.2.4)$$

$$e = 1 - \frac{r_p}{a} \qquad (9.2.5)$$

$$e = \frac{r_a}{a} - 1 \qquad (9.2.6)$$

$$e = \frac{r_a - r_p}{r_a + r_p}. \qquad (9.2.7)$$

The semi-major axis of an ellipse is $a$ in figure 9.3. Equations for finding the length of the semi-major axis include:

$$a = \frac{c}{e} \tag{9.2.8}$$

$$a = \frac{r_p + r_a}{2} \tag{9.2.9}$$

$$a = \sqrt{c^2 + b^2} \tag{9.2.10}$$

$$a = \frac{\rho}{1 - e^2}. \tag{9.2.11}$$

The semi-minor axis is $b$ in figure 9.3. Equations for finding the length of the semi-minor axis include:

$$b = a\sqrt{1 - e^2} \tag{9.2.12}$$

$$b = \sqrt{a^2 - c^2}. \tag{9.2.13}$$

The linear eccentricity is $c$ in the figure. Equations for finding its length are:

$$c = ae \tag{9.2.14}$$

$$c = \frac{(r_p + r_a)e}{2} \tag{9.2.15}$$

$$c = \sqrt{a^2 - b^2}. \tag{9.2.16}$$

Equations for finding the semi-latus rectum $\rho$ are:

$$\rho = a(1 - e^2) \tag{9.2.17}$$

$$\rho = \frac{2 r_p r_a}{r_p + r_a} \tag{9.2.18}$$

$$\rho = \frac{b^2}{a}. \tag{9.2.19}$$

Equations for finding perigee and apogee heights (apogee height is not shown) are

$$h_p = r_p - r_e \tag{9.2.20}$$

$$h_a = r_a - r_e. \tag{9.2.21}$$

The radius (i.e., distance from the center of the Earth) to the satellite can be calculated from the semi-latus rectum and true anomaly by these equations:

$$r = \frac{\rho}{1 + e \cos \upsilon} \tag{9.2.22}$$

$$r = \frac{a(1 - e^2)}{1 + e \cos \upsilon}. \tag{9.2.23}$$

Note that at apogee, $r_a = r$ while at perigee $r_p = r$. From the definition of the true anomaly, $\upsilon = 0°$ at perigee and $\upsilon = 180°$ at apogee. We can make use of these facts about perigee and apogee and apply them to equations 9.2.22 and 9.2.23, while also making use of equations 9.2.5 and 9.2.6, to derive equations for the apogee radius and perigee radius. The apogee radius is given by

$$r_a = \frac{p}{1-e} \tag{9.2.24}$$

$$r_a = a(1+e) \tag{9.2.25}$$

$$r_a = a+c. \tag{9.2.26}$$

Similarly, the perigee radius is given by

$$r_p = \frac{p}{1+e} \tag{9.2.27}$$

$$r_p = a(1-e) \tag{9.2.28}$$

$$r_p = a-c. \tag{9.2.29}$$

Chapter 4 described how to calculate the true anomaly by either solving the equation of the center or Kepler's equation. That discussion applies here as well. However, if we know the radius to the satellite, the orbital eccentricity, and either the semi-major axis or the semi-latus rectum, then the true anomaly can be easily calculated from equations 9.2.22 and 9.2.23. If we also know the apogee radius and perigee radius, we can use equation 9.2.18 to determine the true anomaly and thereby derive these equations:

$$\upsilon = \cos^{-1}\left(\frac{p-r}{re}\right) \tag{9.2.30}$$

$$\upsilon = \cos^{-1}\left[\frac{a(1-e^2)-r}{re}\right] \tag{9.2.31}$$

$$\upsilon = \cos^{-1}\left(\sqrt{\frac{2r_p r_a - r(r_p + r_a)}{r(r_a - r_p)}}\right). \tag{9.2.32}$$

Equations such as these that allow us to determine orbital information from satellite distances are particularly useful because radar and other types of sensors can be used to determine the distance (range) to a satellite. Of course, a ranging sensor will typically give the distance from the sensor to the satellite rather than the distance from the center of the Earth. Thus, the range reported by the sensor must be adjusted to obtain the proper value for $r$. Equation 9.2.1 cannot be applied unless the satellite's range was measured when the satellite was at the sensor's zenith.

We can determine a satellite's true anomaly without resorting to the computational complexity involved with the equation of the center and Kepler's equation in a few other special cases. In these additional special cases, we can take advantage of what we know about the true anomaly at specific points in a satellite's orbit. When a satellite is at perigee, $v = 0°$ while $v = 180°$ when it is at apogee. As a satellite travels from perigee toward apogee and reaches the semi-latus rectum shown as $\rho$ in figure 9.3, we have $v = 90°$. Another semi-latus rectum extends downward from the occupied focus to the ellipse, which is the semi-latus rectum that is reached as a satellite travels from apogee toward perigee. When a satellite reaches that semi-latus rectum, we have $v = 270°$.

The true anomaly can also be determined when a satellite reaches the semi-minor axis point in its orbit. When it reaches the semi-minor axis point traveling from perigee toward apogee, the true anomaly is

$$v = \cos^{-1}(-e). \tag{9.2.33}$$

When a satellite reaches the semi-minor axis point while traveling from apogee toward perigee, the true anomaly is

$$v = 360° - \cos^{-1}(-e). \tag{9.2.34}$$

Equations in this section are useful because they allow us to determine the essential elements of an ellipse from the elements we know. This is important because we may be able to determine only some of the orbital elements from observing an object over a period of time, such as its distance at perigee/perihelion and apogee/aphelion. However, we can then use that information to calculate other orbital information, such as eccentricity and length of the semi-major axis. Moreover, we can apply Kepler's laws to deduce other orbital characteristics, as we did in subsection 8.12.3 to calculate a planet's period from its semi-major axis. We can apply exactly the same techniques to satellites to determine, for instance, a satellite's period from its semi-major axis.

## 9.3 Geocentric and Topocentric Coordinates

Before looking at a satellite's orbital elements and other attributes, we need a coordinate system for describing a satellite's location that meets 4 criteria. First, Earth should be the coordinate system origin to simplify calculations. Second, the system should be independent of any specific observer's location. Third, a location described in such a system should be fixed regardless of the

# Satellites

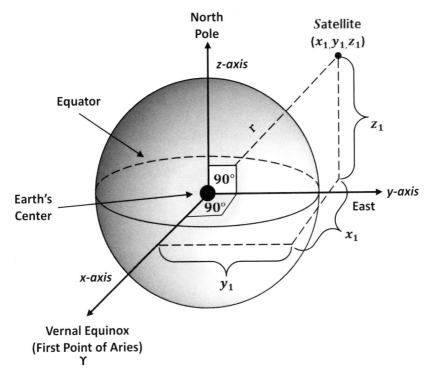

**Figure 9.5** Cartesian Coordinate System
A geocentric Cartesian coordinate system can be used to specify a satellite's location.

time zone. Fourth, it should be simple to convert a location in such a coordinate system to an observer's horizon-based coordinate system so that an observer knows where to point a telescope or sensor to view the satellite.

Criteria 2 and 3 are of practical importance. Without them, if we wanted to publish a satellite's position, we would have to publish it for different places around the globe and for different time zones. A coordinate system that requires publishing so many different coordinates for the same satellite to account for different locations and time zones would make publishing a catalog of satellite positions impractical.

Figure 9.5 shows a geocentric Cartesian coordinate system that meets all 4 criteria. Its $x$-axis lies in the plane of the Earth's equator and extends from the center of the Earth toward the vernal equinox. The $y$-axis also lies in the plane of the equator and extends eastward from the center of the Earth. The $z$-axis is perpendicular to the plane of the equator and extends northward from the center of the Earth through Earth's North Pole. The location of any object in

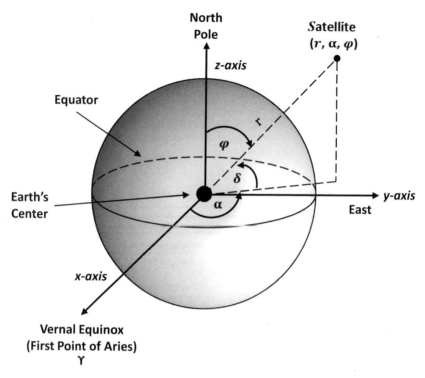

**Figure 9.6** Spherical Coordinate System
This figure shows a geocentric coordinate system in spherical coordinates rather than Cartesian.

the sky, such as a satellite, can be uniquely described by stating its Cartesian coordinates, as shown in the figure. The length $r$ is the distance from the center of the Earth to the object (e.g., satellite). We will see shortly how to calculate that distance directly from an object's Cartesian coordinates.

The coordinate system shown in figure 9.5 is technically known as the Earth Centered Inertial (ECI) coordinate system.[2] It is a fixed geocentric coordinate system in which coordinates do not change as the Earth rotates because it is oriented (via the $x$-axis) to a fixed point in the sky—namely, the vernal equinox.[3]

---

2. It is also known as the IJK coordinate system and as the Conventional Inertial System (CIS) coordinate system.

3. This is not quite true. Precession causes the location of the vernal equinox to change over time. To account for this, a standard epoch, such as J2000, is used to state the location of the vernal equinox. Nutation also causes the orientation of Earth's axis of rotation to change over time. We will ignore these practical considerations, but they must be accounted for when high accuracy is required.

# Satellites

The issue we must now deal with is satisfying our fourth criterion; that is, how can we convert between the just-introduced ECI coordinate system and a horizon-based coordinate system? To answer that question, suppose instead of defining ECI as a Cartesian coordinate system, we define it as a spherical coordinate system that meets the same criteria that motivated defining the ECI coordinate system in the first place. The result is shown in figure 9.6.[4] It, too, can be used to uniquely describe the location of any object in the sky. It is exactly the same as figure 9.5 except that figure 9.6 describes a location in terms of 2 angles and a distance rather than in terms of 3 distances, as the Cartesian coordinate system does.

As we saw in section 5.3, converting Cartesian to spherical coordinates requires 3 equations:

$$r = \sqrt{x^2 + y^2 + z^2} \tag{9.3.1}$$

$$\alpha = \tan^{-1}\left(\frac{y}{x}\right) \tag{9.3.2}$$

$$\varphi = \cos^{-1}\left(\frac{z}{r}\right). \tag{9.3.3}$$

Equation 9.3.1 gives the distance from the center of the Earth to the satellite shown in figure 9.6; it is simply the magnitude of a vector that captures the satellite's position in Cartesian coordinates (compare equations 9.3.1 and 9.1.1). Because equation 9.3.2 involves the arctangent, we must use the method presented in section 4.1 to remove the ambiguity of the arctangent function to ensure that $\alpha$ is in the range [0°, 360°].

Converting spherical to Cartesian coordinates is simple and requires 3 equations:

$$x = r \cos\alpha \sin\varphi \tag{9.3.4}$$

$$y = r \sin\alpha \sin\varphi \tag{9.3.5}$$

$$z = r \cos\varphi. \tag{9.3.6}$$

What have we accomplished by changing from a Cartesian coordinate system to a spherical one? Well, let us make 1 minor change to figure 9.6. Instead

---

4. The Greek letters used to label the angles in the figure are unimportant. Although mathematicians have historically used $\theta$ instead of $\alpha$ when describing the spherical coordinate system (which we also did in section 5.3), the reason we choose to use $\alpha$ instead of $\theta$ will become apparent shortly. Also note that in describing a spherical coordinate system, physicists typically use the same Greek letters for the angles that mathematicians use, but historically they have reversed which Greek letter is used for which angle. This can be *very* confusing when looking at figures and formulas from physicists versus mathematicians!

of using the angle $\varphi$, which is measured from the $z$-axis downward toward the $xy$ plane, let us use the angle $\delta$ and measure it upward from the $xy$ plane to the radius $r$. The relationship between these 2 angles is simply

$$\delta = 90° - \varphi. \tag{9.3.7}$$

If we make this minor change, then figure 9.6 is the same as the equatorial coordinate system defined in section 4.6 with $\delta$ being the declination and $\alpha$ being the right ascension! This simple change means we can easily convert a satellite location in the ECI coordinate system to the horizon coordinate system presented in section 4.7. The process required is as follows: convert the satellite's Cartesian coordinates to a spherical coordinate system, apply equation 9.3.7 to convert the angle $\varphi$ to the declination $\delta$, and then convert equatorial coordinates $(\alpha, \delta)$ to horizon coordinates through the methods presented in chapter 4. To convert horizon coordinates to the ECI coordinate system, we merely reverse this process.

Do not confuse the horizon coordinate system's altitude $h$ with distance to the satellite $r$! Altitude in the horizon coordinate system is just an angle that indicates how far above the horizon one must look to see an object while $r$ is the actual distance (from the center of the Earth) to the object.

Consider an example. Assume that at precisely $18^h$ UT on February 15, 2016, the Cartesian coordinates, expressed in km, for a satellite are

$$\vec{R} = [15{,}300 \ \ 24{,}600 \ \ -18{,}000].$$

An observer at sea level located at 38° N latitude, 78° W longitude in the Eastern Standard Time zone was stargazing at that same time (i.e., $13^h$ LCT). For that observer, what were the satellite's horizon coordinates and what was the satellite's distance from the center of the Earth? Assume that the observer was not on daylight saving time.

1. Convert the UT to its LCT, LST, and GST times for the observer.
   (Ans: $LCT = 13^h$, $UT = 18^h$, $GST = 3.678934^h$, $LST = 22.478934^h$, date = 2/15/2016.)
2. Convert the Cartesian coordinates to spherical coordinates. Remember to use the sign of $y$ and $x$ to adjust $\alpha$ if necessary to put it into the correct quadrant.
   (Ans: $r = 34{,}106.45100271$ km, $\alpha = 58.12040315°$, $\varphi = 121.85419091°$.)
3. Use equation 9.3.7 to convert $\varphi$ to declination.
   (Ans: $\delta = -31.85419091°$.)
4. Convert $\alpha$ from degrees to decimal hours (i.e., divide by 15).
   (Ans: $\alpha_t = 3.87469354^h$.)

5. Convert the equatorial coordinates ($\alpha_t, \delta$) to horizon coordinates.
(Ans: $h = -12°40'42.72''$, $A = 120°42'40.55''$.)

At the stated time, the satellite was 34,106.451 km above the center of the Earth, and it was below the horizon.

The preceding example has a significant problem. The steps just shown produce *geocentric* horizon coordinates; the results are valid *only* for an observer at the Earth's center. We failed to take into account an observer's distance from the center of the Earth. When calculating horizon coordinates in the preceding chapters, an observer's distance from the center of the Earth was negligible because that distance was insignificant compared to the distance from Earth to the object. Or, to state it differently, when calculating distances in the preceding chapters we could essentially treat Earth as a point rather than a sphere.

For satellites we need a horizon-based system whose origin is located at the observer's location on the surface of the Earth rather than at the center of the Earth. A coordinate system whose origin is on the Earth's surface is called a topocentric coordinate system and can be expressed in Cartesian or spherical coordinates. In a topocentric system, the $x$-axis points south, the $y$-axis points east, and the $z$-axis points toward the observer's zenith. A topocentric spherical coordinate system is the same as the horizon coordinate system presented in chapter 4 except for where the origin is located: observer's location (topocentric) or center of the Earth (geocentric).

Given a satellite's Cartesian coordinates expressed in km, how do we convert them to topocentric coordinates?[5] Assume an observer at latitude $\phi$, longitude $\psi$ is $h_{sea}$ meters above sea level. We first convert the observer's location to Cartesian ECI coordinates and rotate the result to account for how far away the observer is from the vernal equinox. A rotation is necessary because the $x$-axis in the ECI coordinate system points toward the vernal equinox while the $x$-axis direction in an observer's topocentric coordinate system depends upon the observer's longitude. The LST is how far an observer is from the vernal equinox and, when converted from hours to degrees, it is the angle by which the coordinates must be rotated. The necessary equations are

$$r'_e = r_e + (h_{sea}/1000) \tag{9.3.8}$$

$$r_{eq} = r'_e * \cos\phi \tag{9.3.9}$$

$$x_{obs} = r_{eq} * \cos(LST_d) \tag{9.3.10}$$

---

5. The approach here is adapted from Kelso's articles on orbital coordinate systems referenced in section 10.7.

$$y_{obs} = r_{eq} * \sin(LST_d) \qquad (9.3.11)$$

$$z_{obs} = r'_e * \sin\phi, \qquad (9.3.12)$$

where $r_e$ is the Earth's radius in km, $r'_e$ combines the Earth's radius and the observer's distance above sea level, $r_{eq}$ is a projection of the observer's latitude onto the plane of the equator, and $LST_d$ is the observer's LST multiplied by 15 to convert it to degrees.

We must now adjust the satellite's coordinates so that its distance is with respect to the observer's location instead of the center of the Earth. This is done by calculating the range vector, which is just the difference between $\vec{R}$ (the satellite's position) and the observer's location. That is, the range vector is

$$\vec{R}' = [x' \ y' \ z'] = [x - x_{obs} \ y - y_{obs} \ z - z_{obs}]. \qquad (9.3.13)$$

The range vector's length is the satellite's distance from the observer's location.

The next step is to rotate the range vector by $LST_d$ degrees about the $z$-axis and by $\phi$ around the $y$-axis so that it is aligned with the observer's location. The equations required are:

$$x'' = x' * \sin\phi * \cos(LST_d) + y' * \sin\phi * \sin(LST_d) - z' * \cos\phi \qquad (9.3.14)$$

$$y'' = -x' * \sin(LST_d) + y' * \cos(LST_d) \qquad (9.3.15)$$

$$z'' = x' * \cos\phi * \cos(LST_d) + y' * \cos\phi * \sin(LST_d) + z' * \sin\phi. \qquad (9.3.16)$$

The final step is to convert the rotated range vector to topocentric coordinates to get a topocentric azimuth and altitude. The required equations are:

$$r_{dist} = \sqrt{x''^2 + y''^2 + z''^2} \qquad (9.3.17)$$

$$A_{topo} = \tan^{-1}(-y''/x'') \qquad (9.3.18)$$

$$h_{topo} = \sin^{-1}(z''/r_{dist}). \qquad (9.3.19)$$

The subscript *topo* is used to reinforce the fact that the altitude and azimuth are topocentric coordinates, not geocentric. Because it involves an arctangent, the azimuth computed by equation 9.3.18 may have to be adjusted to put it into the correct quadrant.

The astute reader may have wondered why an observer's longitude does not seem to appear in any of the equations because obviously an observer's longitude must be considered. In fact, the observer's longitude *is* accounted for in the calculations that include the LST because calculating the LST requires a longitude.

**Satellites** 283

To demonstrate the process, let us compute the topocentric coordinates for the example whose geocentric horizon coordinates we just calculated.

1. Convert the UT to its LCT, LST, and GST times for the observer.
  (Ans: $LCT = 13^h$, $UT = 18^h$, $GST = 3.678934^h$, $LST = 22.478934^h$, date = 2/15/2016.)
2. Convert the observer's LST to an angle.
  (Ans: $LST_d = 337.184015°$.)
3. Convert the observer's location to ECI coordinates.
  (Ans: $r'_e = 6378.135$ km, $r_{eq} = 5026.03896796$ km, $x_{obs} = 4632.77655106$, $y_{obs} = -1948.96103999$, $z_{obs} = 3926.77200393$.)
4. Compute the range vector.
  (Ans: $x' = 10{,}667.22344894$, $y' = 26{,}548.96103999$, $z' = -21{,}926.77200393$.)
5. Rotate the range vector by $LST_d$ and $\phi$.
  (Ans: $x'' = 16{,}993.85177868$, $y'' = 28{,}608.09635272$, $z'' = -13{,}863.84303584$.)
6. Compute the rotated range vector's magnitude.
  (Ans: $r_{dist} = 36{,}047.47312815$ km.)
7. Compute the topocentric azimuth. Adjust if necessary to put it in the correct quadrant.
  (Ans: $A_{topo} = 300.711263°$.)
8. Compute the topocentric altitude.
  (Ans: $h_{topo} = -22.618884°$.)
9. Convert the altitude and azimuth to DMS format.
  (Ans: $h_{topo} = -22°37'7.98''$, $A_{topo} = 300°42'40.55''$.)

The topocentric coordinates are *significantly* different from the geocentric horizon coordinates computed for this example!

Practitioners sometimes collectively refer to a satellite's altitude and azimuth as the look angle. Moreover, the word "elevation" will sometimes be used instead of "altitude," but it has the same meaning as the topocentric altitude. We will refer to altitude and azimuth instead of look angle as a reminder of their similarity to the horizon coordinate system used in the preceding chapters.

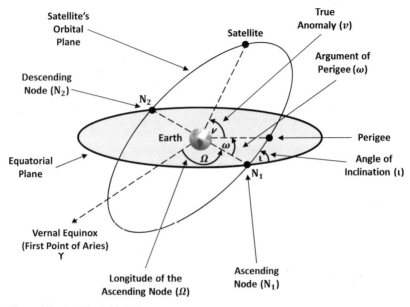

**Figure 9.7**  Satellite Orbital Elements
A satellite's orbital elements are the same as for a planet or other celestial object. The major difference is what plane (equatorial or ecliptic) is used as a reference plane.

## 9.4  Satellite Orbital Elements

Section 9.2 presented various components of an ellipse to describe a satellite's orbit in 2 dimensions. Of course, satellites travel through space in 3 dimensions. Therefore, just as we did in chapter 4 for celestial objects, we must extend the discussion to consider orbital motion in 3-dimensional space. In this case, we must describe how a satellite is oriented with respect to Earth.

We begin with the orbital elements shown in figure 9.7. This figure is nearly identical to figure 4.12 (subsection 4.5.6), which describes a planet's orbital elements. Because we are dealing here with objects orbiting Earth, we use perigee/apogee rather than perihelion/aphelion to describe the points at which a satellite is closest to/farthest away from Earth. Some authors prefer the more generic terms "periapsis" and "apoapsis," but they are the same as perigee and apogee when dealing with objects orbiting Earth.

Note that a satellite's angle of inclination $\iota$ is measured with respect to the plane of the Earth's equator, *not* the ecliptic plane. An object's angle of inclination is always stated relative to a reference plane, which by convention is the ecliptic plane for objects orbiting the Sun but is the Earth's equatorial plane for satellites orbiting Earth. If $\iota$ is in the range $[0°, 90°]$, the satellite is orbiting in an easterly direction. If in the range $[90°, 180°]$, the satellite is orbiting in a westerly direction and is said to have a retrograde orbit.

Orbital elements in figure 9.7 have the same meaning as for the planets except that the frame of reference is different (equatorial versus ecliptic plane). So, $N_1$ and $N_2$, the ascending and descending nodes, are the points at which a satellite goes above ($N_1$) or below ($N_2$) the equatorial plane. A satellite's longitude of the ascending node ($\Omega$) is typically called the right ascension of the ascending node and abbreviated RAAN.

Before leaving this discussion of orbital elements, we note that problems arise with circular (i.e., $e=0$) and equatorial orbits (i.e., $\iota=0°$ or $\iota=180°$). Some orbital elements are undefined for such orbits because the reference points used to define them do not exist. Refer to figure 9.7. If an orbit lies within the equatorial plane ($\iota=0°$ or $\iota=180°$), there are no ascending or descending nodes ($N_1$ and $N_2$) for defining the RAAN ($\Omega$). Similarly, if the orbit is a true circle ($e=0$), the object is always at a constant distance from the occupied focus so there is no perigee or apogee from which to measure the argument of perigee ($\omega$), nor from which to measure the true anomaly ($\upsilon$). To handle circular and equatorial orbits, 3 additional angles are defined:

- $u$, argument of latitude for circular orbits that are not in the equatorial plane
- $\lambda_t$, the true longitude at the epoch for circular orbits in the equatorial plane
- $\omega_t$, true longitude of perigee for elliptical orbits in the equatorial plane

The equations for $u$, $\lambda_t$, and $\omega_t$ are given later in subsection 9.4.2.

Four cases must be considered to handle circular and equatorial orbits:

1. $e=0$, and $\iota=0°$ or $\iota=180°$: These are circular orbits in the equatorial plane. $\Omega$ and $\omega$ are undefined, and both are given the value $0°$. $\upsilon$ is also undefined and given the value $\lambda_t$.

2. $e>0$, and $\iota=0°$ or $\iota=180°$: Such orbits are elliptical and are in the equatorial plane. $\Omega$ is undefined, so it is given the value $0°$. $\omega$ is also undefined, so $\omega_t$ is used instead.

3. $e=0$ and $\iota>0°$: These orbits are circular but are inclined with respect to the equatorial plane. $\omega$ is undefined and given the value $0°$. $\upsilon$ is also undefined and given the value $u$.

4. $e>0$ and $\iota>0°$: These elliptical orbits are inclined with respect to the equatorial plane. All orbital elements in figure 9.7 are defined and have their usual meaning.

We did not need to worry about different orbit types before now because all objects in prior chapters have elliptical orbits inclined with respect to their reference orbital plane.

For the algorithms ahead, we will sometimes need to know what type of orbit is being analyzed. For convenience, define $O_{type}$ to be an integer whose value (1, 2, 3, or 4) corresponds to 1 of the 4 orbit types in the preceding list. Thus, if $O_{type} = 3$, we are dealing with a circular orbit that is inclined with respect to the equatorial plane. We will usually be concerned with orbit type 4.

### 9.4.1 Which Orbital Elements?

The labeled items in figures 9.3, 9.4, and 9.7 provide a lot of information about an object's orbit. How many of those items do we actually need to completely specify an object's orbital characteristics? It turns out that only 6 measurements are needed to completely and uniquely describe the location and orientation of any orbiting object. This is because, regardless of the coordinate system being used, 3 measurements are required to uniquely describe an object's position in 3-dimensional space, and 3 more are required to uniquely describe its orientation. In astronomy these 6 elements have historically been called Keplerian elements in honor of Johannes Kepler, who first discovered that heavenly bodies move in elliptical orbits. More precisely, the historical Keplerian elements are:

1. Inclination ($\iota$)
2. Eccentricity ($e$)
3. Length of the semi-major axis ($a$)
4. Longitude of the ascending node ($\Omega$)
5. Argument of perigee/perihelion ($\omega$)
6. Mean anomaly at the epoch ($M_0$)

The length of the semi-major axis specifies how big an orbit is while the eccentricity specifies an orbit's shape. The inclination, longitude of the ascending node, and argument of perigee specify an orbit's orientation with respect to a reference plane. The only remaining item required is to indicate where an object is in its orbit, which in the case of satellites is done by specifying the mean anomaly with respect to an instant in time (i.e., an epoch).

Sometimes there are variations in this basic set, such as giving the true anomaly instead of the mean anomaly, or giving some attribute other than the semi-major axis. Recall from chapter 4 that the true anomaly can be determined from the mean anomaly and vice versa. Likewise, equations such as those presented in section 9.2 provide flexibility in computing the attributes of an ellipse when given one attribute but another is desired. Thus, these 6 Keplerian elements can always be derived even if some other set of orbital elements are provided, as long as the orbital elements given are sufficient to completely describe an object's location and orientation.

Let us pause for a moment to be more precise with terminology. The phrases *orbital elements*, *Keplerian elements*, *mean orbital elements*, and *osculating orbital elements* are often used interchangeably. However, there are subtle differences that should be understood even though practitioners often neglect to distinguish between these phrases. One way to understand the distinctions is as follows:

- **Orbital elements** refers to the elements shown in figure 9.7, plus the orbital eccentricity and length of the semi-major axis. When used generically, as this phrase often is, the elements that define an orbit are not referenced to any particular point in time.
- **Keplerian elements** are the 6 items previously listed plus an epoch as a reference point in time. Keplerian elements assume an ideal elliptical orbit that is unaffected by real-world effects such as precession and gravitational forces.
- **Mean orbital elements** are the same as Keplerian elements except that mean orbital elements are "averaged out" over an entire orbit so that various perturbation effects are accounted for through the averaging process. That is, although an ideal elliptical orbit is assumed, orbital elements have been averaged out over the object's real, perturbed orbit. Moreover, for satellites mean orbital elements are typically given with respect to a stated model of the Earth (spherical, ellipsoid, etc.) and with respect to some model for representing the orbit's various perturbations.
- **Osculating orbital elements** are the same as Keplerian and mean elements except that they are instantaneous values instead of mean values. These elements reflect a real orbit undergoing "real world" perturbations from various effects (solar weather, nonspherical Earth, solar and lunar gravitational effects, etc.).

We will not concern ourselves with the subtle differences between these descriptions of orbital characteristics. We will generally refer to them simply as orbital elements or Keplerian elements, in keeping with common practice.

### 9.4.2 Keplerian Elements and State Vectors

If we do not know a satellite's orbital elements, it is necessary to locate the satellite at 3 different points in its orbit to determine its orbital elements. Once derived, they are typically stated as Keplerian elements (the set

[$\iota$ $e$ $a$ $\Omega$ $\omega$ $M_0$]). However, Keplerian elements are not the only way to describe a satellite's position. Another way is to use Euler angles, which Leonhard Euler (1707–1783) developed to describe the orientation of objects in 3-dimensional space. We will not discuss Euler angles, but you may see them in other references on satellite orbits.

State vectors are frequently encountered as a way to specify a satellite's position and orientation. A state vector expresses position and orientation in terms of two 3-element vectors: a position vector $\vec{R}$ and a velocity vector $\vec{V}$, which we will express as

$$\vec{R} = \begin{bmatrix} x & y & z \end{bmatrix}$$

$$\vec{V} = \begin{bmatrix} V_x & V_y & V_z \end{bmatrix}.$$

$V_x$ is the object's velocity in the $x$ direction, $V_y$ is the velocity in the $y$ direction, and $V_z$ is the velocity in the $z$ direction. $\vec{R}$ and $\vec{V}$ combined completely describe an object's position and orientation.

An object's position and orientation in 3-dimensional space can be described by Keplerian elements or state vectors because the two are mathematically equivalent. The choice of one over the other is largely a matter of preference. Although osculating orbital elements are often expressed as state vectors, Keplerian elements are more widely used in satellite catalogs, and especially when expressed in the Two-Line Elements (TLE) format, which we will address in subsection 9.4.3.

Significant mathematical expertise is required to derive a process for converting between Keplerian elements and state vectors. We will show how to do the conversions, but we will not attempt to explain why the process works because doing so requires calculus and linear algebra, which are beyond the scope of this book. Readers who wish to understand the mathematics involved can find a list of helpful resources in chapter 10, such as *Fundamentals of Astrodynamics* by Bate, Mueller, and White and *Fundamentals of Astrodynamics and Applications* by Vallado. The conversion techniques presented here were derived by combining methods presented in both those references.

To convert between Keplerian elements and state vectors, we will need to determine the argument of latitude ($u$), true longitude at the epoch ($\lambda_t$), and true longitude of perigee ($\omega_t$) for certain orbit types. The required equations are

$$u = \cos^{-1}\left(\frac{xN_x + yN_y}{N_{len}R_{len}}\right) \qquad (9.4.1)$$

$$\lambda_t = \cos^{-1}\left(\frac{x}{R_{len}}\right) \qquad (9.4.2)$$

# Satellites

$$\omega_t = \cos^{-1}\left(\frac{e_x}{e}\right), \qquad (9.4.3)$$

where $x$ and $y$ are elements in the positional vector $\vec{R}$, $R_{len}$ is the length of the positional vector, $N_x$ and $N_y$ are elements in the node vector (see the following algorithm), $N_{len}$ is the length of the node vector, $e$ is the orbital eccentricity, and $e_x$ is a component of the eccentricity vector (see the following algorithm). We will address only satellites with circular ($e=0$) or elliptical ($0 < e < 1$) orbits. The algorithms presented here will *not* work for parabolic ($e=1$) or hyperbolic ($e > 1$) orbits.

Let us first convert a state vector to its equivalent Keplerian elements. On November 19, 1996, at $20^h$ UT (the date and time will not be needed), the positional components in the state vector for space shuttle STS-80 were

$$\vec{R} = [x \ y \ z] = [-6260.434 \ 2221.183 \ 1094.143],$$

where all units are in km. STS-80's velocity components were

$$\vec{V} = [V_x \ V_y \ V_z] = [-2.675419 \ -5.812131 \ -3.449434],$$

where all units are in km/s. What were STS-80's Keplerian elements for this state vector, how far was it above the center of the Earth, and how fast was it going?

The steps required to answer these questions are as follows:

1. Compute the length (magnitude) of the position vector (use equation 9.1.1).
   (Ans: $\overline{R} = 6732.29802462$ km, which is how far STS-80 was above Earth's center.)

2. Compute the length of the velocity vector.
   (Ans: $\overline{V} = 7.26892898$ km/s, which is how fast STS-80 was traveling.)

3. Compute the angular momentum vector $\vec{H} = [H_x \ H_y \ H_z]$ by the equations

$$H_x = yV_z - zV_y,$$
$$H_y = zV_x - xV_z,$$
$$H_z = xV_y - yV_x.$$

   (Ans: $H_x = -1302.52171169$, $H_y = -24{,}522.24486527$, $H_z = 42{,}329.05772553$.)

4. Compute the length of the angular momentum vector.
   (Ans: $\overline{H} = 48{,}936.55263679$.)

**Figure 9.8** Space Shuttle *Columbia*
This is space shuttle *Columbia* (STS-80) being launched November 19, 1996 from the Kennedy Launch Complex in Florida. With a mission lasting 17 days, 15 hours, and 53 minutes, STS-80 was the longest of all the space shuttle missions. (Image courtesy of NASA)

**Satellites**

5. Compute the node vector $\vec{N} = [N_x \; N_y \; N_z]$ by the equations

$N_x = -H_y,$

$N_y = H_x,$

$N_z = 0.0.$

(Ans: $N_x = 24{,}522.24486527$, $N_y = -1302.52171169$, $N_z = 0.0.$)

6. Compute the length of the node vector.
(Ans: $\overline{N} = 24{,}556.81282337.$)

7. Calculate 2 temporary variables, $A$ and $B$, defined as

$A = V_{len}^2 - \dfrac{\mu}{R_{len}}$

$B = xV_x + yV_y + zV_z,$

where $\mu$ is the Earth's standard gravitational parameter (we use a more precise value for $\mu$ than given in section 8.6 to provide more accuracy).

(Ans: $\mu = 398{,}600.4418 \text{ km}^3/\text{s}^2$, $A = -6.36986053$, $B = 65.30343581.$)

8. Compute the eccentricity vector via the equations

$e_x = (xA - BV_x)/\mu,$

$e_y = (yA - BV_y)/\mu,$

$e_z = (zA - BV_z)/\mu.$

(Ans: $e_x = 0.10048360$, $e_y = -0.03454355$, $e_z = -0.01691990.$)

9. Compute the length of the eccentricity vector to obtain the orbital eccentricity.
(Ans: $e = 0.10759411.$)

10. Compute the semi-latus rectum via the equation

$\rho = (\overline{H})^2/\mu.$

(Ans: $\rho = 6007.98677783$ km.)

11. Use the semi-latus rectum and eccentricity to determine the length of the semi-major axis (see equation 9.2.11).
(Ans: $a = 6078.35278389$ km.)

12. Compute the orbital inclination from the equation

$\iota = \cos^{-1}\left(\dfrac{H_z}{H_{len}}\right).$

(Ans: $\iota = 30.11976874°.$)

13. Determine the orbit type.
    (Ans: $O_{type} = 4$.)
14. If $O_{type} = 1$ or $O_{type} = 2$, the RAAN is undefined, so set $\Omega = 0°$. Otherwise, calculate the RAAN from the equation

$$\Omega = \cos^{-1}\left(\frac{-H_y}{N_{len}}\right).$$

If $H_x < 0$, then subtract $\Omega$ from 360° to put the result in the correct quadrant.
    (Ans: $\Omega = 356.95953889°$.)
15. If $O_{type} = 1$ or $O_{type} = 3$, the argument of perigee ($\omega$) is undefined, so set $\omega = 0°$ and skip to step 18.
    (Ans: must perform the next step to calculate $\omega$.)
16. If $O_{type} = 2$, set $\omega$ to $\omega_t$ and skip to step 18. $\omega_t$ is given by

$$\omega_t = \cos^{-1}\left(\frac{e_x}{e}\right).$$

If $e_y < 0$, then subtract $\omega_t$ from 360° to put the result in the correct quadrant.
    (Ans: must perform the next step to calculate $\omega$, but $\omega_t = 339.0533209°$.)
17. For all other orbit types, calculate the argument of perigee from the equation

$$\omega = \cos^{-1}\left(\frac{e_y H_x - e_x H_y}{e N_{len}}\right).$$

If $e_z < 0$, then subtract $\omega$ from 360° to put the result in the correct quadrant.
    (Ans: $\omega = 341.73698644°$.)
18. If $O_{type} = 1$, set $\upsilon = \lambda_t$ and skip to step 21. $\lambda_t$ is given by the equation

$$\lambda_t = \cos^{-1}\left(\frac{x}{R_{len}}\right).$$

If $y < 0$, then subtract $\lambda_t$ from 360° to put the result in the correct quadrant.
    (Ans: must perform the next step to calculate $\upsilon$, but $\lambda_t = 158.42085347°$.)
19. If $O_{type} = 3$, set $\upsilon = u$ and skip to step 21. $u$ is given by the equation

$$u = \cos^{-1}\left(\frac{x N_x + y N_y}{N_{len} R_{len}}\right).$$

If $z < 0$, then subtract $u$ from 360° to put the result in the correct quadrant.
    (Ans: must perform the next step to calculate $\upsilon$, but $u = 161.10280826°$.)
20. For all other orbit types, compute the true anomaly from the equation

$$\upsilon = \cos^{-1}\left(\frac{e_x x + e_y y + e_z z}{e R_{len}}\right).$$

If $B < 0$, then subtract $\upsilon$ from $360°$ to put the result in the correct quadrant.

(Ans: $\upsilon = 179.36582182°$.)

21. Use the true anomaly to calculate the eccentric anomaly. We will use the equation

$$E = \cos^{-1}\left(\frac{e + \cos \upsilon}{1 + e \cos \upsilon}\right).$$

This form of the eccentric anomaly differs from that given in subsection 4.5.4, but it is easier to use in this situation. If $\upsilon > 180°$, then subtract $E$ from $360°$ to put it in the correct quadrant.

(Ans: $E = 179.29348834°$.)

22. Use the eccentric anomaly to compute the mean anomaly. From Kepler's equation,

$$M = E - e \sin E,$$

where $E$ *must* be in radians and the result is in radians. Multiply $E$ by $\pi/180$ to convert it to radians before applying this equation.

(Ans: $M = 3.12793499$ radians.)

23. Multiply the mean anomaly by $180/\pi$ to convert it to degrees.

(Ans: $M_0 = 179.21747378°$.)

Thus, STS-80 was traveling at 7.269 km/s (4.517 miles/s) at a distance of 6732.298 km (4183.256 miles) above the center of the Earth. Its Keplerian elements were

$\iota = 30.11976874°$,

$e = 0.10759411$,

$a = 6078.35278389$ km,

$\Omega = 356.95953889°$,

$\omega = 341.73698644°$,

$M_0 = 179.21747378°$.

Converting Keplerian elements to a state vector is similarly complex. The strategy for doing so is to use yet another coordinate system, the perifocal coordinate system, for which the state vector is easy to derive. We will not describe the perifocal coordinate system beyond noting that it is defined

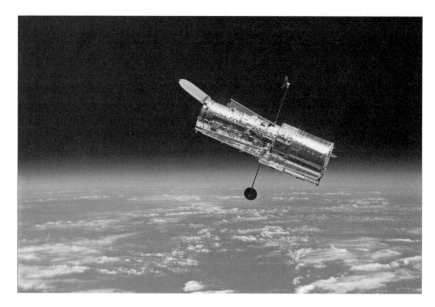

**Figure 9.9**  Hubble Space Telescope
The Hubble Space Telescope (HST) as photographed in 1997 from space shuttle *Discovery* (STS-82). The HST gives astronomers views of distant objects in the universe that are impossible to obtain from Earth-based telescopes. (Image courtesy of NASA)

relative to the object whose orbital elements we know rather than relative to Earth. Once the state vector is in the perifocal coordinate system, a series of rotations about coordinate system axes will transform it to the ECI coordinate system. The details need not concern us too much, but understanding the strategy will help in following the logic behind the algorithm presented below.

Let us do an example. On February 18, 2016, at $22^h26^m51^s$ UT (as for the previous example, the date and time will not be needed) the Hubble Space Telescope (HST) had the following Keplerian elements:

$\iota = 28.47°$,

$e = 0.000284$,

$a = 6919.90 \, \text{km}$,

$\Omega = 326.13°$,

$\omega = 11.39°$,

$M_0 = 83.01°$.

Determine the HST's state vector, how fast it was traveling, and how far it was above the center of the Earth.

To properly handle circular and equatorial orbits, the following algorithm assumes that if $O_{type} = 1$, the mean anomaly at the epoch ($M_0$) we are given was derived from the true longitude at the epoch ($\lambda_t$). If $O_{type} = 3$, it is assumed that the given mean anomaly was derived from the argument of latitude ($u$). Similarly, if $O_{type} = 2$, it is assumed that the argument of perigee ($\omega$) is actually the true longitude of perigee ($\omega_t$). Based on these assumptions, the steps required to determine the HST's state vector are:

1. Compute the semi-latus rectum from the length of the semi-major axis and the eccentricity (use equation 9.2.17).
   (Ans: $\rho = 6919.89944187$ km.)

2. Solve Kepler's equation to get the eccentric anomaly. (We will use the simple iterative method from subsection 4.5.5 for 2 iterations. So few iterations are required because the orbital eccentricity is so close to a circle in this example.)
   (Ans: $E = 83.02615162°$.)

3. Apply equation 4.5.8 to the eccentric anomaly and eccentricity to get the true anomaly.
   (Ans: $\upsilon = 83.04230351°$.)

4. Apply equation 9.2.22 to the true anomaly, eccentricity, and semi-latus rectum to get the length of the positional vector.
   (Ans: $\overline{R} = 6919.6613642$ km, which is how far the HST was above the center of the Earth.)

5. Compute the positional vector $\vec{R}' = [x'\ y'\ z']$ in the perifocal coordinate system using the equations

$x' = R_{len} \cos \upsilon,$

$y' = R_{len} \sin \upsilon,$

$z' = 0.0.$

   (Ans: $x' = 838.22341038,\ y' = 6868.70404203,\ z' = 0.0$.)

6. Calculate a temporary variable $A = \sqrt{\mu/\rho}$, where $\mu$ is the Earth's standard gravitational parameter.
   (Ans: $A = 7.58960190$.)

7. Compute the velocity vector $\vec{V}' = [V'_x\ V'_y\ V'_z]$ in the perifocal coordinate system using the equations

$V'_x = -A \sin \upsilon,$

$V'_y = A(e + \cos \upsilon),$

$V'_z = 0.0.$

(Ans: $V'_x = -7.53371102$, $V'_y = 0.92153309$, $V'_z = 0.0$.)

8. Compute the length of the velocity vector $\vec{V}'$.

(Ans: Since a rotation does not change a vector's length, the length of $\vec{V}'$ will be the length of the state vector's velocity vector. $\overline{V'} = 7.58986330 \, \text{km/s}$, which is how fast the HST was traveling.)

9. Determine the orbit type.

(Ans: $O_{type} = 4$.)

10. If $O_{type} = 1$ or $O_{type} = 2$, set $\Omega' = 0°$. Otherwise, set $\Omega' = \Omega$.
(Ans: $\Omega' = 326.13°$.)

11. If $O_{type} = 1$ or $O_{type} = 3$, set $\omega' = 0°$. Otherwise, set $\omega' = \omega$.
(Ans: $\omega' = 11.39°$.)

12. Use the $g$ family of functions to rotate $\vec{R}'$ by $-\omega'$ degrees about the $z$-axis.
(Ans: $x' = -534.75943697$, $y' = 6898.96702755$, $z' = 0.0$.)

13. Use the $f$ family of functions to rotate the new $\vec{R}'$ vector from the previous step by $-\iota$ degrees about the $x$-axis.
(Ans: $x' = -534.75943697$, $y' = 6064.65308848$, $z' = 3288.72755997$.)

14. Use the $g$ family of functions to rotate the new $\vec{R}'$ vector from the previous step by $-\Omega'$ degrees about the $z$-axis. The result will be the positional vector component $\vec{R}$ of the state vector.

(Ans: $x = 2935.88146621$, $y = 5333.53398371$, $z = 3288.7275600$.)

15. Use the $g$ family of functions to rotate the $\vec{V}'$ vector by $-\omega'$ degrees about the $z$-axis.
(Ans: $V'_x = -7.56732964$, $V'_y = -0.58442022$, $V'_z = 0.0$.)

16. Use the $f$ family of functions to rotate the new $\vec{V}'$ vector from the previous step by $-\iota$ degrees about the $x$-axis.
(Ans: $V'_x = -7.56732964$, $V'_y = -0.51374443$, $V'_z = -0.27859227$.)

17. Use the $g$ family of functions to rotate the new $\vec{V}'$ vector from the previous step by $-\Omega'$ degrees about the $z$-axis. The result will be the velocity vector component $\vec{V}$ of the state vector.

(Ans: $V_x = -6.56950077$, $V_y = 3.79078765$, $V_z = -0.27859227$.)

Thus, the HST's positional component of the state vector at the stated time was

$\vec{R} = [2935.881 \; 5333.534 \; 3288.738]$,

where all units are in km. The velocity component of the state vector was

$$\vec{V} = [-6.570 \ 3.791 \ -0.279],$$

where all units are in km/s. From the lengths of these vectors, the HST was 6919.661 km (4299.678 miles) above the center of the Earth and traveling at 7.590 km/s (4.716 miles/s).

### 9.4.3 Satellite Catalogs

A satellite catalog is similar to a star catalog, with some important differences:

- A star catalog provides the location of stars and other natural objects. A satellite catalog provides the location of man-made objects orbiting Earth (e.g., satellite, International Space Station [ISS], expended rocket bodies).
- Star catalogs provide an object's coordinates (typically equatorial). Satellite catalogs provide orbital elements from which a satellite's coordinates can be calculated.
- Star catalogs are referenced to a standard epoch (e.g., J2000) and may not be updated to a new epoch for several years. Satellite catalogs are updated daily, weekly, or monthly, depending on how frequently a satellite catalog provider publishes their catalog.
- All objects in a star catalog are referenced to the same standard epoch. By contrast, each object in a satellite catalog will likely be referenced to a different epoch.

Satellite catalogs can be published in several different formats, including a format unique to a specific satellite or space mission. However, the most widely used format is the Two-Line Element (TLE) format (shown in table 9.1), which is a fixed format that dates back to the 1960s when punched cards were the standard method for entering data into a computer program. We will not need the classification, piece of the launch, mean motion derivatives, drag, ephemeris type, element number, or checksum.

Table 9.2 shows several examples of TLE-encoded orbital data extracted over a period of time from multiple satellite catalogs. The first 2 lines in the table are not part of a satellite catalog but are provided to help delineate the columns in the catalog data fields. The first line after the column markers is the name of the orbiting object to which the immediately following 2 lines pertain. This name is not part of the TLE data format either, but it is provided by some catalog publishers to make it easier to find objects of interest. In this example table, the first set of TLE data is for an Atlas Centaur rocket body (R/B is an abbreviation for "rocket body"), which is the remains of a rocket that was

**Table 9.1** TLE Data Fields
This table lists the columns in which TLE data is located and what the data means.

| | TLE Data Line 1 |
|---|---|
| Column | Description |
| 1 | Always the number 1 |
| 3–7 | Satellite catalog number |
| 8 | Classification |
| 10–11 | Last 2 digits of the launch year |
| 12–14 | Launch number for the launch year |
| 15–17 | Piece of the launch (for multiple payloads) |
| 19–20 | Last 2 digits of the epoch year |
| 21–32 | Epoch day of the year and fractional part of the day |
| 34–43 | 1st derivative of the mean motion |
| 45–52 | 2nd derivative of the mean motion |
| 54–61 | Drag term |
| 63 | Ephemeris type |
| 65–68 | Element number |
| 69 | Checksum |
| | TLE Data Line 2 |
| Column | Description |
| 1 | Always the number 2 |
| 3–7 | Satellite catalog number (same as TLE data line 1) |
| 9–16 | Orbital inclination in degrees |
| 18–25 | RAAN in degrees |
| 27–33 | Orbital eccentricity (a leading decimal point is assumed) |
| 35–42 | Argument of perigee in degrees |
| 44-51 | Mean anomaly in degrees |
| 53–63 | Mean motion in revolutions per day |
| 64–68 | Number of orbits as of the epoch |
| 69 | Checksum |

used to launch some object into space. The first set of TLE data shown is an example of space debris!

Using the Atlas Centaur rocket body as an example, let us decode the information in its 2 TLE lines of data. Column 1 in each data line shows whether the data is in the format for data line 1 or data line 2. Then starting in column 3 of either card, the next 5 digits are the standard satellite catalog number for the object. This must be the same on both data lines, and it is a way to ensure that the 2 TLE lines of data refer to the same object. In this example, the authoritative

**Table 9.2** Sample TLE-Encoded Orbital Data
The first 2 rows at the top of this table indicate columns. The first line in each group of 3 rows that follow indicates the object (e.g., Hubble) to which the next 2 rows of TLE data pertain.

```
         1         2         3         4         5         6
1234567890123456789012345678901234567890123456789012345678901234567890123456789

ATLAS CENTAUR R/B
1 06155U 72065B   16111.89078645  .00000269  00000-0  46396-4 0  9993
2 06155  35.0043  52.6400 0037491 110.2876 250.1858 14.71170973329733
DIRECTV 7S
1 28238U 04016A   16110.82212388 -.00000015  00000-0  00000+0 0  9993
2 28238   0.0368 299.3000 0003158  92.6354 353.2969  1.00272514 43861
HUBBLE
1 20580U 90037B   16105.15775463  .00001419  00000-0  77939-4 0  9999
2 20580  28.4706 321.5331 0002637 239.4290  47.7753 15.08299214225439
HST
1 20580U 90037B   16112.45538286  .00000992  00000-0  51061-4 0  9993
2 20580  28.4700 273.3474 0002697 322.5891  68.4694 15.08311977226539
ISS (ZARYA)
1 25544U 98067A   04130.36064403  .00008712  00000-0  77799-4 0  7409
2 25544  51.6265 163.6249 0010999 114.4982 338.6494 15.69280476312220
ISS (ZARYA)
1 25544U 98067A   04130.48651632  .00010015  00000-0  88635-4 0  7419
2 25544  51.6262 162.9814 0011098 115.5233 329.2179 15.69287238312240
ISS (ZARYA)
1 25544U 98067A   04130.68807870  .00013239  00000-0  11546-3 0  7426
2 25544  51.6263 161.9566 0011096 116.0931  28.1298 15.69296145312270
ISS
1 25544U 98067A   16123.57970227  .00005319  00000-0  86668-4 0  9998
2 25544  51.6443 285.5413 0001902  85.2256   2.3282 15.54406887997889
NOAA-15
1 25338U 98030A   16105.49980982  .00000088  00000-0  55868-4 0  9995
2 25338  98.7836 109.5213 0010891  17.3827 342.7723 14.25716987931913
NOAA-18
1 28654U 05018A   16105.51343707  .00000068  00000-0  62268-4 0  9994
2 28654  99.1983 110.7876 0014120 189.3160 170.7751 14.12278716561742
NOAA-19
1 33591U 09005A   16105.52875582  .00000154  00000-0  10855-3 0  9992
2 33591  99.0337  61.8026 0014786  41.5380 318.6912 14.12061607370158
STS-134
1 37577U 11020A   11136.56805556  .00002509  11310-4  93195-5 0    17
2 37577  51.6414 323.1388 0070233 179.6223   5.9150 15.98772296    01
STS-134
1 37577U 11020A   11152.08121728 -.00033187  00000-0 -20329-3 0   367
2 37577  51.6547 243.2564 0005364 343.6536  83.5992 15.76245694  2438
STS-134
1 37577U 11020A   11152.21025809 -.00393293  00000-0 -25707-2 0   377
2 37577  51.6547 242.5934 0003887 349.9801  90.0149 15.76147802  2453
```

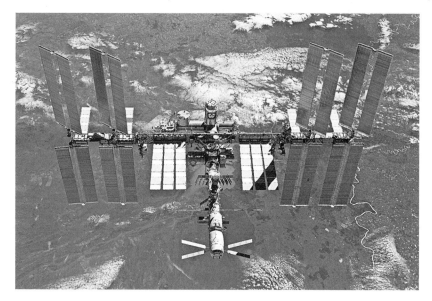

**Figure 9.10**  International Space Station
This photograph of the International Space Station (ISS) was taken in 2011 by the crew of space shuttle *Discovery* (STS-133). The ISS allows long-term space missions to study the Earth and the universe. (Image courtesy of NASA)

international identifier for this particular Atlas Centaur rocket body is 06155. The standard satellite catalog number for the Hubble Space Telescope is 20580 while the satellite catalog number for the International Space Station is 25544.

The satellite catalog number should always be used to reference an object because the catalog number is the only authoritative identifier and will be the same across all catalogs.[6] For example, you can see from the table that the Hubble Space Telescope is referred to as HST in one catalog but as Hubble in another. Similarly, the International Space Station is known as ISS in one catalog but as ISS (ZARYA) in another. In both cases, the satellite catalog numbers 20580 (HST) and 25544 (ISS) ensure that there is no confusion about what object is being referred to regardless of the satellite catalog being used.

Columns 10–14 on the first data line provide information about when the object was launched. The launch year is represented by the 2 digits in columns 10 and 11 while columns 12–14 indicate in what sequence this object was launched in that year. Only the last 2 digits of the year are encoded in

---

6. Satellite catalog number 00001 is the rocket body that launched Sputnik 1 into space while 00002 is the satellite itself. Neither object is still in orbit. The oldest object still in orbit is Vanguard 1, which was launched in March 1958 and whose catalog number is 00005.

the TLE data format. If the 2 digits are greater than or equal to 57, then the year is assumed to be in the 1900s; otherwise, the year is in the 2000s.[7] In this example, the Alpha Centaur rocket was the 65th launch in 1972.

Columns 19–32 in data line 1 provide the epoch for when this Atlas Centaur observation was made. Columns 19 and 20 are the last 2 digits of the year, which for this example was the year 2016. The date in 2016 on which this data was captured is provided in columns 21–32 (111.89078645). The first 3 digits (111) are how many days into the year, while the fractional portion (0.89078645) indicates the UT time for the observation. Applying the algorithm from section 3.7, 111 days into the year occurred on April 20, 2016. Converting the fractional part of the day to UT time (see section 3.6) gives

$0.89078645 * 24 = 21.3788748^h$,

which is $21^h 22^m 43.95^s$. Thus, the epoch at which this Alpha Centaur rocket body data was captured was April 20, 2016, at $21^h 22^m 43.95^s$ UT.

The Keplerian orbital elements are obtained from columns 9–63 on the second data line. Following the format indicated in table 9.1 for the second data line, we have

$\iota = 35.0043°$ (columns 9–16),

$\Omega = 52.6400°$ (columns 18–25),

$e = 0.0037491$ (columns 27–33),

$\omega = 110.2876°$ (columns 35–42),

$M_0 = 250.1858°$ (columns 44–51).

Note that a leading decimal point is assumed for the eccentricity's value.

One Keplerian element is missing in the TLE data: the length of the semimajor axis ($a$). It must be computed from the mean motion (columns 53–63). Data line 2 indicates that the Alpha Centaur rocket body's mean motion at the epoch was 14.71170973 revolutions per day, meaning that the Alpha Centaur rocket body orbits Earth about 14.7 times per day. Let $n$ be the mean motion expressed in revolutions per second rather than per day. That is,

$$n = \frac{\text{Mean Motion}}{86{,}400}, \qquad (9.4.4)$$

where the scaling factor 86,400 is the number of seconds in a day and is used to convert the mean motion from orbits per day to orbits per second. Applying

---

7. It should be obvious why 57 was chosen as the demarcation point between the 1900s and 2000s. The first object launched into space, Sputnik 1, was launched in 1957.

equation 9.4.4 to the Alpha Centaur rocket body, we obtain $n = 0.00017027$ revolutions/second. The length of the semi-major axis is then given by

$$a = \sqrt[3]{\frac{\mu}{(2\pi n)^2}}, \qquad (9.4.5)$$

where $\mu$ is the standard gravitational parameter for the Earth (398,600.4418 km$^3$/s$^2$). Applying equation 9.4.5, the length of the Alpha Centaur rocket body's semi-major axis is $a = 7035.46936588$ km. This, then, gives us all 6 of the standard Keplerian elements.

It is sometimes useful to convert the length of the semi-major axis to its equivalent mean motion, such as to encode an object's Keplerian elements as TLE data. This can be easily done by reversing the process for converting mean motion to the length of an object's semi-major axis. Equation 9.4.5 can be rewritten as

$$n = \left(\frac{1}{2\pi a}\right)\sqrt{\frac{\mu}{a}}, \qquad (9.4.6)$$

which gives the mean motion in revolutions per second. This can be converted to revolutions per day by the equation

Mean Motion $= 86{,}400n$. $\qquad (9.4.7)$

If we take the semi-major axis that we just computed for the Alpha Centaur rocket body (7035.46936588 km), apply equation 9.4.6 and then equation 9.4.7, we get the expected result

Mean Motion $= 14.71170973$ rev/day.

The last piece of information we will consider from the TLE data is the number of orbits as of the epoch, obtained from columns 64–68 of the second line of TLE data. For the Alpha Centaur rocket body, the value given is 32,973. This means that as of the epoch at which this TLE data is referenced, the Alpha Centaur rocket body had orbited Earth 32,973 times since its launch in 1972! While interesting, be aware that this TLE data field is not always kept up to date and hence may be grossly inaccurate.

## 9.5 Categorizing Satellite Orbits

An interesting aspect of satellites is that we are free, at least theoretically, to choose any orbit we want for a satellite and launch it into that orbit or maneuver it into that orbit at some time after launch. Satellite orbits can be highly elliptical, moderately elliptical, or even circular in nature, as appropriate for a satellite's mission. Satellites can be placed at different heights above Earth, and assuming that they have fuel on board, they can be maneuvered to change their orbital characteristics. In fact, they are often placed in temporary parking orbits from where they will be maneuvered into their final orbits. For example, a spare satellite may be placed in a parking orbit so that it is readily available to replace a satellite that has failed or reached the end of its useful life. There is even a "graveyard" orbit where expended satellites and other space junk are intentionally placed so that they will not pose a danger to other objects orbiting Earth. In this section we briefly mention some of the orbital regimes into which satellites are placed. The characteristics of a particular orbital regime directly impact what missions a satellite can or cannot perform, and they are important factors in determining from where a satellite is launched.

The location from which a satellite is launched is important because it directly relates to the orbital inclination that the satellite will achieve. The launch site and desired orbital inclination also determine the amount of energy required to achieve orbit, which of course determines the size of the rocket required for the launch, the amount of fuel required, and consequently the cost. A satellite's orbital inclination determines the maximum and minimum latitude that a satellite can reach with respect to the Earth. For example, if a satellite is traveling eastward with an orbital inclination of 40°, then that satellite will never appear over the Earth at a latitude greater than 40° N or at a latitude lower than 40° S. Along with a satellite's altitude, this determines what areas of the Earth the satellite can see as well as the locations on Earth that can see the satellite. If a satellite is traveling in a westward direction (i.e., retrograde orbit), the maximum/minimum latitude that the satellite can reach is $180° - \iota$.

The orbital inclination that a satellite will achieve is related to the latitude of the launch site by the equation

$$\iota = \cos^{-1}(\sin A \cos \phi), \tag{9.5.1}$$

where $A$ is the azimuthal direction in which the satellite is launched and $\phi$ is the latitude of the launch site. To illustrate, assume a satellite is launched from Cape Canaveral with an azimuthal direction of 45°. Table 9.3 shows that the Cape Canaveral launch site is located at 28.46675° N latitude. Using these values with equation 9.5.1, we find that the resulting orbital inclination for our satellite will be

$$\iota = \cos^{-1}(\sin 45° \cos 28.46675°) \approx 51.57°.$$

Of course, a satellite's inclination and other orbital characteristics can be changed once it is in orbit, but that takes fuel to accomplish.

**Table 9.3**  US Launch Sites
This table gives the latitude of several launch sites in the United States.

| Site | Latitude |
| --- | --- |
| Pacific Missile Range, Hawaii | 22.02278° N |
| Cape Canaveral, Florida | 28.46675° N |
| Kennedy Space Center, Florida | 28.60820° N |
| White Sands, New Mexico | 32.56460° N |
| Vandenberg, California | 34.77204° N |
| Wallops, Virginia | 37.84621° N |
| Kodiak, Alaska | 57.43533° N |

Instead of launching at azimuth 45°, suppose our satellite is launched at azimuth 100° from Cape Canaveral. Then equation 9.5.1 indicates that the new orbital inclination will be about 30.03°. If the satellite were launched with an azimuth greater than 180°, it would be placed into a retrograde orbit.

Equation 9.5.1 can be rewritten in the form

$$A = \sin^{-1}\left(\frac{\cos \iota}{\cos \phi}\right), \tag{9.5.2}$$

which allows us to determine the azimuth required to achieve a specific orbital inclination from a given launch site. Thus, to achieve an orbital inclination of 32° from Cape Canaveral requires launching with an azimuth of

$$A = \sin^{-1}\left(\frac{\cos 32°}{\cos 28.46675°}\right) \approx 74.73°.$$

Equation 9.5.2 provides an important piece of information that must be kept in mind when selecting a launch site. The equation tells us that the orbital inclination must be greater than the launch latitude, or else the target orbital inclination cannot be achieved without post-launch maneuvers once in orbit. This is because if the orbital inclination is less than the launch latitude, the term $(\cos \iota)/(\cos \phi)$ in equation 9.5.2 will be greater than 1.0 or less than $-1.0$. The inverse sine function is undefined for values outside the range $\pm 1$, so equation 9.5.2 cannot be solved for such a combination of orbital inclination and launch latitude.

Satellites are often launched in an easterly direction because that allows them to take advantage of the "sling shot" effect of Earth's rotational speed to reduce the energy a rocket must supply to achieve a desired launch velocity. The least amount of energy required is when the orbital inclination is equal to the launch site's latitude and the launch azimuth is 90°. From the perspective of reducing the size of the rocket required, the optimum situation is to launch from an equatorial site at a launch azimuth of 90°. Here's why. At the equator, Earth rotates at 1669.81 km/hour, which means that any object launched due east from the equator is already traveling at 1669.81 km/hour. Now Earth's rotational speed depends upon latitude, so the velocity obtained for "free" due to Earth's rotation at a given latitude is approximately

$$V = 1669.81 \cos \phi, \qquad (9.5.3)$$

where the result is in km/hour.

To illustrate, assume a satellite to be launched from Cape Canaveral needs a velocity of 28,000 km/hour to achieve the desired orbit. Then applying equation 9.5.3, the velocity we get for free is

$$V = 1669.81 \cos 28.46675° \approx 1468 \, \text{km/hr}.$$

This means that the launch rocket must supply enough energy to achieve

$$28{,}000 - 1468 = 26{,}532 \, \text{km/hour}.$$

If launched from the equator ($\phi = 0°$), the rocket must supply only an additional 26,330 km/hour. By contrast, if launched from Kodiak, Alaska, the rocket must supply an additional 27,101 km/hour, or nearly all of the required velocity to achieve orbit.

Satellite orbits are classified in multiple ways, the most common of which is by their distance above the Earth. Orbital regimes based on distance above the Earth are Low Earth Orbit (LEO), Medium Earth Orbit (MEO), Geosynchronous Earth Orbit (GEO), and, although not based purely on height, Highly Elliptical Orbit (HEO). One must carefully note how an author measures distance to a satellite: from the *center* of the Earth or from the *surface* of the Earth.[8] The difference is important!

In addition to height-based orbital regimes, satellite orbits are also classified by their orbital eccentricity (circular, elliptical, highly elliptical) and inclination (equatorial, inclined, highly inclined, polar, retrograde). These various methods for classifying orbits provide useful information about the orbital characteristics of a particular satellite and what its limitations are. In this section, we will primarily be interested in height-based classifications, starting with GEO orbits.

---

8. A quick way to determine how an author is measuring distance is to look at the author's definition for LEO. If the author states that LEO extends up to 2,000 km, then clearly the author is measuring distances from the surface of the Earth because the radius of the Earth (over 6,300 km) is greater than 2,000 km. If the author were measuring distance from the center of the Earth, then LEO orbits would be inside the Earth!

In his calculations for placing spaceborne communications relays into orbit, Arthur C. Clarke deduced that a satellite with a circular equatorial orbit ($e = 0$, $\iota = 0°$) located 42,164 km above the center of the Earth would have an orbital period equal to Earth's rotational period. (We'll see how he arrived at that distance in subsection 9.9.1.) Such a satellite would appear to be motionless from the perspective of Earth and would therefore be ideal as a spaceborne communications relay.

Objects in a circular orbit at an altitude of 42,164 km above the center of the Earth are in a GEO orbit, and they have an orbital period equal to Earth's rotational period (approximately 24 hours). If a GEO object is in the equatorial plane ($\iota = 0°$), the orbit is also geostationary, which means that the object will appear to remain in the same place in the sky relative to Earth. Geostationary orbits are sometimes called Clarke orbits in honor of Arthur C. Clarke. Note that while all geostationary orbits are GEO, not all GEOs are geostationary because a circular orbit can be inclined with respect to the equatorial plane. A GEO, non-geostationary orbit ($\iota > 0°$) means that the satellite will return to exactly the same point relative to Earth each orbit, but it will not appear stationary. The key things to remember are that GEO objects are in a circular orbit, they have an orbital period equal to Earth's rotational period, and they are geostationary if their orbit is also equatorial.

Geostationary orbits are useful for communications satellites and weather satellites because they are always over the same spot on Earth. From their great height they can see from about 75° S latitude to about 75° N latitude, which allows them to cover about a third of the Earth's surface. GEO, and particularly geostationary orbits, are ideal from the perspective of always viewing fixed portions of the Earth. However, GEO orbits are so far above the Earth that it is difficult to make sensors with sufficient resolution to provide highly detailed information about the Earth's surface.

With orbits less than 8,400 km above the center of the Earth, LEO satellites are much closer to the Earth than they would be in a GEO orbit. Therefore, LEO orbits are useful for remote sensing missions because they place spaceborne sensors closer to the Earth than higher orbits would. Because they are closer to the Earth, LEO satellites require less power to transmit data to, or receive commands from, Earth. However, a significant disadvantage of LEO orbits is that they are affected by atmospheric drag and require periodic station-keeping maneuvers to overcome those effects.

LEO objects, depending upon their exact height, will make 1 complete orbit around Earth in about 90 minutes. Most human spaceflight missions, including the space shuttle missions and the International Space Station, take place in LEO orbits. If you refer back to the example problems in subsection 9.4.2, you will see that indeed STS-80 (at 6,732 km above the Earth) and the HST (at 6920 km above the Earth) were both in LEO obits.

MEO orbits are above LEO but lower than GEO (i.e., in the range 8,400–42,164 km above the center of the Earth). The orbital period for MEO objects varies considerably because there is such a wide range of heights in the MEO orbital regime. At the lower end of orbital heights, MEO objects may have an orbital period as short as 2 hours while at the higher end their orbital period approaches 24 hours.

The greater the distance from Earth, the more of Earth a satellite can see, which makes MEO orbits prime candidates for navigation and communications satellites. For example, Global Positioning System (GPS) satellites are typically at a distance of about 26,600 km above the center of the Earth and have an orbital period of 12 hours. The greater distances from Earth afforded by MEO also mean that fewer satellites are required to provide worldwide coverage than if the same satellites were positioned in LEO orbits.

LEO, MEO, and GEO orbits are typically circular or near-circular orbits, which brings us to HEO orbits. Although some authors refer to HEO orbits as High Earth Orbits and define them as orbits above GEO, the acronym usually refers to an orbit's eccentricity rather than its height above the Earth. We will use HEO in that context and note that HEO orbits typically vary from about 7,400 km to about 45,000 km above the center of the Earth.

HEO orbits have multiple uses. They are used as a temporary parking orbit for transferring satellites into a GEO orbit or as a parking orbit before sending a space vehicle to the Moon, planets, or outer reaches of space. The higher a satellite is above the Earth, the greater its field of view and the longer it can view the Earth, particularly when at apogee. This makes HEO orbits particularly useful for applications, such as reconnaissance and surveillance, that require extended periods of time over a place on the Earth.

## 9.6 Locating a Satellite

We now have all the preliminary information necessary to locate where a satellite will appear for a given observer. The basic process can be summarized in the following steps:

- Obtain the Keplerian orbital elements for the satellite of interest. The orbital elements will typically be given as TLE data and should be referenced to an epoch as close to the desired viewing date as possible.

- Using an appropriate propagation model, adjust the orbital elements from the reference epoch to the desired viewing date and time.

- Convert the propagated orbital elements to a state vector (subsection 9.4.2). This gives the satellite's location in the ECI coordinate system.
- Convert ECI coordinates to topocentric coordinates (section 9.3).

The propagation model mentioned in the second bullet refers to the process of taking an object's position at one instant in time and predicting where that object will be at a future point in time according to some mathematical model of the object's orbit. In general, the shorter the difference between the epoch and the desired future time, the more accurate propagation tends to be because there is less time for various perturbations to affect an orbit. Moreover, the shorter the time period, the less important the differences between underlying mathematical models tend to be. When dealing with orbital elements to perform precise orbital calculations, it is important to know what propagation model is being used. For one reason, comparing the future predicted position of 2 satellites, such as to predict whether they will collide, is meaningful only if the same propagation model is used to propagate the orbit for both satellites.

To illustrate the process just summarized in the bulleted list, we will use a very simple propagation model that adjusts the mean anomaly at the epoch ($M_0$) to the mean anomaly at the time we wish to observe the satellite ($M_t$). Let $t_0$ be the epoch time obtained from the TLE data and let $t$ be the time at which we wish to observe the satellite. The difference between the 2 times is

$$\Delta t = t - t_0 \qquad (9.6.1)$$

and is expressed in days. The TLE data gives us a satellite's mean motion, which is how many times per day the satellite orbits Earth. If we multiply mean motion by $\Delta t$, that will tell us how many times the satellite has orbited Earth between the epoch to which the TLE data is referenced and the time at which we wish to observe the satellite.

Recall that the mean anomaly is how far an object has gone along a mean orbit from some stated reference point. In the case of the planets, that reference point is perihelion. In the case of a satellite, the reference point is perigee. So, $M_0$ is how far (measured in degrees) a satellite has gone around the Earth in a perfectly circular orbit since the moment of perigee. Since there are 360° in a circle, the distance the satellite traveled in time $\Delta t$ is

$$360° * \Delta t * (\text{Mean Motion})$$

so that our propagated mean anomaly is

$$M_t = M_0 + 360° * \Delta t * (\text{Mean Motion}). \qquad (9.6.2)$$

**Satellites**   309

$M_t$ will likely need to be adjusted after applying this equation to ensure that it is in the range [0°, 360°]. Once $M_t$ has been obtained, we have the Keplerian elements at the desired viewing time and must then convert the elements to topocentric coordinates.

Let us now work through a complete example. Assume an observer at sea level is located at 38° N latitude, 78° W longitude in the Eastern Standard Time zone is on daylight saving time. What were the topocentric coordinates for the International Space Station (ISS) for that observer on May 9, 2004, at $7^h 40^m 35^s$ LCT? Use the first TLE data set for ISS from table 9.2 to obtain the Keplerian orbital elements.

1. Extract the Keplerian elements and epoch date from the TLE data. The TLE epoch date is $t_0$ for equation 9.6.1.

    (Ans: Epoch date is 5/9/2004 at $8^h 39^m 19.64^s$ UT, $\iota = 51.6265°$, $e = 0.0010999$, $\Omega = 163.6249°$, $\omega = 114.4982°$, $M_0 = 338.6494°$, Mean Motion = 15.69280476 rev/day.)

2. Convert mean motion to the length of the semi-major axis.

    (Ans: $a = 6739.09316404$ km.)

3. Convert the given date and time to UT and LST times for the observer. The UT time is $t$ for equation 9.6.1.

    (Ans: $LCT = 7.676389^h$, $UT = 11.676389^h$ on the same day, $GST = 2.851474^h$, $LST = 21.651474^h$.)

4. Compute the Julian day number for the epoch at which the Keplerian elements were captured in step 1.

    (Ans: $JD_e = 2,453,134.86064$.)

5. Compute the Julian day number for the desired date, using the Greenwich date and UT from the prior step.

    (Ans: $JD = 2,453,134.98652$.)

6. Compute total elapsed days, including fractional days, by subtracting $JD$ from $JD_e$.

    (Ans: $\Delta t = 0.12587217$ days.)

7. Apply $\Delta t$ and equation 9.6.2 to get the propagated mean anomaly.

    (Ans: $M_t = 1049.75288141°$.)

8. Adjust $M_t$ if necessary to place it in the range [0°, 360°].

    (Ans: $M_t = 329.75288141°$.)

9. Using the mean anomaly $M_t$ in place of $M_0$, convert the Keplerian orbital elements to a state vector. Use the Newton/Raphson method to find the eccentric anomaly.

(Ans: $\vec{R} = [-1826.444 \ -3797.174 \ 5251.163]$, $\vec{V} = [7.213$
$-2.621 \ 0.608]$, velocity is 7.698 km/s, distance is 6732.692 km.)
10. Using the method explained in section 9.3, convert the positional vector (i.e., Cartesian coordinates) to topocentric coordinates for the stated observer. (Ans: $h_{topo} = -24°58'56.42''$, $A_{topo} = 131°44'39.19''$.)

The desired observation time (May 9, 2004, at $7^h40^m35^s$ LCT, which is UT = $11.676389^h$) for this example, is essentially the epoch time for the TLE data in table 9.2 immediately following the TLE data we used for this example. This example was intentionally chosen so that the desired observation time coincides with the next ISS TLE data. This allows us to gauge the accuracy of our simple propagation model. If the propagation model is accurate, the adjusted mean anomaly $M_t$ should be very close to the mean anomaly at the epoch in the second TLE data set.

The mean anomaly at the epoch for the second ISS TLE data is $329.2179°$, which differs from our propagated mean anomaly ($M_t = 329.75288141°$) by about $0.5°$. For such a simple propagation model, the difference is relatively small. However, the difference in time between the 2 TLE data sets was only about 3 hours. The farther away the desired observation time is from the epoch at which the orbital elements were captured, the less accurate this simple propagation model will be.

If you compare the Keplerian elements from the first 2 sets of TLE data for the ISS, you will notice that all the orbital elements, including the orbital eccentricity, have changed over only about a span of 3 hours! Table 9.4 summarizes the differences. The point is that since the Keplerian elements do not stay constant over time, an accurate propagation model must adjust *all* the orbital elements and not just the mean anomaly. The simple propagation model presented here that adjusts only the mean anomaly (equation 9.6.2) is not very accurate, but it is sufficient for illustrating what a propagator does.

There are several reasons why a satellite's orbital elements do not remain constant over time. The most important factor is Earth's gravity, which is not uniform because the Earth is not a perfect sphere whose center of mass is located precisely at the Earth's center. This means that Earth's gravitational force acting upon a satellite changes depending on where the satellite is located, the shape of the Earth underneath the satellite, and the precise location of Earth's center of mass. Besides Earth's gravity, Earth's atmosphere creates drag that impacts a satellite's velocity as it moves through the atmosphere. The change in velocity directly affects all the satellite's orbital elements. Of less impact than drag but still important for highly accurate orbital predictions, solar radiation and the gravitational effect of the Sun, Moon, and planets act

# Satellites

**Table 9.4** Effects of Orbital Perturbations
This table shows how orbital perturbations can affect a satellite's orbital elements over a very short period of time.

| Orbital Element | TLE Set 1 (ISS) | TLE Set 2 (ISS) | Δ |
|---|---|---|---|
| $t_0$ | 130.36044403 | 130.48651632 | $3.03^h$ |
| $\iota$ | 51.6265° | 51.6262° | 1.1″ |
| $e$ | 0.0010999 | 0.0011098 | 0.0000099 |
| $\Omega$ | 163.6249° | 162.9814° | 38′37″ |
| $\omega$ | 114.4982° | 115.5233° | 1°01′30″ |
| $M_0$ | 338.6494° | 329.2179° | 9°25′53″ |
| Mean Motion | 15.69280476 | 15.69287238 | 0.00006762 |

*Note*: The data in this table is for the International Space Station and represents 2 snapshots that are only $3.03^h$ apart.

cumulatively over time to noticeably alter a satellite's orbit. Even the effects of relativity must be considered when extremely accurate orbital predictions are required.

Because of these types of orbital perturbations, satellite professionals require a far more comprehensive and accurate propagation model than the simple technique presented here. A widely used model is the Simplified General Perturbation version 4 (SGP4) model, which uses state vectors and a complex orbital model to propagate satellite orbits. SGP4 works well—unless an object burns a thruster to alter its orbit, which then requires new observations to be made to determine the object's new orbital characteristics. We will not describe SGP4, but more information about the model, including source code for its implementation, can be found through the references given in section 10.7.

## 9.7 Satellite Rise and Set Times

It is natural to ask, When will a satellite be in view for a given location on Earth? This is analogous to determining the rising and setting times for the stars, planets, Sun, and Moon. Because the stars and planets are so far away, we effectively considered them as stationary so that simple formulas could be applied to approximate their rising and setting times. The situation for the Sun and Moon is more complicated because they cannot be considered as stationary with respect to the stars over a $12^h$–$24^h$ period of time. Even so, in both cases we employed a 3-step process. First, we calculated the position of the

Sun or Moon at midnight on the date prior to the desired viewing time and then calculated the object's position $12^h$ later (for the Moon) or $24^h$ later (for the Sun). Second, we used those 2 positions to compute 2 rising and setting times. Finally, we interpolated the 2 sets of rising and setting times to obtain an approximate time at which the Sun/Moon will rise and set.

Unfortunately, except for geostationary orbits, satellites orbit Earth in time frames measured in minutes or hours and thus are far from stationary. Therefore, an interpolation scheme as used for the Sun and Moon is impractical. Deriving an accurate equation for predicting rise and set times for satellites is very difficult because there are several complicating factors to consider.

For example, suppose a satellite orbits Earth every 90 minutes. Assuming that the satellite always goes over the same location on Earth during each orbit, a satellite with a 90-minute orbital period will rise and set over a given location 16 times during a 24-hour interval of time. However, because Earth is also rotating on its axis as the satellite orbits, except for very specific types of orbits, a satellite will not return to the exact same spot over the Earth at the end of its next orbit. This means that where a satellite will rise and set relative to locations on Earth may change with every orbit.

Other complicating factors must be addressed if one desires to view an object, such as the International Space Station, through a telescope. The object must come into view when the glare of the Sun or Moon will not obscure the object. Even if the Moon is ignored, limiting the search for rise and set times to nighttime hours is not sufficient. The satellite must be in a position relative to the Sun so that even though the Sun may not be visible, sunlight reflects off the satellite so that the satellite is visible. Of course, if the intent is to see a satellite with an antenna in order to communicate with it, daytime hours are also candidates for rise and set times because the position of the Sun is likely irrelevant. Another complication, for reasons discussed in section 9.6, is that an accurate propagation model must be used to predict the position of a satellite at future points in time.

Complicated algorithms for determining when a satellite will be visible to an observer or ground station have been developed that address all these complications. A good reference for such an algorithm is Vallado's *Fundamentals of Astrodynamics and Applications*. The Heavens Above website (http://heavens-above.com/), as well as a general Internet search for "satellite visibility passes" or "satellite pass predictors," may also yield useful sources.

This chapter's program can compute a rough approximation for when a satellite *might* be visible. It takes as input a satellite's Keplerian elements, an observer's location and LCT time, the number of hours over which to do a rise/set time estimation, and a time increment. With the observer's stated LCT as a starting point, the program computes the topocentric coordinates for where the satellite will be at that initial LCT time. Then the simple propagation model presented earlier is used to propagate the satellite's location into the future by the requested time increment, and the topocentric coordinates for the newly propagated position are calculated. This iterative process of propagating forward by a given time increment and computing the new topocentric coordinates is continued until the number of hours requested for the estimation has been completed. If the altitude between successive iterations changes from being above or below the observer's horizon, an interpolation is done to estimate when that transition occurred.

A satellite *may* or *may not* be visible whenever the altitude computed by this process is positive (i.e., the satellite is above the observer's horizon) because of all the previously discussed complications. Even so, the program for this chapter should give a reasonable estimate of a time interval during which a satellite will be above an observer's horizon and a "ball park" approximation of the LCT time, altitude, and azimuth at which the satellite will appear. Keep in mind that to view a satellite with a telescope, the satellite must be illuminated by the Sun and is therefore visible only when it is above the horizon *and* the Sun is not too far below the horizon. This means that the best times to view a satellite with a telescope are generally for a few hours after sundown or a few hours before sunrise.

## 9.8 Satellite Distance

Determining the distance of a satellite from the center of the Earth is a simple calculation that can be accomplished in a variety of ways. One obvious technique is to convert the satellite's Keplerian elements to a state vector and then apply equation 9.3.1 to determine its distance from the resulting positional vector. However, it is much easier to determine the distance directly from the properties of an ellipse than going through all the calculations required to convert Keplerian elements to a state vector.

In particular, equations 9.2.25 and 9.2.28 allow us to compute a satellite's apogee and perigee distances directly from its orbital eccentricity and semi-major axis. For example, assume a satellite is circling Earth in an elliptical orbit whose eccentricity is 0.5 and whose semi-major axis is 40,000 km in length. The maximum distance the satellite will be from Earth (the apogee radius from equation 9.2.25) is

$$r_a = a(1+e) = 40{,}000(1+0.5) = 60{,}000 \text{ km}.$$

Similarly, the minimum distance (perigee radius) as given by equation 9.2.28 is

$r_p = a(1 - e) = 40{,}000(1 - 0.5) = 20{,}000$ km.

If we need to know a satellite's distance at an arbitrary point in its orbit, that can be obtained by applying equation 9.2.23 to the satellite's orbital eccentricity, semi-major axis, and true anomaly at the time in which we are interested. As we have seen, the true anomaly can be obtained from the mean anomaly by solving Kepler's equation.

As an example, assume that the satellite for which we just determined the minimum and maximum distance is at a point in its orbit at which the true anomaly is 45°. Then applying equation 9.2.23 yields

$$r = \frac{40{,}000(1 - 0.5^2)}{1 + 0.5 \cos 45°} \approx 22{,}163.88 \text{ km.}$$

This is the same problem we solved in subsection 4.5.1, which further demonstrates that satellites obey the same laws of physics as any other object orbiting another.

Recall from section 9.2 that the true anomaly for an orbit is 0° at perigee and 180° at apogee. These values can be used with equation 9.2.23 to calculate a satellite's distance at $r_p$ and $r_a$. In fact, since $\cos 0° = 1$, equation 9.2.23 simplifies to equation 9.2.28. In like fashion, because $\cos 180° = -1$, equation 9.2.3 simplifies to equation 9.2.25.

As we pointed out in the introduction to this section, the distance from the center of the Earth to a satellite can be computed by converting the satellite's Keplerian elements to a state vector and then finding the length of the resulting positional vector. That approach requires a significant amount of work. However, a close examination of the algorithm in subsection 9.4.2 shows that the process can be greatly simplified so that only 3 of the 6 Keplerian elements (eccentricity, semi-major axis, and mean anomaly) are required along with the epoch date and mean motion.

To illustrate this streamlined approach, let us estimate the distance from the center of the Earth to the Atlas Centaur rocket body whose TLE data lines are given in table 9.2. Let us find the distance on 4/21/2016 at $10^h$ LCT for an observer at 40° N latitude, 80° W longitude in the Eastern Standard Time zone who is not on daylight saving time. The required steps are:

1. Calculate the length of the semi-major axis from the mean motion.
   (Ans: $a = 7035.46936588$ km.)

2. Convert the observer's stated LCT to UT, GST, and LST times.
   (Ans: $LCT = 10^h$, $UT = 15^h$, $GST = 5.007567^h$, $LST = 23.674234^h$, date = 4/21/2016.)

3. Compute the total number of elapsed days, including fractional days, since the TLE data's epoch (4/20/2016 at $21^h22^m43.95^s$ UT).
   (Ans: $\Delta t = 0.73421355$ days.)
4. Use equation 9.6.2 to propagate the mean anomaly by $\Delta t$.
   (Ans: $M_t = 4138.73898646°$.)
5. If necessary, adjust $M_t$ to be in the range $[0°, 360°]$.
   (Ans: $M_t = 178.73898646°$.)
6. Use the simple iteration method to solve Kepler's equation to get the eccentric anomaly from $M_t$.
   (Ans: $E = 249.98396744°$.)
7. Use equation 4.5.8 to compute the true anomaly from the eccentric anomaly.
   (Ans: $\upsilon = -110.2177362°$.)
8. Use the true anomaly, eccentricity, length of the semi-major axis, and equation 9.2.23 to get the distance from the center of the Earth.
   (Ans: $r = 7044.49765652$ km.)

All these steps are required to convert Keplerian elements to their equivalent state vector, but eliminating those steps that are not required when all we want is the distance significantly reduces the computations required.

## 9.9  Other Flight Dynamics

The preceding sections introduced various topics that are important for locating and observing satellites. We hope this short introduction has convincingly demonstrated that what has been learned about the orbits of the planets and other celestial objects applies equally well to the thousands of man-made objects that now encircle our home planet. We will conclude this brief introduction to satellites by examining 3 topics related to the flight dynamics of orbiting satellites. These include determining a satellite's orbital period, ways to determine how fast a satellite is moving and how that relates to its orbit, and how much of the Earth's surface a satellite can see.

### 9.9.1  Orbital Period

The orbital period for any object is the time that it takes for the object to complete 1 full revolution around whatever is at the orbit's occupied focus. In the case of the planets, the Sun lies at the occupied focus so a planet's orbital period is how long it takes for the planet to complete 1 trip around the Sun. This length of time is typically expressed in years. By contrast, the orbital period for a satellite is how long it takes to complete 1 trip around Earth, and this length of time is typically expressed in minutes or hours.

A satellite's orbital period is a fundamental piece of information that helps characterize its orbit. Among other things, a satellite's orbital period tells a mission planner how often a satellite will revisit a particular spot on Earth and therefore how frequently sensors can provide information about that location. For example, if a weather satellite's orbital period is 10 hours, then weather data for a given location on Earth can only be obtained from that satellite every 10 hours.[9] If updates are required every 30 minutes (perhaps to study an impending storm), then a meteorologist has 3 choices: (a) use an appropriately located geostationary satellite, (b) use a satellite whose orbital period is 30 minutes or less, or (c) obtain data from multiple satellites that in the aggregate revisit a target area every 30 minutes. We will reconsider this problem at the end of this subsection.

A satellite's orbital period is given by the equation

$$\tau = 2\pi \sqrt{\frac{a^3}{\mu}}, \tag{9.9.1}$$

where $\mu$ is the standard gravitational parameter for Earth. Since $2\pi/\sqrt{\mu}$ is a constant, it can be calculated and inserted into the equation to give

$$\tau \approx 0.00995201\sqrt{a^3},$$

which is the orbital period around Earth in seconds. Dividing by 60 gives the period in minutes, which is the equation

$$\tau \approx 0.00016587\sqrt{a^3}. \tag{9.9.2}$$

Note that equation 9.9.1 is general and applies to any orbit; it only requires that the correct standard gravitational parameter be used. In fact, you can see that equation 9.9.2 is very similar to equation 8.12.11. The 2 equations differ because the Sun's standard gravitational parameter was used to derive equation 8.12.11 for objects orbiting the Sun while Earth's standard gravitational parameter was used to derive equation 9.9.2 for objects orbiting Earth. A second difference between the 2 equations is that the scaling factor applied to equation 8.12.11 gives the orbital period in years, while the scaling factor applied to equation 9.9.2 gives the orbital period in minutes.

For a couple of examples, suppose space shuttle STS-8 was in an orbit whose semi-major axis was 6,675 km. Then the shuttle's orbital period was

$$\tau \approx 0.00016587\sqrt{(6{,}675)^3} \approx 90.5 \text{ minutes.}$$

---

9. A real weather satellite would more likely be geostationary so that revisit times are not an issue.

# Satellites

In subsection 9.4.3, we used TLE data to determine that the length of the semi-major axis for the Alpha Centaur rocket body was $a \approx 7035.469$ km. What was the rocket body's orbital period? Again applying equation 9.9.2, we have

$$\tau \approx 0.00016587\sqrt{(7035.469)^3} \approx 97.9 \text{ minutes.}$$

After a moment's thought, it should be clear that orbital period is fundamentally the same thing as mean motion. The only difference is that the orbital period given by equation 9.9.2 is expressed in minutes whereas mean motion is expressed in revolutions per day. So, orbital period and mean motion are related by the simple equation

$$\text{Mean Motion} = [24/(\tau/60)] = \frac{1,440}{\tau}, \tag{9.9.3}$$

where mean motion is in revolutions/day and $\tau$ is in minutes. If you perform the necessary computations, you will find that the mean motion for the Alpha Centaur rocket body given by this equation appears to be different from what is captured in the TLE data in table 9.2. (Using an orbital period of 97.9 minutes and equation 9.9.3 yields a mean motion of 14.70888662 versus 14.71170973 as captured in the TLE data.) The discrepancy is due to round-off errors introduced in going from equation 9.9.1, which is exact, to equation 9.9.2, which is an approximation.

A careful examination of equation 9.9.1 reveals how Arthur C. Clarke concluded that a satellite with a circular orbit 42,164 km above the center of the Earth would be geosynchronous. We start by noting that equation 9.9.1 can be rewritten in the form

$$a = \sqrt[3]{\frac{\mu \tau^2}{4\pi^2}},$$

where $\mu$ is the Earth's standard gravitational parameter. Recognizing that $\sqrt[3]{\mu/(4\pi^2)}$ is a constant, this equation can be simplified to

$$a \approx 21.613545\sqrt[3]{\tau^2},$$

where $\tau$ is in seconds and $a$ is in kilometers. To allow $\tau$ to be expressed in minutes, the scaling factor must be adjusted to

$$a \approx 331.253274\sqrt[3]{\tau^2}, \tag{9.9.4}$$

For a satellite to be in a geosynchronous orbit, its orbital period must be the same as the time it takes for Earth to rotate once on its axis with respect to the stars. That is, a satellite must be at a distance so that its orbital period is precisely 1 sidereal day. A sidereal day for Earth is $23^h56^m04.1^s$, or 1436.0683333 minutes. Using this value as the desired orbital period and applying equation 9.9.4, we arrive at $a \approx 42{,}164.17$ km, which is the same conclusion that Clarke reached. A satellite in a circular orbit 42,164 km above the center of the Earth will orbit Earth in the same time that it takes for Earth to rotate once on its axis. If the satellite is also in an equatorial orbit, it will be geostationary.

Let us apply this information to consider the weather example proposed at the beginning of this subsection. Suppose we want to place a satellite in a circular orbit such that it orbits Earth every 30 minutes gathering weather data. How far above the center of the Earth must the orbit be? Using 30 minutes as our desired orbital period, equation 9.9.4 yields

$$a \approx 331.253274\sqrt[3]{(30)^2} \approx 3198.215 \text{ km}.$$

Unfortunately, Earth's radius is 6378.135 km, so it is impossible for an unpowered satellite to orbit Earth in 30 minutes because it would have to be orbiting *inside* the Earth!

Let us consider an alternative approach. Assume that as a safety margin, we require our satellite to be at least 100 km above sea level. What is the fastest orbital period possible within that constraint? This time apply equation 9.9.2, where the desired length of the orbital semi-major axis is $a = 6378.135 + 100 = 6478.135$ km. The result is

$$\tau \approx 0.00016587\sqrt{(6478.135)^3} \approx 86.5 \text{ minutes}.$$

The only way to achieve a 30-minute revisit rate is to utilize a geostationary satellite that views the part of the Earth we are interested in, or use 3 satellites properly spaced so that they are in orbits that are $86.5/3 \approx 8.8$ minutes apart.

### 9.9.2 Velocity

We have already seen that a satellite's orbital velocity can be obtained from the velocity vector component of the state vector. However, there are simpler ways to determine velocity directly from the geometry of a satellite's orbit, just as we did in subsection 8.12.4 for the planets. If a satellite is in a circular orbit, the velocity is very simple to calculate directly from the satellite's orbital period and distance from Earth. From the equation for the circumference of a circle, the total distance around a circle of radius $r$ is $d = 2\pi r$, where $d$ is in the same units (e.g., kilometers) as $r$. A satellite's orbital period $\tau$ is the time it takes the satellite to travel that distance, so the velocity for a circular orbit is

$$V = \frac{2\pi r}{\tau}.$$

## Satellites

Since we have been expressing a satellite's period in minutes, this equation will be in km/minute. It must be multiplied by 60 to convert it to km/hour, or divided by 60 to convert it to km/s. We will choose the latter, so the equation for the velocity of a satellite traveling in a circular orbit is

$$V = \frac{2\pi r}{60\tau} = \frac{\pi r}{30\tau}, \tag{9.9.5}$$

where $r$ is in kilometers, $\tau$ is in minutes, and the resulting $V$ is in km/s.

As an example, assume a space shuttle in a circular orbit is at an altitude of 350 km above sea level. What is the shuttle's orbital period and how fast is it moving? Since the problem statement is given in terms of an altitude and a circular orbit, we can apply equation 9.2.1 to determine how far the space shuttle is above the center of the Earth. That is,

$$r = 6378.135 \text{ km} + 350 \text{ km} = 6728.135 \text{ km}.$$

For a circular orbit, the semi-major axis is the same as the distance above the center of the Earth ($a = r$) because the occupied focus and geometric center (see figure 9.3) are the same point. Thus, equation 9.9.2 gives the shuttle's orbital period as

$$\tau \approx 0.00016587\sqrt{(6728.135)^3} \approx 91.5 \text{ minutes}.$$

Finally, equation 9.9.5 gives the velocity as

$$V = \frac{\pi(6728.135)}{30(91.5)} \approx 7.70 \text{ km/s}.$$

The equation for a noncircular orbit is only slightly more complicated. The velocity in km/s is given by

$$V = \sqrt{\mu\left(\frac{2}{r} - \frac{1}{a}\right)}, \tag{9.9.6}$$

where $\mu$ is Earth's standard gravitational parameter, and $r$ and $a$ are in kilometers. This equation, called the vis-viva law, is derived directly from Kepler's third law and Newton's law of universal gravitation. Since $\mu$ is a constant, the vis-viva law can be simplified (for satellites orbiting Earth) by extracting $\sqrt{\mu}$, resulting in

$$V \approx 631.348115\sqrt{\left(\frac{2}{r} - \frac{1}{a}\right)}. \tag{9.9.7}$$

As noted earlier, $a = r$ for a circular orbit, so the vis-viva law for a circular orbit around Earth reduces to

$$V \approx 631.348115\sqrt{\frac{1}{r}}. \tag{9.9.8}$$

The vis-viva law relates a satellite's velocity to its orbital parameters, specifically the relationship between velocity, distance above the center of the Earth, and the length of the orbital semi-major axis. This is important information for determining how large a rocket is required to put a satellite into orbit and how much fuel an orbiting vehicle must expend to change its orbit.

To illustrate, assume that the space shuttle from the previous example is 350 km above sea level, but it is in an elliptical orbit with $e = 0.029849$ and has an orbital semi-major axis of 6,700.5 km. What is the space shuttle's velocity? The answer, which comes directly from equation 9.9.7, is

$$V = 631.348115\sqrt{\left(\frac{2}{6,728.135} - \frac{1}{6,700.5}\right)} \approx 7.68 \text{ km/s}.$$

If we apply equations 9.2.25 and 9.2.28, we find that the shuttle's apogee distance is 6,900.5 km while its perigee distance is 6,500.5 km. Applying equation 9.9.7 to these 2 distances, we find that the shuttle's apogee and perigee speeds are

$$V_{apogee} = 7.49 \text{ km/s}, \quad V_{perigee} = 7.95 \text{ km/s}.$$

A maneuver known as the Hohmann transfer orbit takes advantage of the vis-viva law to determine the velocity increments required to transition a satellite from a circular orbit of radius $r_1$ to another circular orbit of radius $r_2$ that is at a greater height but is in the same orbital plane. The maneuver is shown in figure 9.11. First, an elliptical transfer orbit, whose perigee distance is equal to the radius $r_1$ of the orbit that the satellite is currently in and whose apogee distance is equal to the radius $r_2$ of the desired new circular orbit, is designed. When the satellite reaches perigee, a thruster is ignited to change the satellite's velocity by $\Delta V_1$. This places the satellite into the elliptical Hohmann transfer orbit. When the satellite reaches apogee in this elliptical orbit, a thruster is again ignited to change the satellite's velocity by $\Delta V_2$. This then places the satellite in the desired circular orbit.

To illustrate, assume a space probe is to be sent to Venus. The mission plan initially places the probe into a circular LEO orbit at a height of 6,900 km. After confirming that the probe is fully functional, it is to be placed into a circular HEO orbit at a height of 40,000 km. At the appropriate point and time in the HEO orbit, the probe will be "slingshotted" to travel to Venus. What are the velocity changes required ($\Delta V_1$ and $\Delta V_2$) to transfer the space probe from the circular LEO orbit to the circular HEO orbit?

# Satellites

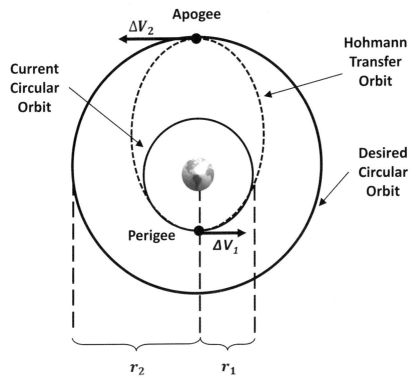

**Figure 9.11** Hohmann Transfer Orbit
A Hohmann transfer orbit is a maneuver used to move an object from one circular orbit to another higher circular orbit in the same orbital plane.

First, we need to design an elliptical transfer orbit. Since the perigee distance needs to be the radius of the LEO orbit ($r_1$) and the apogee distance needs to be the radius of the HEO orbit ($r_2$), equation 9.2.9 gives the length of the required semi-major axis as

$$a = \frac{r_1 + r_2}{2} = \frac{6{,}900 + 40{,}000}{2} = 23{,}450 \text{ km}.$$

Next, we need to determine the orbital velocity for the current circular LEO orbit, the desired circular HEO orbit, and the velocities at perigee and apogee in the Hohmann transfer orbit. Since the LEO and HEO orbits are circular, equation 9.9.8 gives their velocities:

$$V_1 \approx 631.348115 \sqrt{\frac{1}{6{,}900}} \approx 7.60 \text{ km/s},$$

$$V_2 \approx 631.348115 \sqrt{\frac{1}{40,000}} \approx 3.16 \, \text{km/s}.$$

The Hohmann transfer orbit is an elliptical orbit, so equation 9.9.7 gives the apogee and perigee velocities as

$$V_{apogee} \approx 631.348115 \sqrt{\frac{2}{40,000} - \frac{1}{23,450}} \approx 1.71 \, \text{km/s},$$

$$V_{perigee} \approx 631.348115 \sqrt{\frac{2}{6,900} - \frac{1}{23,450}} \approx 9.93 \, \text{km/s}.$$

The last step is to compute $\Delta V_1$, which is the velocity change required to transfer from the circular LEO orbit to the elliptical orbit at perigee, and $\Delta V_2$, which is the velocity change required to transfer from the elliptical orbit at apogee to the circular HEO orbit. Thus,

$$\Delta V_1 = V_{perigee} - V_1 = 9.93 - 7.60 = 2.33 \, \text{km/s},$$

$$\Delta V_2 = V_2 - V_{apogee} = 3.16 - 1.71 = 1.45 \, \text{km/s}.$$

The total velocity change required to accomplish the Hohmann transfer orbit maneuver is

$$\Delta V = \Delta V_1 + \Delta V_2 = 3.78 \, \text{km/s}.$$

### 9.9.3 Ground Tracks and Footprint

In this last subsection we consider the problem of determining how much of Earth a satellite can see. The higher a satellite is above Earth, the more of the Earth's surface it can see. Similarly, the higher a satellite is above Earth, the greater the area on the Earth from which telescopes, antennas, and sensors can view a passing satellite. To describe how much of Earth a satellite can see, or conversely from how large an area on the Earth a satellite can be seen, imagine drawing a line from the center of the satellite to the center of the Earth, as shown in figure 9.12. The point at which the line intersects the Earth's surface is a projection of the satellite's position onto the Earth. If an observer is located at that projection point, then the satellite is at the observer's zenith.

The circular area on the surface of the Earth that a satellite can see is called the satellite's footprint. The center of the footprint is always the projection point of the satellite onto the surface of the Earth. Obviously, the location of the

# Satellites

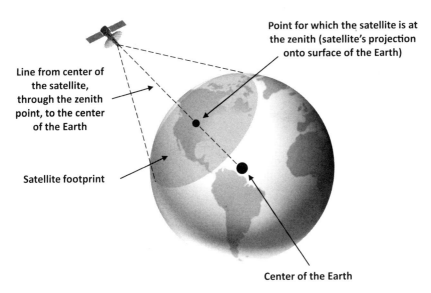

**Figure 9.12** Satellite Footprint
A satellite's footprint is how much of Earth's surface the satellite can see. The center of the footprint is the observation point for which the satellite is at the zenith.

footprint changes as the satellite orbits Earth and as Earth itself rotates. Moreover, the size of the footprint will also change if the distance from the satellite to Earth changes as the satellite follows its orbit. The size of the footprint will be constant only for truly circular orbits (i.e., $e = 0$).

If one plots the center of a satellite's footprint as the satellite orbits Earth, the result is a path that moves across the surface of the Earth. This path is called the satellite's ground track, an example of which is shown in figure 9.13. If the Earth were not rotating, then as a satellite orbits Earth it would trace out a great circle inclined with respect to the Earth at an angle equal to the satellite's orbital inclination. However, the Earth is rotating so a satellite will typically trace out paths that are slightly offset each orbital pass that the satellite makes. The precise pattern that a satellite's ground track makes depends on the satellite's orbital characteristics. Some ground tracks form a figure-eight pattern, some form a circle, some are periodic wave patterns as shown in figure 9.13, and some may even exhibit retrograde motion.

Plotting a satellite ground track is not particularly difficult, but it does require an accurate orbit propagator. The general process is to first convert a satellite's position to equatorial coordinates at some instant in time, such as

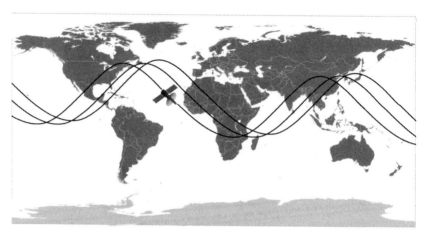

**Figure 9.13** Satellite Ground Tracks
The path a satellite traces as it orbits Earth is the satellite's ground track.

the epoch at which the satellite's orbital elements are referenced. The declination is the geocentric latitude for the satellite's position at that epoch. That is,

$$\phi = \delta. \tag{9.9.9}$$

The satellite's right ascension (expressed in degrees) is its longitude, but adjusted for how much Earth has rotated since the Greenwich prime meridian was aligned with the vernal equinox. This is simply the GST expressed as degrees rather than units of time (multiply time by 15 to convert to degrees). So,

$$\psi = \alpha - GST, \tag{9.9.10}$$

where both $\alpha$ and GST are expressed in degrees rather than time. Instead of computing GST through the method given in section 3.10, one can instead use the equation

$$GST = 280.4606° + 360.9856473°d, \tag{9.9.11}$$

where $d$ is the number of days (including fractional days) since the epoch J2000.

After converting position to latitude and longitude (equations 9.9.9 and 9.9.10), plot a point on a map at that location. Propagate the orbit by some time increment, use the propagated orbital elements to compute the satellite's equatorial coordinates, and then convert them to latitude/longitude to plot a new point. Continue this process for at least 1 orbit to create the satellite's ground track.

Calculating the size of a satellite's footprint is simpler than calculating its ground track. The equation required is

$$r_{fp} = \left(\frac{\pi r_e}{180}\right) \cos^{-1}\left(\frac{r_e}{r}\right) \tag{9.9.12}$$

where $r_e$ is the radius of the Earth, $r$ is the distance from the center of the Earth to the satellite, and $r_{fp}$ is the maximum radius of the satellite's footprint.[10] The fraction of the Earth's surface (expressed as a percentage) that equation 9.9.12 represents is

$$F = \frac{50(r - r_e)}{r}. \tag{9.9.13}$$

For example, assume a satellite is 35,000 km above the center of the Earth. Then the maximum radius of the satellite's footprint is

$$r_{fp} = \left(\frac{6378.135\pi}{180}\right) \cos^{-1}\left(\frac{6378.135}{35,000}\right) \approx 8850 \text{ km},$$

which is

$$F = \frac{50(35,000 - 6378.135)}{35,000} \approx 41\%$$

of the Earth's surface. If the satellite were only 10,000 km above the center of the Earth, the results would be

$$r_{fp} \approx 5607 \text{ km}, \quad F \approx 18\%.$$

If a satellite's velocity $V$ is known in km/s, then the total time in minutes that the satellite will be visible within a fixed footprint whose radius is $r_{fp}$ is

$$FP_t = \frac{2r_{fp}}{60V} = \frac{r_{fp}}{30V}. \tag{9.9.14}$$

To illustrate, suppose the space shuttle was 350 km above sea level. What was its footprint, and how long was it visible within that footprint? From subsection 9.9.2 we calculated

$$r = 6728.135 \text{ km}, \quad V \approx 7.70 \text{ km/s}.$$

---

10. The factor $\frac{\pi}{180}$ is in the equation because we are assuming that the inverse cosine returns a result in degrees.

The shuttle's footprint was therefore

$$r_{fp} = \left(\frac{6378.135\pi}{180}\right) \cos^{-1}\left(\frac{6378.135}{6728.135}\right) \approx 2066 \text{ km},$$

and it remained within that footprint for

$$FP_t = \frac{2{,}066 \text{ km}}{30(7.70 \text{ km/s})} \approx 9.0 \text{ minutes}.$$

Equation 9.9.14 is really suitable only for orbits, such as LEO, in which Earth does not rotate very far during the time that the satellite makes a single pass around Earth. Otherwise, the motion of the Earth must also be considered to obtain an accurate estimate. For instance, assume a satellite 35,000 km above the center of the Earth is traveling at 1.25 km/s. Then equation 9.9.14 indicates that the satellite will remain in its footprint for about 236 minutes (3.9 hours). During that time, Earth will have rotated by 58.5°! This drastically affects calculations for how long the satellite will remain in its footprint.

## 9.10 Program Notes

The program for this chapter allows orbital elements to be entered from the keyboard or read from a data file. Two sample TLE data files are provided, which can be viewed with a simple text editor to understand their format. The data in table 9.2 is in the file TLE-Table9.dat while the data file TLEs.dat contains more example satellite data that can be used with this chapter's program. As mentioned earlier, the name of the object to which TLE data lines refer is not really part of the TLE data format. However, both data files provided with this book include object names to make it easier to find an object of interest. You can download TLE data from publicly available sources to create your own TLE data files. For data files that you create, you may include or exclude object names as you see fit. The data file Launch-Sites.dat has the launch site data from table 9.3. You can modify this file to add your own launch sites, or create your own launch sites data file for use with this chapter's program.

The purpose of this chapter was to briefly introduce artificial satellites and demonstrate that their orbits are governed by the same laws of physics that govern all objects (Sun, Moon, planets) that orbit another. Because that is this chapter's primary focus, be forewarned that the calculations to locate a satellite or to predict when a satellite may be visible will not be very accurate for more than a short time after a TLE's stated epoch. Use Keplerian elements with an

**Satellites**

epoch as close as possible to the time at which you wish to locate an object to obtain the most accurate results.

Although this chapter's program includes a simple propagator (section 9.6), a more accurate propagator, such as SGP4, is required to obtain better results. Source code for the SGP4 algorithm can be found in several of the resources listed in section 10.7. You may wish to modify this chapter's program to include an SGP4 propagator because incorporating a more accurate propagator is the most significant thing that can be done to improve results.

Unfortunately, even when Keplerian elements close to the desired time are used and a better propagator is incorporated, the results may still prove disappointing. This is because the accuracy and precision with which angles and times are computed or entered have a significant effect when dealing with objects that move through space as rapidly as those that orbit Earth. Even the precision with which a programming language's trigonometric and other functions (e.g., square root) are computed greatly impacts the accuracy of the resulting solution.

## 9.11 Exercises

1. If a satellite has Cartesian coordinates $x = 23,300$ km, $y = -14,600$ km, $z = 16,000$ km, what are its spherical coordinates?
   (Ans: $r = 31,812.733$ km, $\alpha = 327.92835339°$, $\varphi = 59.80508327°$.)

2. If the spherical coordinates for an object are $r = 34,106.451$ km, $\alpha = 58.12040315°$, $\varphi = 121.85419091°$, what are the object's Cartesian coordinates?
   (Ans: $x = 15,300$ km, $y = 24,600$ km, $z = -18,000$ km.)

3. Suppose the Cartesian coordinates in problem 1 were determined on June 5, 2015, at $15^h$ UT. If an observer at sea level in the Pacific Standard Time zone at 40° N latitude, 120° W was on daylight saving time, what would the horizon coordinates be for the observer? (Note: the LCT time for the observer was $8^h$.)
   (Ans: $h = 63°07'57.21''$, $A = 258°15'25.45''$.)

4. What would the topocentric coordinates for the previous problem be?
   (Ans: $h_{topo} = 56°50'9.38''$, $A_{topo} = 78°15'25.45''$.)

5. If the state vector for a satellite was

$\vec{R} = [-1880.723 \ 2173.591 \ 1473.570]$,

$\vec{V} = [-8.816 \ -6.582 \ -1.531]$,

what were the object's Keplerian elements? How far was the object above the center of the Earth, and how fast was it moving?

(Ans: $\iota = 28.46631111°$, $e = 0.0005148$, $a = 3229.58950374$ km, $\Omega = 21.8733225°$, $\omega = 1.82234382°$, $M_0 = 104.95712948°$, distance 3230.019 km, speed 11.108 km/s.)

6. If the Keplerian elements for an object were $\iota = 30.11976874°$, $e = 0.10759411$, $a = 6078.35278389$ km, $\Omega = 356.95953889°$, $\omega = 341.73698644°$, $M_0 = 179.21747378°$, what was the object's state vector? How far was the object above the center of the Earth, and how fast was it moving? Use the simple iteration method to solve Kepler's equation.

(Ans: $\vec{R} = [-6260.434\ 2221.184\ 1094.144]$, $\vec{V} = [-2.675\ -5.812\ -3.449]$, distance 6732.298 km, speed 7.269 km/s.)

7. If an object orbiting Earth has a mean motion of 2.5 revolutions per day, what is the length of the object's orbital semi-major axis?

(Ans: 22,931.995 km.)

8. If an object orbiting Earth has a semi-major axis length of 35,000 km, what is its mean motion in revolutions per day?

(Ans: 1.326 rev/day.)

9. From table 9.2, what was the epoch at which the NOAA-18 TLE data was obtained? What were the Keplerian elements for the NOAA-18 satellite at that epoch?

(Ans: Epoch date was 4/14/2016 at $12^h 19^m 20.96^s$ UT, $\iota = 99.1983°$, $e = 0.001412$, $a = 7229.721$ km, $\Omega = 110.7876°$, $\omega = 189.316°$, $M_0 = 170.7751°$.)

10. Convert the Keplerian elements from the previous problem to their equivalent state vector. How far was the NOAA-18 above the center of the Earth, and how fast was it traveling?

(Ans: $\vec{R} = [-2567.223\ 6769.330\ 14.593]$, $\vec{V} = [1.113\ 0.408\ 7.319]$, distance 7239.798 km, speed 7.415 km/s.)

11. A satellite was launched from Wallops, Virginia, with a launch azimuth of 88°. What was the resulting orbital inclination for the satellite?

(Ans: $\iota = 37.89°$.)

12. What would the orbital inclination be if the satellite in the previous problem was launched from the Kennedy Space Center in Florida with the same launch azimuth?

(Ans: $\iota = 28.67°$.)

13. A satellite is being planned that must achieve an orbital inclination of 38.45° and an orbital velocity of 18,500 km/hr. If launched from Wallops, Virginia, what will the launch azimuth need to be? How much additional velocity

must the launch rocket provide?

(Ans: $A = 82.64°$, $\Delta V = 17,181.42$ km/hr, or 10,676.04 mph.)

14. If the launch site for the previous problem was the Pacific Missile Range in Hawaii, what would the launch azimuth need to be, and how much additional velocity must the launch rocket provide?

(Ans: $A = 57.65°$, $\Delta V = 16,952.03$ km/hr, or 10,533.50 mph.)

15. From table 9.2, the orbital inclination for the International Space Station at the epoch was $51.6443°$. If a space shuttle was launched from Cape Canaveral at that time to rendezvous with the Space Station, at what launch azimuth must it be launched to achieve the same orbital inclination as the Space Station?

(Ans: $\iota = 44.90°$.)

16. Using the last set of TLE data from table 9.2 (STS-134), what were the Keplerian elements at the epoch for the STS-134 mission? What were the topocentric coordinates 3 hours after the epoch for an observer at sea level at 45° N, 78.5° W in the Eastern Standard Time zone who is not on daylight saving time? Use the simple iteration method to solve Kepler's equation. (Hint: the LCT time 3 hours later for the observer is $3^h 02^m 46.30^s$.)

(Ans: Epoch date was 6/1/2011 at $5^h 02^m 46.30^s$ UT, $\iota = 51.6547°$, $e = 0.0003887$, $a = 6719.5040$ km, $\Omega = 242.5934°$, $\omega = 349.9801°$, $M_0 = 90.0149°$, $h_{topo} = 19°32'1.60''$, $A_{topo} = 248°19'25.21''$.)

17. Assume that the Keplerian elements for a satellite were captured on 3/12/2016 at $19^h$ UT and have the following values: $\iota = 30.9297°$, $e = 0.076012$, $\Omega = 52.6400°$, $\omega = 284.2856°$, $M_0 = 119.1380°$, and $a = 8200.05$ km. For a satellite ground station at sea level located at 28.5° N latitude, 74.2° W longitude in the Eastern Standard Time zone that is not on daylight saving time, at what time after $15^h$ LCT is the estimated first opportunity for the ground station's antenna to see the satellite after the stated epoch? What are the approximate topocentric coordinates at that time? (Hint: propagate forward from the epoch for at least 2 hours with a time increment of 1 minute to get an approximate rising time, then calculate the satellite's topocentric coordinates at the estimated rising time.)

(Ans: LCT $= 15^h 41^m 30^s$, $h_{topo} = 0°52'58.47''$, $A_{topo} = 8°45'42.46''$.)

18. A satellite in a circular LEO orbit is 8,000 km above the center of the Earth. What is the satellite's orbital period?

(Ans: $\tau = 118.6869$ minutes, or 1.9781 hours.)

19. A satellite in a circular orbit has a period of $26^h 30^m$. How far above the center of the Earth is the satellite? (Hint: be sure to convert the period to decimal minutes!)

(Ans: $a \approx 45,125.8$ km.)

20. If a satellite is in an orbit with $e = 0.760122$ and $a = 27,169.5436$ km, what are the satellite's perigee and apogee distances from the center of the Earth?

(Ans: $r_p = 6517.3758$ km, $r_a = 47,821.7114$ km.)

21. For the previous problem, what is the satellite's orbital period, speed at perigee, and speed at apogee?

(Ans: $\tau = 742.8333$ minutes, $V_{perigee} = 10.38$ km/s, $V_{apogee} = 1.41$ km/s.)

22. If a satellite is traveling at 10.25 km/s in a circular orbit 6517.5 km above the center of the Earth, how much of the Earth can the satellite see? How large is the satellite's ground footprint?

(Ans: $r_{fp} = 1321.36$ km, $F = 1.1\%$, $FP_t = 4.29$ minutes.)

# 10  Astronomical Aids

A major emphasis of this book has been to demonstrate the calculations needed to predict the location of various objects. Such calculations are typically long and tedious, but with the aid of a computer the tedium and potential for mistakes are greatly reduced. Even so, sometimes it is more convenient to use published resources than to calculate things yourself. This is especially true when great accuracy is required, or if you do not have the equations at hand to calculate the desired information. This chapter describes some common observational aids and resources that are both useful and time saving.

A number of observational aids are in the form of charts and tables to help locate an object of interest. Some are published monthly or annually while others are referenced to a standard epoch and cover a much longer period of time. Most of these aids provide the location of an object, usually in equatorial coordinates, but they may also provide other information, such as an object's visual magnitude and distance from Earth. Unless great accuracy is needed, you may find that data provided in monthly periodicals, such as *Sky & Telescope*, or *Astronomy*, are exactly what you need. This is especially true for general sightseeing when all you want is a basic idea of where to point a telescope. Also, both *Sky & Telescope* and *Astronomy* contain regular feature articles that amateur astronomers and nonscientists may find to be quite readable.

When using the various astronomical aids, you will find that they are unlikely to be adjusted for your exact location or time zone. Some aids assume a location in the middle of the United States, but more accurate ones typically assume $0^h$ UT at $0°$ latitude, $0°$ longitude. In any case, once the equatorial coordinates of an object are determined, the methods presented in chapters 3 and 4 can be used to calculate an object's horizon coordinates for your location and time zone. Rising and setting times can be approximated by applying the methods from chapter 5 (for distant celestial objects), chapter 6 (the Sun), chapter 7 (the Moon), or chapter 8 (the planets).

To avoid confusion before proceeding, some definitions are in order. The terms "star chart" and "star map" refer to a pictorial representation of the relative position of celestial objects. The representation may be for the entire night sky as seen from some vantage point, or it may only be an expanded view of a small section of the sky. A star atlas is typically a bound collection of several star charts. The key thing to remember is that star charts, maps, and atlases are pictorial representations. Even though they are called star charts, such observational aids may include galaxies, nebulae, and even planets. A set of star charts (such as Wil Tirion's *Sky Atlas 2000.0* or the *Uranometria 2000.0* by Tirion, Rappaport, and Remaklus) is a good addition to your observatory, and they should be an early purchase because they are so useful for planning a night of observations.

The terms "star catalogs," "star atlases," "ephemerides," "astronomical tables," and "almanacs" are often used interchangeably. The main feature of these aids is that they provide tables giving the coordinates of celestial objects. A star catalog will often include only objects that lie outside our Solar System. An ephemeris (a book containing ephemerides), astronomical table, or almanac generally lists celestial objects that lie within our Solar System. Moreover, these aids typically give the position of an object for several instants during the year, perhaps even an object's daily position. The main point is that aids such as star catalogs and ephemerides give numeric coordinates for objects rather than providing only a pictorial representation of relative locations.

## 10.1 Recommended Authors

There are many good astronomy books written by both professionals and amateurs. Of special note in the area of computational astronomy are the classic works *Astronomical Formulae for Calculators*[1] by Meeus and *Practical Astronomy with Your Calculator* by Duffett-Smith.[2] These books were significant resources for many of the algorithms presented in this book, and they contain additional algorithms that may be of interest to an amateur astronomer.

---

1. Meeus's most recent edition is *Astronomical Algorithms*. He has also authored *Mathematical Astronomy Morsels*, a 5-volume set that is a treasure trove of formulas, explanations, and diagrams of various phenomena.

2. The most recent edition of Duffett-Smith's book, *Practical Astronomy with Your Calculator or Spreadsheet*, has been updated to include spreadsheets that implement the various algorithms he presents.

The algorithms in both books are suitable for implementation on a personal computer. Duffett-Smith's book is easier to read than the one by Meeus, but Meeus's book is an excellent value and contains a more rigorous mathematical development of the topics than Duffett-Smith's book.

For the mathematically inclined, van Brummelen's *Heavenly Mathematics* is an excellent treatise on the rarely taught topic of spherical trigonometry, which is the foundational area of mathematics required to understand coordinate systems and their conversions. *Fundamentals of Astrodynamics* by Bate, Mueller, and White is also a recommended classic for studying celestial motion, but be aware that it requires a substantial level of mathematical expertise. In that same vein, Vallado's *Fundamentals of Astrodynamics and Applications* is an outstanding reference that covers a wide range of topics related to orbits, particularly for satellites. However, be forewarned that his work is also highly technical in nature and requires a considerable amount of mathematical expertise, particularly in calculus and vector spaces.

Fairman's *3D Astronomy with Java* is a good choice for those whose primary interest is implementing satellite algorithms. Boulet's *Methods of Orbit Determination for the Micro Computer* strikes a reasonable balance between presenting the mathematics required for determining orbits and the details necessary to implement various algorithms on a computer. As with Vallado's book, Boulet's book requires a significant degree of mathematical expertise to understand the subject matter.

## 10.2 Star Charts

While some star charts are more detailed than others, their essential idea is to graphically represent what the night sky looks like from some particular viewpoint. Star charts published monthly in popular astronomy periodicals may already be adjusted to account for precession, but the resolution of star charts in such periodicals is generally such that it makes little practical difference whether precession has been accounted for or not.

To use a star chart, align the chart's north direction (typically the top) with magnetic north. When looking at a star chart, carefully note the location of east and west. On terrestrial maps depicting roadways and cities, we are accustomed to west being on the left of the map and east on the right. However, star charts are frequently reversed so that the star chart presents a view of the sky as if you were lying on your back and looking up into the night sky. After noting the compass directions and aligning the star chart with magnetic north, select an easily recognizable constellation on the chart, such as the Big Dipper, as a reference point, and then try to find that constellation in the night sky. The chart's publisher may have drawn lines on the chart to connect the stars in a constellation to make recognizing them easier. Star charts will often give a relative indication of a star's magnitude by drawing brighter stars larger than fainter ones, which also aids in locating specific stars in the night sky.

You may not be able to see all the stars shown on a star chart. Charts printed on a monthly basis will often include horizon circles as a guide for determining what stars you will be able to see. For example, if your latitude is greater than 30°, you will not be able to see any stars that lie outside a 30° horizon circle printed on the chart. Besides providing horizon circles, star charts will often have a grid to provide a rough indication of the equatorial coordinates of objects within the grid. The sides of such grids will usually provide right ascension and declination. Right ascension may be expressed as an hour angle since most equatorial-mount telescopes have hour circles as aids for pointing the telescope.

Star charts, such as those from *Sky Atlas 2000.0*, are far more detailed than those that have been described so far. At first you may find that the added detail makes the charts more difficult to use, but they are worth the effort because they provide more information than the less detailed charts found in most periodicals.

A star atlas will be referenced to a standard epoch. This may be of little practical importance since coordinates read off from a start chart may be accurate to only within a few arcminutes, depending upon the grid spacing used to measure the equatorial coordinates. You will need to apply the methods from chapter 4 to determine where a section of interest from a star chart will appear for your location. That is, pick a sample equatorial coordinate from the chart, calculate the corresponding horizon coordinates for your location, and then you will be able to tell where in the sky to look for objects on that section of the star chart. Moreover, you will need to apply the results of chapter 5 to determine if that section of the sky will even be visible for your location at the time you choose for observation.

Star charts are useful for locating objects other than stars. For example, suppose you know Saturn will be in the constellation Libra at some particular time. By using a star chart as a guide, you can locate the constellation of Libra and probably spot Saturn very easily. Saturn will be the bright "star" that does not appear on your star chart.

Periodicals such as *Astronomy* and *Sky & Telescope* include monthly star charts similar to the one shown in figure 10.1. Besides constellations and bright stars, such star charts may include the location of the Moon and planets. Additionally, there are free resources that you can use to create your own

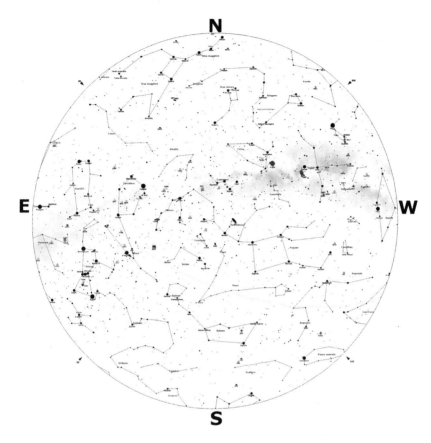

**Figure 10.1** A Monthly Star Chart
This star chart shows the November sky at $23^h$ LMT for observers at 42° N latitude. Star charts are useful for finding constellations and other objects in the sky. (Image courtesy of Roberto Mura)

customized star chart for any desired time and location. Stellarium is a free software "planetarium" that can create a star chart for the night sky. Its users can select from objects such as galaxies, nebulae, and other deep sky objects for inclusion in a customized star chart.

Instead of downloading and installing software, the *Astronomy* magazine publishers provide a resource accessible from an Internet browser (http://www.astronomy.com/stardome.aspx) that can create star charts. Another online resource that offers similar functionality is John Walker's *Your Sky* website (http://www.fourmilab.ch/yoursky). Finally, the *Sky & Telescope* publishers provide an article showing how to create a star wheel (http://www.skyandtelescope.com/observing/make-a-star-wheel/), which is a round star chart that can be rotated to account for the seasons as well as the time of day.

## 10.3 Star Catalogs

If you need arcsecond accuracy for the location of an object outside the Solar System, a star catalog is probably your best bet. Besides providing equatorial coordinates, star catalogs often contain other useful information, such as a star's magnitude, distance, and classification (red giant, white dwarf, etc.). Unfortunately, to use a star catalog you must already know the name of the object you are looking for, which is not necessarily as easy as it sounds. Perhaps the best way to determine an object's name is first to find it on a star chart. A star chart may in fact have multiple names for an object. For example, if you look at figure 10.2 and find the bright star in the lower right corner of the constellation Orion, you will see that the star is called both Rigel and $\beta$. However, if you look at the neighboring constellation of Taurus shown in figure 10.3, you will see that the star Elnath is *also* called $\beta$. Even more perplexing, if you refer back to figure 10.2, you will see that the 6 stars in the constellation of Orion that form Orion's shield are all named $\pi$! This may seem confusing because multiple stars can have the same name whether they are in different constellations, or even within the same constellation. Alas, it also turns out that besides using Greek letters to name stars, there are several *different* ways to use our usual Arabic numbers (1, 2, 3, etc.) to name stars and other deep space objects! Let's look at some of the more common methods for naming celestial objects that you are likely to encounter.

The key to finding a star is to first determine what constellation it is in. Roughly speaking (and there are exceptions), the brightest star in a constellation is labeled $\alpha$, the first lowercase letter in the Greek alphabet. The second brightest is labeled $\beta$, the second lowercase letter in the Greek alphabet, and so on. Some of the brighter stars in a constellation may also have a proper name, such as the star Rigel already mentioned. Thus, because Rigel is labeled $\beta$, it is the second brightest star in the Orion constellation. The brightest star in the Orion constellation is Betelgeuse, and hence it is labeled $\alpha$ in addition to being called Betelgeuse. To find Betelgeuse in a star catalog, go to the section in the catalog that lists the stars for Orion and you will find it designated as both Betelgeuse and $\alpha$.

The first person to use Greek letters to systematically name the stars in this fashion was the German astronomer Johann Bayer, who listed 1564 stars in

# Astronomical Aids

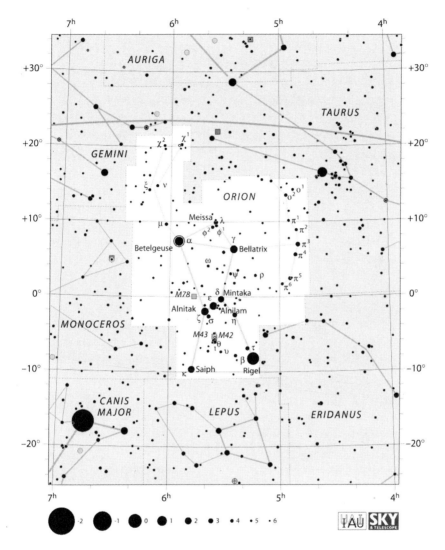

**Figure 10.2** Naming Stars in the Orion Constellation
Using Bayer's method for naming stars, the brightest star in a constellation is $\alpha$, the second brightest is $\beta$, etc. For Orion, Betelgeuse is also called $\alpha$ Orionis (or abbreviated as $\alpha$ Ori) while Rigel is also called $\beta$ Orionis (or $\beta$ Ori). (Image courtesy of the IAU and *Sky & Telescope*, Roger Sinnott & Rick Fienberg, released under CC BY 3.0, see http://creativecommons.org/licenses/by/3.0/)

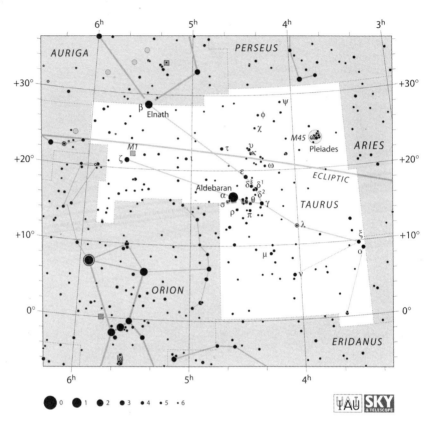

**Figure 10.3** The Taurus Constellation
The brightest star in the constellation of Taurus is Aldebaran ($\alpha$ Tau) while the second brightest is Elnath ($\beta$ Tau). (Image courtesy of the IAU and *Sky & Telescope*, Roger Sinnott & Rick Fienberg, released under CC BY 3.0, see http://creativecommons.org/licenses/by/3.0/)

the star atlas *Uranometria Omnium Asterismorum* (usually referenced as just *Uranometria*) published in 1603.[3] Bayer named a star with a lower case Greek letter indicating the star's brightness relative to the other stars in the constellation followed by the Latin name of the constellation the star is in. So, he listed Betelgeuse in his atlas as $\alpha$ Orionis and Rigel as $\beta$ Orionis. Similarly, since Sirius is the brightest star in the constellation Canis Major, Bayer listed it in his atlas as $\alpha$ Canis Majoris. Because Bayer's method for naming stars always

---

3. Uranometry is the science of measuring the position or distance of celestial objects. A uranographer is a celestial cartographer—that is, one who produces charts of the heavens.

# Astronomical Aids

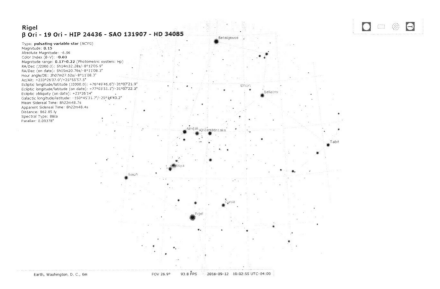

**Figure 10.4** Rigel in the Orion Constellation
The author used the Stellarium program to create this snapshot of the Orion constellation to demonstrate that stars have multiple "names." In this case, the star Rigel (located at the bottom of the Orion constellation) has the following designations: β Ori (Bayer's designation), 19 Ori (Flamsteed's number), HIP 24436 (Hipparcus main catalog entry), SAO 131907 (Smithsonian Astrophysical Observatory star catalog entry), and HD 34085 (Henry Draper star catalog entry).

includes a Greek letter *and* the name of the constellation in which it falls, Rigel (β Ori) and Elnath (β Tau) are uniquely named and will not be confused.

In modern star charts and catalogs, 2 stars in the same constellation may be labeled with the same Greek letter but differentiated by a superscript. For example, look just above and to the right of Betelgeuse in figure 10.2. You will find 2 stars, one labeled $\phi^1$ and the other $\phi^2$. As astronomers more systematically cataloged the sky, many "new" stars were discovered. Some were so close in location to "old" already lettered stars and of about the same magnitude that the expedient of superscripts was adopted to avoid confusion.

So far so good, but what happens when a constellation contains more stars than there are letters in the Greek alphabet? The Greek alphabet has 24 letters in it, and *every* constellation has many more stars than that! When Bayer ran out of Greek letters, he used a sequence of uppercase and lowercase Roman letters, which was adequate for the relatively few stars contained in his atlas. Although still used today, the so-called Bayer designation is far from adequate for handling the millions of stars that have been named, even when the stars are grouped by constellation and brightness.

The quest to catalog more stars and celestial objects soon required a more advanced method for labeling celestial objects. Using a French edition of John Flamsteed's *Atlas Coelestis*, Joseph Jerome de Lalande took up the task of labeling the stars. His Flamsteed numbers are simply integers assigned to each star by constellation in order of right ascension. For Rigel, the Flamsteed number is 19. Thus, the star Rigel might be listed in a catalog in several different ways: as Rigel, as $\beta$ Ori, and as 19 Ori!

The situation becomes even more discouraging when identifying nonstellar and deep-sky objects. These objects are labeled with a number that depends upon which catalog you are using! Six commonly encountered catalogs are the *New General Catalog* (NGC), the *Index Catalog I and II* (IC), the *Messier Catalog* (M), the *Hipparcus* catalog (HIP), the *Smithsonian Astrophysical Observatory* catalog (SAO), and the *Henry Draper* catalog (HD). In Wil Tirion's *Sky Atlas 2000.0*, numbers without a letter prefix preceding them are NGC designations, those with the prefix "M" are Messier catalog designations, and those with an "I" prefix are IC designations. Proper names may also exist for an object, as in the case of the Great Orion Nebula. The Great Orion Nebula appears in the *Sky Atlas 2000.0* as 1972-M42, indicating that its NGC catalog designation is 1972 while its Messier catalog designation is 42. Many star charts label the Great Orion Nebula as NGC-1972 to make it clear which catalog numbering system is being used.

## 10.4 Ephemerides and Almanacs

An ephemeris or astronomical almanac contains tables giving the location of Solar System objects over some period of time, which may be a month, a week, or even for various times throughout the day. In addition to location, an ephemeris will often give other useful information, such as rising and setting times, times for eclipses, times of the equinoxes and solstices, and times for perihelion and aphelion. Equatorial coordinates must be referenced to a particular date and time, with the time most often being $0^h$ UT. Data such as rise and set times provided in an ephemeris must also be given for a specific location, which is usually 0° latitude (the equator), 0° longitude (the Greenwich Prime Meridian). In some almanacs the tables may be calculated for several latitudes in addition to the equator. The time conversions in chapter 3 can be used to convert whatever an almanac or ephemeris provides to your particular LCT.

Using an ephemeris to locate a planet is more complicated than using a star catalog to locate a star. The reason is that because of the rapid motion of Solar System objects with respect to Earth, frequent corrections (monthly, daily, hourly, etc., depending on the object and accuracy required) must be made to

## Astronomical Aids

**Table 10.1** Lunar Polynomial Coefficients
This table gives the polynomial coefficients for computing the Moon's equatorial coordinates on June 5, 2015.

| Coefficient | Right Ascension | Declination |
| --- | --- | --- |
| $a_0$ | 283.8189020 | −17.7747834 |
| $a_1$ | 14.3661042 | 1.1912985 |
| $a_2$ | 0.0013541 | 0.5299011 |
| $a_3$ | −0.0340516 | −0.0082501 |
| $a_4$ | 0.0019849 | −0.0039707 |
| $a_5$ | 0.0003718 | 0.0001560 |

*Note*: Data taken from the US Naval Observatory's website for *The Astronomical Almanac Online*.

the equatorial coordinates. We will describe 2 methods frequently used to make such corrections. Both methods are based on evaluating a polynomial, but the polynomial to evaluate will differ depending on the exact resource referenced, the method used, and the celestial object under consideration.

For example, the technique used in *The Astronomical Almanac* to compute a highly accurate position for the Moon is to evaluate a pair of polynomial functions that take time as input and produce the Moon's equatorial coordinates as output. *The Astronomical Almanac* provides one function for calculating the Moon's right ascension and another function for calculating the declination. Both polynomial functions are of the form

$$f(p) = a_0 + a_1 p + a_2 p^2 + a_3 p^3 + a_4 p^4 + a_5 p^5, \quad (10.4.1)$$

where $p$ is the desired UT expressed as a fractional part of a day.[4] Coefficients $a_0$ through $a_5$ in equation 10.4.1 differ for right ascension and declination, and differ for each day of the year.

To illustrate, calculate the Moon's equatorial coordinates for June 5, 2015, at $22^\text{h}33^\text{m}$ UT. Table 10.1 gives the polynomial coefficients for June 5, 2015, for both right ascension and declination. The following steps are required:

1. Convert the time of day to decimal format.
   (Ans: 22.55).

2. Convert the time of day to a fractional part of a day by dividing the results of step 1 by 24.
   (Ans: $p = 0.93958333$.)

---

[4]. *The Astronomical Almanac* polynomials require that Universal Time 1 (UT1) be converted to Terrestrial Time (TT). For this section we will assume that $UT1 = TT = UT$ and ignore converting UT1 to TT.

3. Use the coefficients in the second column of table 10.1 to form the polynomial shown in equation 10.4.1, and evaluate for the value of $p$ calculated in step 2. This gives the Moon's right ascension in degrees.
   (Ans: $\alpha_{deg} = 297.2918236°$.)
4. Convert the result from step 3 from degrees to hours by dividing by 15.
   (Ans: $\alpha = 19.81945491^h$.)
5. Use the coefficients in the third column of table 10.1 to form the polynomial shown in equation 10.4.1, and evaluate for the value of $p$ calculated in step 2. This gives the Moon's declination in degrees.
   (Ans: $\delta = -16.19747725°$.)
6. If desired, convert $\alpha$ to HMS format and $\delta$ to DMS format.
   (Ans: $\alpha = 19^h 49^m 10.04^s$, $\delta = -16°11'50.92''$.)

Although the polynomials provided in *The Astronomical Almanac* provide highly accurate equatorial coordinates (within $\pm 0.003^s$ for the Moon's right ascension and $\pm 0.003''$ for declination) for Solar System objects, they have a significant disadvantage. The pair of polynomials to be evaluated are different for every object, for every day of the year, and for every year. A software program based on such techniques must contain a database of polynomial coefficients so that multiple dates can be evaluated, and it must be periodically updated as coefficients are developed for new dates.

A simpler but significantly less accurate method can be used with the ephemerides from popular periodicals such as *Sky & Telescope* and *Astronomy*. The main idea behind this simpler method is that if positions are given for an object at 2 different times, an interpolation can be done to determine where the object is at other times within that time interval. That is, suppose an object's position is $A$ at day $D_a$ and position $B$ at day $D_b$. Let $d$ be the date of interest, noting that $d$ must lie between $D_a$ and $D_b$. Armed with this information, we first calculate the number of days from $D_a$ to $d$ via the equation

$$t = d - D_a. \tag{10.4.2}$$

Then we evaluate the polynomial

$$f(t) = \frac{t(B - A)}{D_b - D_a} + A. \tag{10.4.3}$$

As an example, calculate the equatorial coordinates for Mercury on January 15, 2016, at $0^h$ UT. Table 10.2 shows the equatorial coordinates for Mercury at selected times, as published by *Sky & Telescope*. The date we are interested in (January 15) falls between the table entries for January 11 and

**Table 10.2** Equatorial Coordinates for Mercury
This table gives equatorial coordinates for Mercury at $0^h$ UT for epoch J2000.

| January | Right Ascension | Declination |
|---|---|---|
| 1  | $20^h 05.5^m$ | $-21°06'$ |
| 11 | $19^h 58.4^m$ | $-18°29'$ |
| 21 | $19^h 09.6^m$ | $-19°03'$ |
| 31 | $19^h 09.4^m$ | $-20°31'$ |

*Note*: This data was extracted from the January 2016 issue of *Sky & Telescope*.

January 21. Using the data from those 2 rows of table 10.2, the following steps are required:

1. Obtain $D_a$ and $D_b$ from the table and $d$ from the problem statement.
   (Ans: $D_a = 11$, $D_b = 21$, $d = 15$.)
2. Use equation 10.4.2 to compute $t$.
   (Ans: $t = 4$.)
3. Use the table to determine the positions $A$ and $B$, and convert the positions to decimal format. Note that the positions $A$ and $B$ consist of 2 coordinates, 1 for the right ascension and 1 for the declination.
   (Ans: $A_\alpha = 19^h 58.4^m = 19.973333^h$, $A_\delta = -18°29' = -18.483333°$, $B_\alpha = 19^h 09.6^m = 19.160000^h$, $B_\delta = -19°03' = -19.050000°$.)
4. Using the dates from step 1, the value of $t$ from step 2, and the decimal values from step 3, apply equation 10.4.3 to compute the right ascension. That is, compute

$$f(t) = \frac{t(B_\alpha - A_\alpha)}{D_b - D_a} + A_\alpha.$$

   (Ans: $\alpha = f(4) = 19.648000 = 19^h 38^m 53^s$.)
5. Similarly, use the appropriate declination values from step 3 and apply equation 10.4.3 to compute the declination.
   (Ans: $\delta = -18.710000° = -18°42'36''$.)

Because table 10.2 provides only a precision of minutes for right ascension and arcminutes for declination, the results in steps 4 and 5 should also be limited to a precision of minutes and arcminutes. The additional precision given in the previous steps is merely to serve as an aid for readers to check their math against this worked-out example. *The Astronomical Almanac* gives Mercury's equatorial coordinates for this example as

$\alpha = 19^h 38^m 18.785^s$, $\delta = -18°22'08.93''$.

Instead of using the dates in table 10.2 closest to our desired date of January 15, we could have used the first and last rows of the table (i.e., January 1 and January 31). However, accuracy is better in general when the interval in which the desired date falls is as small as possible. Had we used January 1 and January 31 in our calculations, we would have estimated Mercury's coordinates to be

$$\alpha = 19^\text{h}39^\text{m}19^\text{s}, \delta = -20°49'40'',$$

which is clearly a less accurate approximation of Mercury's position.

This interpolation technique is easier than the method used in *The Astronomical Almanac*, but a penalty is paid in terms of accuracy. Still, for manually pointing a backyard telescope, the accuracy obtained from doing a simple interpolation may be sufficient.

## 10.5 Astronomical Calendars

If one is willing to sacrifice some accuracy, a useful aid is an astronomical calendar. One such calendar, the *Skygazer's Almanac*, is published annually in *Sky & Telescope*. Their calendar is readily available and is representative of how astronomical calendars are organized. Besides being easy to use, an astronomical calendar graphically depicts the comings and goings of celestial bodies throughout the year to give a better feel for the rhythmic motions of the Sun, Moon, and planets. An astronomical calendar makes it possible to estimate at a glance events such as the following (among many others):

- Times for sunrise and sunset
- Times for moonrise and moonset
- Phase of the Moon
- Times when planets rise and set
- When planets are in conjunction or opposition
- Transit times for major stars such as Vega
- When meteor showers are most likely to occur

Besides estimating the times for various astronomical events, some calendars provide a quick way to calculate Julian day numbers, sidereal time, and the equation of time.

The *Skygazer's Almanac* lists dates along its vertical axis and time on its horizontal axis. Dates along the vertical axis are 1 week apart while a grid of dots extending vertically from top to bottom represents individual days within a week. Similarly, dots extending left to right each represent 5 minutes of time. The vertical lines formed by the dots are spaced half an hour apart while the horizontal lines are 1 week apart. The intersection of these vertical and horizontal dotted lines with various "event lines" allows the times for events to be estimated with relatively high accuracy, more than might be imagined for such a simple graphical device. An event line indicates that some astronomical event, such as the rising or setting of a planet, is to occur. Events depicted on an astronomical calendar are relative to a specific latitude (40° N is commonly used in the *Skygazer's Almanac*), which means that time corrections must be applied to adjust them for a particular observer's location. Instructions for how to do so are usually included with an astronomical calendar.

Reading an astronomical calendar is simple. Using the *Skygazer's Almanac* as an example, locate the desired date along the vertical axis of the chart, and read across the chart until an event line is reached. Having reached an event line, the LCT time at which the event will occur is found by looking directly above (or below as convenient) the event line to the time marked along the chart's horizontal axis. You must then make the necessary corrections to adjust the LCT time obtained from the chart for your specific location.

## 10.6 Online Resources

The Internet is an indispensable tool that places at our disposal an unbelievably vast amount of information about virtually any topic. In this section we present a small sampling of available online resources that are relevant to astronomy and satellites. The URLs presented here were accurate at the time this section was written, but because the Internet is highly dynamic and constantly changing, you may encounter a URL that is no longer correct. In such a case, you can do a Google search to find where the resource has moved, or find a more recent version of the resource. At the time this section was written, all the resources described here were free of charge.

NASA and JPL maintain multiple websites with up-to-date information about the Earth, Solar System, stars and galaxies, and various NASA and JPL space exploration missions. The HORIZONS system is accessible online and can be used to generate ephemerides for various objects within our Solar System, including planets, asteroids, and comets. The NASA and JPL websites also contain tools for finding the orbital elements for Solar System objects as well as for date/time conversions. The main URLs are:

- http://www.nasa.gov/ website for following NASA projects and obtaining information about upcoming launches.

- http://www.jpl.nasa.gov/ website for information about JPL missions, including photographs and data from missions such as the Mars Rover mission.
- http://ssd.jpl.nasa.gov/ website for information about our Solar System and a variety of astronomical tools relevant to exploring the Solar System. Of particular note, the HORIZONS application can be accessed from this URL.

The United States Naval Observatory (USNO) website provides a wealth of invaluable online resources for navigation and astronomy. In cooperation with Her Majesty's Nautical Almanac Office in the United Kingdom, the USNO annually publishes *The Astronomical Almanac*. Key URLs for the USNO website are:

- http://www.usno.navy.mil/USNO (the main portal for accessing the USNO online resources)
- http://asa.usno.navy.mil/ (the URL for directly accessing *The Astronomical Almanac Online*)
- http://asa.usno.navy.mil/SecE/Section_E.html (a handy URL for getting the geocentric equatorial coordinates for some of the planets on any given date)

The *Astronomy* and *Sky & Telescope* publishers both maintain websites providing a wealth of information about astronomy, tools such as telescopes and cameras, and shops where one can search for and purchase books, astronomical calendars, and maps. Besides informative articles, both publishers also provide access to photographs and videos about celestial objects. The main URLs are:

- http://www.astronomy.com/
- http://www.skyandtelescope.com/

In addition to these online *Astronomy* and *Sky & Telescope* resources, there are a number of websites that provide free star charts, atlases, and astronomical calendars. Notable examples include:

- http://www.midnightkite.com/
- http://oneminuteastronomer.com/free-star-charts/
- http://eyesonthesky.com/StarCharts.aspx
- http://astroclub.tau.ac.il/skymaps/monthly

# Astronomical Aids

- http://www.skymaponline.net/
- http://www.fourmilab.ch/yoursky/

Several websites provide free software to turn a computer into your own personal planetarium. With these programs you can enter your latitude/longitude and generate displays for what the night sky will look like at your location. Many of the programs also provide detailed photographs of planets, galaxies, nebulae, and other deep-space objects. These programs are generally available for Microsoft Windows, Mac OS X, and various flavors of Linux. Examples include:

- http://www.google.com/earth/ (This free program from Google allows you to browse Earth as well as the heavens.)
- http://www.stellarium.org/ (This site includes source code.)
- http://www.shatters.net/celestia/
- http://www.ap-i.net/skychart/en/start
- http://www.astrosurf.com/c2a/english/ (currently available for Microsoft Windows computers only)
- http://www.worldwidetelescope.org/

Several resources are available for those wishing to explore the satellites and other man-made objects that encircle Earth. Of these resources, CelesTrak is perhaps the most useful, as it provides a reasonably up-to-date catalog with the position of numerous satellites and other objects such as the International Space Station. Additionally, the Home Planet URL is a website that provides a free satellite tracking program for Microsoft Windows computers.

- http://celestrak.com/
- http://www.fourmilab.ch/homeplanet/

It is often necessary when dealing with satellites to convert between state vectors and Keplerian elements. The following websites are useful for understanding the mathematics behind the process. The first URL is an online calculator that will perform the conversions. The remaining URLs are resources for understanding the mathematics involved.

- http://orbitsimulator.com/formulas/OrbitalElements.html
- http://www.navipedia.net/index.php/Osculating_Elements

- https://en.wikibooks.org/wiki/Astrodynamics/Classical_Orbit_Elements
- http://www.orbitessera.com/html/state_vectors.html

A good reference for the TLE format can be found at the following CelesTrak URL:

- http://www.celestrak.com/NORAD/documentation/tle-fmt.asp

## 10.7 High-Accuracy Resources

The algorithms and methods presented in this book, and many of the resources listed in the preceding sections, produce results that are accurate only to a few minutes of time or a few minutes of arc. This should be sufficient to meet the needs of most amateur astronomers because few amateurs are likely to have clocks synchronized to standardized atomic clocks, devices capable of making accurate measurements in tenths of an arcsecond, or telescopes that can be precisely positioned down to an arcsecond. Even so, there are times when it is useful to have much more accurate positional data.

Obtaining more accurate results requires a significantly higher level of mathematical expertise and a much greater knowledge of astrophysics than has been assumed for this book. Those who have the necessary background or wish to delve more deeply into these topics can do an Internet search for information on 2 different theoretical models, VSOP87 and DE405, which are widely used within the community of professional astronomers to analyze orbits. The goal of both models is to provide highly accurate data suitable for space navigation and astronomy. We will only briefly touch on these 2 models here.

The VSOP (Variations Séculaires des Orbites Planétaires) model is an analytical model developed and maintained by scientists at the Bureau des Longitudes in Paris, France. The number 87 in "VSOP87" indicates that one is referencing the 1987 version, which is the most current version, of the VSOP model. VSOP87 allows computing highly accurate positions and orbital elements for celestial objects within the Solar System to an accuracy of at least $1''$ for 2,000 years before and after the year 2000 AD for all the planets, and for a longer time period for certain planets. VSOP87 is widely used in professional astronomy circles and in readily available programs such as Celestia.

The following websites are a good place to start for a deeper understanding of the VSOP model. The first website listed is of special note because it includes a code generator that will generate source code to implement the VSOP87 model in a variety of programming languages.

# Astronomical Aids

- http://www.neoprogrammics.com/vsop87
- http://www.caglow.com/info/compute/vsop87
- https://en.wikipedia.org/wiki/VSOP_(planets)

Instead of developing a new analytic model of the Solar System, JPL used various mathematical methods (including interpolation, least squares estimation, and so on) to analyze decades of astronomical observations to develop highly accurate models of the Solar System. The models they created include classical Newtonian gravitational calculations as well as corrections for effects predicted by Einstein's theory of relativity. All their models are highly accurate and are used to support navigation for both robotic and manned spacecraft missions.

The JPL models are designated as DE#, where DE is an acronym for Development Ephemeris, and the # is merely a sequential number; it is not a reference to the year in which the model was created. JPL released DE405 in 1998, and it was used to compute the highly accurate ephemerides in *The Astronomical Almanac* produced for the years 2003–2014. DE405 is also the basis for the software provided with the *Multiyear Interactive Computer Almanac, 1800–2050* (otherwise known as MICA). Released in 2013, DE430 covers the years 1550–2650 AD and is used as the basis for *The Astronomical Almanac* ephemerides from 2015 to the present. The JPL HORIZONS system mentioned earlier is based on DE431.

The following websites are good places to start for more information about the JPL models. The first URL listed includes source code for implementing the DE43x models.

- http://www.projectpluto.com/jpl_eph.htm
- https://en.wikipedia.org/wiki/Jet_Propulsion_Laboratory_Development_Ephemeris
- http://www.cv.nrao.edu/~rfisher/Ephemerides/ephem_descr.html

To obtain the most accurate results possible, it is important to distinguish between various timekeeping systems. For example, the difference between TT and UT varies by over a minute depending on the specific date under consideration. The USNO maintains highly accurate time and can be consulted to find the precise difference between TT and UT for times in the past, and predictions of what the difference will be in the future. The applicable URL is

- http://www.usno.navy.mil/USNO/earth-orientation/eo-products/long-term/long-term

SGP4 is a widely used model for propagating satellite orbits. The following websites describe the model in some detail; some provide source code for the model in a variety of programming languages.

- https://en.wikipedia.org/wiki/Simplified_perturbations_models
- https://pypi.python.org/pypi/sgp4
- https://sourceforge.net/projects/sgp4-j/
- http://celestrak.com/NORAD/documentation/spacetrk.pdf
- http://celestrak.com/software/tskelso-sw.asp

Coordinate system conversions are a critical component of astronomy, and *Heavenly Mathematics* by van Brummelen is an excellent description of the mathematics behind coordinate system conversions. In the context of satellites, Vallado's *Fundamentals of Astrodynamics and Applications* and Bate, Mueller, and White's *Fundamentals of Astrodynamics* also provide useful methods for a variety of coordinate system transformations. Those who want a more immediate "how to" can easily find a number of coordinate system conversion algorithms through a simple Internet search.

The following URLs are of special note. They are articles written by T. S. Kelso for *Satellite Times* on a variety of topics relevant to satellites. The first URL is an index of all of Kelso's articles while the next two are the first part of a series of articles that deal with orbit propagation and orbital coordinate systems. The last URL is a website that steps through the process of taking a satellite's orbital elements, propagating them via the SGP4 model, and computing the satellite's topocentric location.

- http://celestrak.com/columns/
- http://celestrak.com/columns/v01n01/
- http://celestrak.com/columns/v02n01/
- http://www.castor2.ca/04_Propagation/index.html

# Glossary

**Absolute Magnitude** How bright an object would appear to be if it were 10 parsecs away from the viewer. See also *visual magnitude*, *apparent magnitude*, and *magnitude*.

**Altitude** In the horizon coordinate system, how far above or below an object is from an observer's horizon.

**Angle of Inclination** The angle formed by a reference plane (e.g., ecliptic plane) and the orbital plane of an object (e.g., planet).

**Annular Eclipse** A solar eclipse in which the Moon's umbra does not reach Earth. The effect is as if a smaller disk were placed in front of a larger disk.

**Anomalistic Month** A month measured from when the Moon is at perigee until it returns to perigee. It is 27.5546 days in length.

**Anomalistic Year** A year measured from the time that Earth is at perihelion until it is at perihelion again. It is 365.2596 mean solar days in length.

**Anomaly** The angle formed by an orbiting body, its orbital focus, and the orbital major axis. See also *eccentric anomaly*, *mean anomaly*, and *true anomaly*.

**Aphelion** The point in a Solar System object's orbit at which it is farthest away from the Sun.

**Apogee** The point in an orbit about Earth at which an object is farthest away from Earth.

**Apparent Magnitude** How bright an object appears to be. See also *visual magnitude*.

**Apparent Sidereal Time** Sidereal time that has not been corrected for irregularities in Earth's rotation (e.g., nutation). For the purposes of this book, apparent sidereal time and mean sidereal time can be considered to be equal, although they are not. See also *sidereal time* and *mean sidereal time*.

**Apparent Solar Day** A day reckoned by the apparent motion of the Sun as compared to a mean solar day, which is reckoned by the motion of the mean Sun. The qualifier "apparent" is often omitted unless it will be confused with a mean solar day. See also *solar day*.

**Apparent Time** Time as measured by the apparent motion of the Sun. Apparent time is not uniform in length since it is based on a solar day.

**Ascending Node** The point in an orbit at which an object rises above a reference plane. For the Solar System, the ecliptic plane is the reference plane.

# Glossary

**Asteroid**  A large rocky object orbiting our Sun that is too small to be classified as a planet or a dwarf planet. Asteroids are sometimes called minor planets or, when large enough, planetoids. See also *meteoroid*.

**Asteroid Belt**  A debris field of asteroids orbiting the Sun in the region of space between Mars and Jupiter. See also *Kuiper Belt* and *asteroid*.

**Astronomical Almanac**  A publication containing the location of various celestial objects and astronomical events of interest such as sunrise and sunset. Do not confuse with the annually produced *The Astronomical Almanac*, which is only one example of an almanac.

**Astronomical Unit (AU)**  The length of the semi-major axis of Earth's orbit. AUs are typically used to measure distances within the Solar System. One AU is about $9.29 \times 10^7$ miles.

**Autumnal Equinox**  The descending node for the Sun's orbit, assuming the plane of the celestial equator as a reference plane. The Sun's right ascension at the autumnal equinox is $12^h$. This occurs at about September 22.

**Auxiliary Circle**  A circle superimposed on an elliptical orbit whose center is the geometric center of the elliptical orbit and whose diameter is the same as the elliptical orbit's major axis. An auxiliary circle is used to represent a fictitious circular orbit for an object in which the object's orbital speed is *not* constant throughout its orbit. See also *mean orbit* and *eccentric anomaly*.

**Azimuth**  The coordinate in the horizon coordinate system that describes how far around an object is from the north. This angle is measured clockwise from north.

**Barycenter**  The center of mass of a system, such as the Solar System, around which the objects in the system revolve.

**Barycentric Coordinates**  Coordinates that specify the location of an object with respect to the center of mass of the Solar System. See also *geocentric*, *heliocentric*, and *topocentric coordinates*.

**Besselian Year**  A year measured in the same way as the tropical year except that the starting reference point is when the right ascension of the mean Sun is 280°. This makes the Besselian year more closely correspond to the civil year. A Besselian year is about 365.2422 mean solar days in length.

**Celestial Equator**  The great circle formed by the intersection of the plane of Earth's equator and the celestial sphere.

**Celestial Sphere**  An imaginary sphere of infinite size to which the stars and planets are considered to be affixed. Earth is normally considered to be the center of the celestial sphere. See also *celestial equator*.

**Centaur**  A small Solar System body whose orbit around the Sun is between Jupiter and Neptune, and whose unstable orbit causes it to periodically cross the orbits of one or more of the Gas Giants.

**Central Meridian**  The meridian, usually at the longitudinal center of a time zone, chosen as the reference point for establishing the local time for that time zone.

**Civil Month**  Any of the 12 months (January, February, etc.) into which a civil year is subdivided.

**Civil Time**  An international timekeeping system in which the Earth is divided into time zones that are synchronized with the mean time at Greenwich, England. Civil time is based on a mean solar day, which begins at midnight and is exactly 24 hours in length. See also *mean solar day* and *mean time*.

**Civil Year**  A year defined to be either exactly 365 or (in the case of a leap year) 366 days in length.

# Glossary

**Conjunction**  The moment when 2 celestial objects have a common coordinate, such as ecliptic longitude, with the same value as viewed from a particular place, such as Earth. For example, the Moon is in conjunction with the Sun at the time of the New Moon. See also *opposition*.

**Coordinated Universal Time (UTC)**  A timekeeping standard used as the basis for synchronizing civil time around the world. The UTC standard is not adjusted for daylight saving time and is coordinated to always be within about 1 second of mean solar time at 0° longitude. For the purposes of this book, UTC, UT, and GMT are considered to be synonymous.

**Crescent**  The phases of the Moon in which the Moon is illuminated, but it is less than half illuminated as seen from Earth. See also *gibbous*.

**Culminate**  In astronomy, same as *transit*.

**Dawn**  The period of semidarkness just before sunrise. At dawn the Sun is less than 18° below the horizon. See also *twilight*.

**Day**  The time interval between 2 successive transits of a celestial object across an observer's meridian. This period of time is about 24 hours, but it depends upon how it is measured and what celestial object is used as a reference. See also *mean solar day, solar day, sidereal day,* and *mean sidereal day*.

**Day Number**  The number of elapsed days since the beginning of the year. Referred to in this book as days into the year. Do not confuse with *Julian day number*.

**Daylight Saving Time (DST)**  A practice, primarily limited to the Western world, in which an hour is added to the local civil time during the spring and summer months. The adjustment is removed during the fall and winter months.

**DE405/DExx**  See *Development Ephemeris*.

**Decimal Format**  A format for expressing time or angles as a single real number (e.g., $12.567^h$, $355.134°$). See also *DMS format* and *HMS format*.

**Declination**  In the equatorial coordinate system, the angular distance that an object lies away from the celestial equator. Declination is in the range [-90°, 90°]. Positive angles are north of the celestial equator whereas negative angles are south of the celestial equator. Declination is analogous to the concept of terrestrial latitude. See also *right ascension*.

**Delta T ($\Delta T$)**  The difference between TT and UT1 (i.e., $\Delta T = TT - UT1$). Since UT1 is based on the motion of the Earth, its value is determined by historical observations for the past but can only be approximated for the future. See also *Terrestrial Time* and *Universal Time 1*.

**Descending Node**  The point in an orbit at which an object descends below a reference plane. For the Solar System, the ecliptic plane is the reference plane.

**Development Ephemeris (DE)**  Any of a series of Solar System models developed by JPL for very accurate space navigation and astronomy. See also *VSOP87*.

**DMS**  Degrees, Minutes, Seconds.

**DMS Format**  A format in which an angle is expressed in an integer number of degrees, an integer number of minutes, and a fractional number of seconds (e.g., $36°40'39.2''$). See also *decimal format*.

**Draconic Month**  A month measured by using the Moon's ascending node as the reference point. This month is 27.2122 days in length.

**DST**  Daylight saving time.

**Dwarf Planet** A Solar System object that (a) is not a moon orbiting some other body, (b) is massive enough to have become spherical in shape due to gravitational pressures, and (c) has not cleared the neighborhood around its orbit. This classification was introduced by the IAU in 2006 in part to resolve whether Pluto, Eris, and Ceres should be considered as planets. See also *planet*.

**Earth Centered Inertial (ECI)** A fixed (with respect to the stars) geocentric coordinate system whose origin is the center of the Earth, whose $x$-axis lies in the equatorial plane and points toward the First Point of Aries, whose $y$-axis lies in the equatorial plane and points east, and whose $z$-axis goes northward through the Earth's North Pole.

**Eccentric Anomaly** The angle formed by projecting an object's true position onto an auxiliary circle, the geometric center of the object's orbit, and the periapsis for the object's orbit. See also *true anomaly, mean anomaly*, and *Kepler's equation*.

**Eccentricity** A measure of the "flatness" of an ellipse. Mathematically, eccentricity is the ratio of the distance of the focus from the center to the length of the semi-major axis.

**Eclipse** The total or partial obscuring of one body by another caused by one body passing into the shadow of, or in front of, another. For the Sun and Moon these are called solar and lunar eclipses. See also *occultation*.

**Ecliptic** The plane containing the Earth's orbit around the Sun.

**Ecliptic Coordinate System** The coordinate system in which a celestial object's location is stated with respect to the plane of the ecliptic. The ecliptic coordinate system is typically used to locate objects within our Solar System. Ecliptic longitude is measured in degrees from the First Point of Aries (vernal equinox) whereas ecliptic latitude is measured north or south of the ecliptic plane.

**Ecliptic Latitude** In the ecliptic coordinate system, the angular distance in degrees that a celestial object lies north or south of the ecliptic plane.

**Ecliptic Longitude** In the ecliptic coordinate system, the angular distance in degrees that a celestial object lies from the First Point of Aries.

**Ellipse** The collection of all points such that the sum of the distance of each point from the 2 foci is constant. This geometric figure is oval in shape when the 2 foci are distinct points, but is a perfect circle if the 2 foci are the same point. See also *eccentricity*.

**Elongation** The angle formed by the Sun, Earth, and the celestial body in question.

**Ephemeris** A table or publication that provides the coordinates of Solar System objects.

**Epoch** An instant in time used as a standard reference from which time is measured. Typically, an epoch is stated as a reference point for measuring equatorial coordinates.

**Equation of the Center** ($E_c$) The numerical difference between the true anomaly ($v$) and the mean anomaly ($M$), that is, $E_c = v - M$. The equation of the center is typically approximated by a truncated infinite series or by some numerical method. See also *Kepler's equation*.

**Equation of Time** The difference between mean solar time and apparent solar time.

**Equatorial Coordinate System** A coordinate system based on the plane of the celestial equator. See also *right ascension* and *declination*.

**Equinox** Either of 2 points at which the Sun crosses the celestial equator. These 2 points are the ascending and descending nodes for the Sun's orbit with respect to the celestial equator. During the equinoxes, the length of day and the length of night are very close to the same. See also *vernal equinox* and *autumnal equinox*.

**Exoplanet** A planet that orbits a star other than our Sun. See also *planet*.

# Glossary

**Fall** The three-month period of time that begins at the instant of the autumnal equinox.

**Fall Equinox** See *autumnal equinox*.

**First Point of Aries** The First Point of Aries is the location of the vernal equinox. It is used as a reference point in the ecliptic and equatorial coordinate systems. See also *vernal equinox*.

**Galactic Coordinate System** A coordinate system for locating objects within the Milky Way Galaxy. This coordinate system is defined with respect to a plane containing the Sun and the center of the Milky Way Galaxy.

**Galaxy** A large-scale collection of stars and other objects that appears to form more or less a cluster or group with a visible structure. A typical galaxy contains an average of 100 billion stars and measures 1,500 to 300,000 light years across.

**Gas Giant** A giant planet composed almost entirely of gaseous materials rather than rock or other solid matter. A Gas Giant may have a rocky or metallic core, but it does not have a well-defined surface as terrestrial planets do. Gas Giants are not confined to our Solar System. See also *outer planet*, *Jovian planet*, and *terrestrial planet*.

**Geocentric** Centered with respect to the Earth. A geocentric universe assumes Earth as the center of the universe. See also *heliocentric*.

**Geocentric Coordinates** Coordinates that specify the location of an object with respect to the center of the Earth; they may be expressed as Cartesian or spherical coordinates. See also *Earth Centric Inertial*, *heliocentric*, *topocentric*, and *barycentric coordinates*.

**Gibbous** The phases of the Moon in which the Moon is more than half illuminated but less than fully illuminated as seen from Earth. See also *crescent*.

**GMT** Greenwich Mean Time.

**Great Circle** A circle of greatest possible diameter formed by intersecting a plane with a sphere.

**Greenwich Mean Time (GMT)** Historically, the LMT for Greenwich, England. GMT has been superseded by UT. For the purposes of this book, UT, GMT, and UTC are considered to be synonymous.

**Gregorian Calendar** The calendar system introduced by Pope Gregory in 1582 to reduce the errors in the Julian calendar system. This is the most widely used calendar system today.

**Habitable Zone** A region around a star in which orbiting objects large enough to be planets can hold an atmosphere and support liquid water on their surface. See also *planet* and *exoplanet*.

**Heliocentric** Centered with respect to the Sun. A heliocentric Solar System assumes the Sun as the center of the Solar System. See also *geocentric*.

**Heliocentric Coordinates** Coordinates that specify the location of an object with respect to the center of the Sun. See also *geocentric*, *topocentric*, and *barycentric coordinates*.

**HMS** Hours, Minutes, Seconds.

**HMS Format** A format in which time is expressed as an integer number of hours, an integer number of minutes, and a fractional number of seconds (e.g., $12^h 13^m 15.1^s$).

**Horizon Coordinate System** A coordinate system based on the horizon as seen by an observer. See also *azimuth* and *altitude*.

**Hour Angle** The difference between the local sidereal time (LST) and the right ascension ($\alpha$) of a celestial body. In equation form, $H = LST - \alpha$.

**IAU**  International Astronomical Union.

**Inclination**  See *angle of inclination*.

**Inertia**  The tendency of a body to remain at rest or resist a change in velocity.

**Inferior Planet**  A planet whose orbit lies between Earth and the Sun (i.e., Mercury or Venus). The phrase "inferior planet" should not be confused with "inner planet." See also *inner planet, superior planet*, and *outer planet*.

**Inner Planet**  Any of the 4 planets (Mercury, Venus, Earth, and Mars) closest to the Sun. The phrase "inner planet" should not be confused with "inferior planet" because "inferior" refers to a planet's orbit relative to Earth, whereas "inner" refers to the region of space in which a planet lies. Mars is an inner planet but not an inferior planet. All of the inner planets are also terrestrial planets. See also *inferior planet, superior planet, outer planet*, and *terrestrial planet*.

**JD**  Julian Day number.

**Jovian Planet**  Any of the planets Jupiter, Saturn, Uranus, and Neptune. See also *outer planet* and *gas giant*.

**Julian Calendar**  A calendar system introduced by Julius Caesar that divides the year into periods of 365 days except for every fourth year, which contains 366 days. See also the *Gregorian calendar*.

**Julian Century**  A period of 36525 days reckoned from January 0.0.

**Julian Date**  A date in the Julian calendar system. Some authors use Julian date and Julian day number interchangeably. To avoid confusion, this book does not do so because Julian day numbers are unrelated to the calendar system named after Julius Caesar.

**Julian Day Number (JD)**  The number of days that have elapsed since noon UT on January 1, 4713 BC. Do not confuse with *Julian date*.

**Julian Year**  A year defined to be exactly 365.25 days in length, which is the average length of a year in the Julian calendar system.

**KBO**  Kuiper Belt Object.

**Keplerian Elements**  Named in honor of Johannes Kepler, the 6 attributes that precisely define an object's orbit; specifically an orbit's inclination, eccentricity, length of the semi-major axis, longitude of the ascending node, argument of perigee/perihelion, and mean anomaly at the epoch.

**Kepler's Equation**  An equation expressing the mathematical relationship between an object's eccentric anomaly ($E$) and its mean anomaly ($M$). Kepler's equation is given by

$$M = E - e \sin E,$$

where $M$ is expressed in radians, $E$ is expressed in radians, and $e$ is the object's orbital eccentricity. Solving Kepler's equation to find $E$ when given $M$ is usually accomplished by iterative numerical methods. See also *equation of the center*.

**Kuiper Belt**  A debris field in the region of space beyond Neptune to about 50 AUs from the Sun. See also *Asteroid Belt* and *Scattered Disc*.

**Kuiper Belt Object (KBO)**  A celestial object that lies within the Kuiper Belt region of space. Some astronomers consider the Scattered Disc region to be part of the Kuiper Belt. An object within the Kuiper Belt may also be called a KBO. See also *TNO*.

**Lagrange Points**  Any of the 5 places in an orbit (for the 3-body problem) at which the gravitational and centripetal forces acting upon an orbiting object are in balance. See also *trojan*.

# Glossary

**Latitude (ecliptic)**  See *ecliptic latitude*.

**Latitude (terrestrial)**  The distance, measured as an angle, that an object lies from Earth's equator.

**LCT**  Local Civil Time.

**Leap Year**  In the Julian calendar system, any year divisible by 4. In the Gregorian calendar system, any year divisible by 4 except for century years, which must also be divisible by 400.

**Light Year**  The distance that light travels in 1 year. One light year is about $5.87 \times 10^{12}$ miles.

**LMT**  Local Mean Time.

**Local Civil Time (LCT)**  The civil time within a specific time zone. Some authors may refer to this as local mean time or Standard Time. For the purposes of this book, LCT, LMT, and Standard Time are considered to be synonymous. See also *civil time*.

**Local Mean Time (LMT)**  Time measured with respect to the mean Sun transiting the central meridian for an observer. This term has been superseded by local civil time. For the purposes of this book, LCT, LMT, and Standard Time are considered to be synonymous.

**Local Sidereal Time (LST)**  The sidereal time for an observer, taking into account the observer's location on Earth. LST is similar in concept to LCT except that LCT is based on solar time, whereas LST is based upon sidereal time. (Sidereal time can only be local, but LST is used here to distinguish it from GST.)

**Local Time**  Same as local civil time.

**Longitude (ecliptic)**  See *ecliptic longitude*.

**Longitude (terrestrial)**  The distance, measured as an angle, that an object lies from the Greenwich prime meridian.

**Longitude of Ascending Node**  The ecliptic longitude of the ascending node for an object within the Solar System. See also *ascending node* and *ecliptic coordinate system*.

**Longitude of Perihelion**  The ecliptic longitude of the point of perihelion for an object within the Solar System.

**LST**  Local Sidereal Time.

**Lunar Eclipse**  An eclipse of the Moon caused by the Moon passing into Earth's shadow. See also *penumbra* and *umbra*.

**Lunar Libration**  A slow oscillation in the Moon's orbit around its axis of rotation.

**Lunar Month**  A synodic month in which a specific phase of the Moon (e.g., Full Moon) is chosen as the reference point.

**Magnitude**  A measurement, based on a logarithmic scale, of the brightness of a celestial object. The word "magnitude" by itself is ambiguous because it could refer to apparent or absolute magnitude. See also *visual magnitude*, *apparent magnitude*, and *absolute magnitude*.

**Mean Anomaly**  The angle formed by an orbiting body, the center of an assumed constant-speed circular orbit, and the orbital major axis. See also *true anomaly*, *eccentric anomaly*, *equation of the center*, and *Kepler's equation*.

**Mean Orbit**  A fictitious orbit in which an object's orbit is assumed to be a perfect circle and whose orbital speed is constant throughout the orbit. See also *auxiliary circle* and *mean anomaly*.

**Mean Orbital Elements** The 6 Keplerian elements, referenced to a stated epoch, in which the orbital elements are averaged out over the entire orbit to represent an ideal elliptical orbit. See also *orbital elements*, *Keplerian elements*, and *osculating orbital elements*.

**Mean Sidereal Time** Sidereal time that has been corrected for irregularities in Earth's rotation (e.g., nutation). For the purposes of this book, mean sidereal time and apparent sidereal time can be assumed to be equal. See also *apparent sidereal time*.

**Mean Solar Day** A day as measured by the motion of a mean Sun. It is the time interval between 2 successive transits of the mean Sun across an observer's meridian.

**Mean Sun** A fictitious Sun moving in a fictitious circular orbit rather than its true elliptical orbit. The motion of the mean Sun is uniform throughout its entire orbit, which allows a mean solar day to be defined that is uniform in length regardless of the time of year or where the Sun is in its mean orbit.

**Mean Time** Time as measured by the motion of the mean Sun. Mean time is uniform in length because it is based on the motion of a mean Sun. See also *apparent time* and *mean solar day*.

**Meridian** A semicircle that passes through both poles of a sphere.

**Meteor** A meteoroid that enters Earth's atmosphere. Meteors are often called shooting stars. See also *meteorite* and *meteoroid*.

**Meteorite** A meteor that survives vaporization in the Earth's atmosphere and strikes the Earth's surface. See also *meteor* and *meteoroid*.

**Meteoroid** A small rocky object in space thought to be a fragment from an asteroid, comet, or other debris ejected from larger celestial bodies. Meteoroids are generally considered to be the same as asteroids except that they are much smaller in size and mass. See also *asteroid*, *meteor*, and *meteorite*.

**Micromoon** A Full Moon or New Moon that occurs when the Moon is at apogee. See also *supermoon*.

**Modified Julian Day Number (MJD)** The number of Julian days that have passed since November 17.0, 1858. See also *Julian day number* and *Julian date*.

**Month** The time interval required for the Moon to complete 1 orbit from 1 reference point to another. The length of a month depends on the reference point used to measure it. See also *anomalistic*, *draconic*, *nodal*, *sidereal*, and *synodic months*.

**mV** Visual magnitude.

**Nadir** The point on the celestial sphere directly beneath an observer. Nadir is the opposite of zenith.

**Nebula** A diffuse cloud of interstellar gas or dust that is visible as a luminous patch of dark and light areas.

**Nodal Month** See *draconic month*.

**Node** Either of 2 points at which an object goes above or below some reference plane. See also *ascending* and *descending node*.

**Nutation** A small periodic wobble in Earth's axis of rotation.

**Obliquity of the Ecliptic** The angle of inclination (about $23°26'$) that the ecliptic plane makes with respect to the plane containing the celestial equator.

# Glossary

**Observer's Meridian**  The meridian that passes through an observer's zenith and Earth's North and South Poles.

**Occultation**  An astronomical event in which 1 celestial object is hidden by another object that passes between it and an observer. For example, the Moon occults a planet when the Moon is in the line of sight between an Earthbound observer and the planet. See also *eclipse*.

**Oort Cloud**  A theorized region of space located 2,000–5,000 AUs from the Sun and extending 50,000–200,000 AUs from the Sun. The Oort Cloud marks the outermost boundary of our Solar System and is thought to be the source of the long-period comets in our Solar System. See also *Kuiper Belt* and *Scattered Disc*.

**Opposition**  The moment when 2 celestial objects are on opposite sides of the Earth and their ecliptic longitudes are 180° apart. For example, the Moon is in opposition to the Sun when it is a Full Moon. See also *conjunction*.

**Orbital Elements**  The attributes of an orbit in 3-dimensional space (such as inclination, eccentricity, argument of periapsis) that are used to describe the position of an orbiting object. See also *Keplerian elements*, *mean orbital elements*, and *osculating orbital elements*.

**Osculating Orbital Elements**  The instantaneous values for the 6 Keplerian elements at a stated epoch. See also *orbital elements*, *Keplerian elements*, and *mean orbital elements*.

**Outer Planet**  Any of the planets (Jupiter, Saturn, Uranus, and Neptune) whose orbits are farther away from the Sun than the Asteroid Belt. The phrase "outer planet" should not be confused with "superior planet" because "superior" refers to a planet's orbit relative to the Earth whereas "outer" refers to the region of space in which a planet lies. Mars is a superior planet but not an outer planet. See also *inferior planet*, *superior planet*, *Jovian planet*, and *gas giant*.

**Parallax**  The amount by which the apparent position of an object changes as the location of an observer changes.

**Parsec**  A unit of measurement based on the distance from Earth at which the stellar parallax is 1 second of arc. One parsec is about 3.26 light years.

**Penumbra**  The lighter portion of a shadow in which not all light is blocked out. See also *eclipse* and *umbra*.

**Perigee**  The point in an orbit around Earth at which an object is closest to Earth. See also *apogee*.

**Perihelion**  The point in an orbit around the Sun at which an object is closest to the Sun. See also *aphelion*.

**Phase Angle**  The angle between an incident ray striking an object and the light ray reflected back from the object.

**Planet (modern definition)**  A celestial body that orbits the Sun, has sufficient mass and gravity so that the object is nearly round in shape, and has cleared the neighborhood around its orbit. See also *dwarf planet*.

**Planet (old definition)**  A celestial body, other than a comet or a meteoroid, in orbit around a star. Pluto was a planet under this definition. See also *dwarf planet*.

**Precession**  A motion of Earth's axis similar to the motion of the axis of a rapidly spinning top.

**Prime Meridian**  The meridian that passes through Greenwich, England.

**Propagation**  The process of using an object's orbital elements and epoch to predict the position of the object at some future point in time.

**Pulsar** A rotating star that emits a beam of electromagnetic energy. Much like a lighthouse, the emitted beam can be detected only when it is pointing in the same direction as the Earth. See also *quasar*.

**Quadrature** The moment when the angle formed by Earth, a celestial object, and a reference point is 90°. For example, the Moon is in quadrature during First Quarter Moon and Last Quarter Moon.

**Quasar** A distant galaxy that emits waves of electromagnetic energy. See also *pulsar*.

**RAAN** Right Ascension of the Ascending Node.

**Refraction** A bending of light rays caused by a change in medium. For example, light rays are bent as they travel from space through Earth's atmosphere, or when passing from the air through water.

**Right Ascension** In the equatorial coordinate system, the distance that an object lies away from the First Point of Aries. Right ascension is measured in units of time and is in the range $[0^h, 24^h]$. Right ascension is analogous to the concept of terrestrial longitude. See also *declination* and *hour angle*.

**Right Ascension of the Ascending Node (RAAN)** The longitude of the point in a satellite's orbit at which the satellite goes above the equatorial plane. The RAAN is measured from the vernal equinox (First Point of Aries) to the ascending node.

**Roche Limit** The distance at which an orbiting body will disintegrate due to the overpowering tidal forces of the body around which it orbits.

**Saros Cycle** A period of approximately 18 years after which the cycle of lunar and solar eclipses tend to repeat.

**Scattered Disc** A sparsely populated region of space that lies between the Kuiper Belt and Oort Cloud regions of space. See also *Kuiper Belt* and *Oort Cloud*.

**Scattered Disc Object (SDO)** A celestial object whose orbit lies in the Scattered Disc region of space. See also *KBO* and *TNO*.

**Sidereal Day** A day as measured by 2 successive transits of a fixed star across an observer's meridian. A sidereal day is approximately $23^h 56^m$ in length.

**Sidereal Month** A month as measured by using a star as a fixed reference point. A sidereal month is 27.3217 mean solar days in length.

**Sidereal Time** Time measured by Earth's rotation relative to fixed stars instead of the mean Sun.

**Sidereal Year** A year measured as the time it takes for Earth to return to the same position with respect to the stars. A sidereal year is 365.2564 mean solar days in length.

**Solar Day** The time interval between 2 successive transits of the Sun across an observer's meridian. A solar day is not uniform in length. See also *apparent solar day*, *apparent time*, and *mean solar day*.

**Solar Eclipse** An eclipse of the Sun caused by Earth passing into the Moon's shadow. See also *annular eclipse*.

**Solar Time** Time measured by the apparent position of the Sun. A sundial measures solar time and is uneven in length.

**Solstice** Either of the 2 points at which the Sun is farthest away from the plane containing the celestial equator. At the solstices, the Sun has no apparent northward or southward motion. The

# Glossary

longest day of the year occurs at the summer solstice whereas the shortest day of the year occurs at the winter solstice. See also *solstice* and *winter solstice*.

**Spring**  The 3-month period of time that begins at the instant of the vernal equinox.

**Spring Equinox**  See *vernal equinox*.

**Standard Time**  Time at the central meridian for a given time zone. For the purposes of this book, LCT, LMT, and Standard Time are considered to be synonymous.

**Star Atlas**  A book containing a collection of star maps.

**Star Catalog**  A book containing the coordinate locations, usually in the equatorial coordinate system, of stars and other deep sky objects for some stated epoch. A catalog may also contain other related data, such as a star's magnitude.

**Star Chart**  A graphical depiction of the position of the stars with respect to each other.

**Star Map**  See *star chart*.

**Star Time**  An informal way of referring to *sidereal time*.

**State Vector**  The position of an orbiting object described in terms of the object's Cartesian coordinates and velocity along each of the Cartesian axes at the stated epoch. See also *Keplerian elements* and *osculating orbital elements*.

**Station Keeping**  Periodic orbital maneuvers performed to keep a satellite in its assigned orbit.

**Summer**  The 3-month period of time that begins at the instant of the summer solstice.

**Summer Solstice**  The point at which the Sun is farthest north of the plane containing the celestial equator. This occurs on about June 22; it is the longest day of the year in the Northern Hemisphere.

**Superior Planet**  A planet whose orbit lies beyond Earth. The phrase "superior planet" should not be confused with "outer planet." See also *inner planet*, *inferior planet*, and *outer planet*.

**Supermoon**  A Full Moon or New Moon that occurs when the Moon is at perigee. See also *micromoon*.

**Synodic Month**  A month based on the average interval between 2 successive occurrences of the same lunar phase: New Moon to New Moon, Full Moon to Full Moon, etc. The synodic month is 29.5306 mean solar days in length.

**Terrestrial Planet**  A planet that is "Earthlike" in terms of its composition and structure. A terrestrial planet has a solid rocky surface, a molten metallic core, and surface features similar to what is found on Earth (mountains, volcanoes, canyons, etc.). Terrestrial planets are not confined to our Solar System. See also *inner planet* and *gas giant*.

**Terrestrial Time (TT)**  A uniform time system defined to account for irregularities in Earth's rotation. TT is based on the International Atomic Time (TAI) and is given by the equation $TT = TAI + 32.184^s$. See also *Universal Time 1* and *Delta T*.

**Time Zone**  A geographic region in which there is an agreed-upon time standard so that all observers within the region can synchronize with the time standard and thereby agree on the same local time for all observers in the region. A time zone's boundary is typically irregular to account for man-made boundaries, and is typically established with respect to the prime meridian at Greenwich, England.

**TLE**  See *Two-Line Element*.

**TNO**  Trans-Neptunian Object.

**Topocentric Coordinates**  Coordinates that specify the location of an object with respect to the surface of the Earth; they may be expressed as Cartesian or spherical coordinates. See also *geocentric*, *heliocentric*, and *barycentric coordinates*.

**Trans-Neptunian Object (TNO)**  A celestial object within the Solar System whose orbit lies farther away from the Sun than Neptune. See also *KBO*.

**Transit**  In astronomy, the moment at which an object crosses an observer's meridian, or the motion of 1 celestial body across another. Also called culminate.

**Trojan**  An object that shares the same orbit as a planet or a larger moon, and that is in a stable orbit in which it remains in the same position relative to that larger object. Trojans oscillate around Lagrange point L4 (ahead of the object with which it shares its orbit by 60°) or Lagrange point L5 (behind the object with which it shares its orbit by 60°). See also *Lagrange points*.

**Tropical Year**  A year measured by using the vernal equinox as the reference point. A tropical year is the time interval between 2 successive crossings by the Sun of the plane of Earth's equator at the point of the vernal equinox. A tropical year is 365.242191 mean solar days in length.

**True Anomaly**  The angle formed by an orbiting body (e.g., planet), the body around which it orbits (e.g., the Sun), and the orbital major axis. See also *eccentric anomaly*, *mean anomaly*, *equation of the center*, and *Kepler's equation*.

**True Orbit**  The actual elliptical orbit of an object. See also *mean orbit* and *auxiliary circle*.

**TT**  Terrestrial Time.

**Twilight**  The period of semidarkness just after sunset. At twilight the Sun is less than 18° below the horizon. See also *dawn*.

**Two-Line Element (TLE)**  A standard format originally defined by the North American Aerospace Defense Command (NORAD) for encoding an object's Keplerian elements, the epoch, and other items related to a satellite's orbit. See also *Keplerian elements*.

**Umbra**  The darkest portion of a shadow in which all light is blocked out. See also *penumbra*.

**Universal Time (UT)**  The local civil time for the Greenwich time zone (i.e., longitude 0°). UT supersedes the historical term *Greenwich Mean Time*. For the purposes of this book, UT, GMT, UT1, and UTC are considered to be synonymous.

**Universal Time 1 (UT1)**  A timekeeping system that is based on the rotation of Earth, and is therefore irregular. UT1 is related to TT by the equation $UT1 = TT - \Delta T$, where $\Delta T$ is determined by astronomical observations. The value of $\Delta T$ is currently about $65^s$. $\Delta T$ for the past is determined by recorded astronomical observations, but it can only be approximated for the future. See also *Terrestrial Time* and *Delta T*.

**UTC**  An abbreviation for Coordinated Universal Time, which in French is *Temps Universel Coordonné*. The acronym UTC is a compromise between English-speaking (for whom the acronym would have been CUT) and French-speaking (for whom the acronym would have been TUC) peoples. See also *coordinated universal time*.

**Vernal Equinox**  The ascending node for the Sun's orbit assuming the plane of the celestial equator as a reference plane. The vernal equinox is also known as the First Point of Aries. The Sun's right ascension at the vernal equinox is $0^h$, which occurs at about March 21. See also *equinox*.

**Visual Magnitude (mV)**  An object's apparent brightness as measured by considering only wavelengths of light that are in the visible spectrum. Visual magnitude is also called apparent

# Glossary

magnitude and indicates how bright an object appears to be to a viewer. See also *magnitude* and *absolute magnitude*.

**VSOP87** (Variations Séculaires des Orbites Planétaires 1987) An analytic theoretical model of the Solar System developed in 1987 by scientists at the Bureau des Longitudes in Paris, France. VSOP87 is used to compute highly accurate positions and orbital elements for objects within the Solar System. See also JPL's *Development Ephemeris*.

**Vulcan** A hypothesized planet in orbit between the Sun and Mercury that was proposed as a way to explain perturbations in Mercury's orbit. All attempts to find Vulcan have been unsuccessful while Einstein's *Theory of General Relativity* can explain Mercury's perturbations. This leads most astronomers to doubt that Vulcan exists. See also *vulcanoid*.

**Vulcanoid** An asteroid from the hypothesized group of asteroids that some believe may exist in a stable orbit between the Sun and Mercury. Named after the hypothetical planet Vulcan, no vulcanoids have yet been discovered. See also *Vulcan*.

**Waning** The phases of the Moon in which the amount of illumination is decreasing from Full Moon to New Moon.

**Waxing** The phases of the Moon in which the amount of illumination is increasing from New Moon to Full Moon.

**Winter** The 3-month period of time that begins at the instant of the winter solstice.

**Winter Solstice** The point at which the Sun's position is farthest south of the plane containing the celestial equator. This occurs on about December 22; it is the shortest day of the year in the Northern Hemisphere.

**Year** The time interval between 2 successive passages of the Earth or Sun past some reference point. The length of a year varies depending on the reference point used.

**Zenith** The point on the celestial sphere directly overhead an observer. See also *nadir*.

**Zulu Time** Same as GMT and UT.

# Index

ABS, 7
absolute magnitude, 252
Adams, John C., 217
age of the Moon, 173, 175
Aitken Basin, 157, 158
Aldrin, Edwin, 159
*Almagest, The*, 214
Alpha Centauri, 131
altitude, 88, 112
Amor subgroup, 228
angle of inclination, 83, 284
angular diameter, 145
angular size, *see* angular diameter
annular solar eclipse, 182
anomalistic year, 27
Anthe, 213
aphelion, 68, 71, 74, 145, 249
Aphrodite Terra, 198
apoapsis, 68, 250
  distance from focus, 250
apogee, 68, 250
  height, 274
  radius, 275
Apollo program, 155, 159
Apollo subgroup, 228
apparent sidereal time, 36
apparent size, *see* angular diameter
apsides, 250
apsis, 66
Arandas Crater, 201
argument of latitude, 285, 288
argument of perihelion, 83
Argyre Planitia, 201
Armstrong, Neil, 159
Arsia Mons, 201
ascending node, 83, 97, 165
Ascraeus Mons, 202
Asteroid Belt, 3, 187, 223, 226, 229
asteroids, 227
  25143 Itokawa, 228
  433 Eros, 228
  Amor subgroup, 228
  Apollo subgroup, 228
  Aten subgroup, 228
  Ceres, 3, 223, 226
  Chariklo, 226
  Hayabusa, 228
  Juno, 226
  naming scheme, 227
  near Earth, 226
  NEAR Shoemaker, 228
  Pallas, 226
  spectral reflectivity, 228
  Vesta, 188, 226
*Astronomical Algorithms*, 138, 143
*Astronomical Almanac*, 138
astronomical almanac, 332, 340
astronomical calendar, 344, 345
*Astronomical Formulae
  for Calculators*, 142
astronomical unit, *see* AU
Aten subgroup, 228
atmospheric refraction, 101
AU, 14
autumnal equinox, 26, 141
auxiliary circle, 76
average orbital velocity, 260
azimuth, 88, 91, 112
azimuthal angle, 119

Bamberg Crater, 201
barycenter, 130
Batygin, Konstantin, 191
Bayer, Johann, 336
Bell, Jocelyn, 192
BepiColombo mission, 195, 198
Besselian year, 27
Betelgeuse, 115, 336, 338
binary star system, 131
binary system, 131
Brown, Michael, 191, 225
Burney, Venetia, 220

Cabeus crater, 158
calendar, 38
  astronomical, 344
  Gregorian, 39
  Julian, 38, 40
Callisto, 207, 208
Caloris Basin, 195
Calypso, 213
Canis Major, 129, 338
Canopus, 131
Cartesian coordinate system, 277
Cartesian coordinates, 119
Cassini, Domenico, 210
Cassini-Huygens, 207, 210, 211, 214
Caucasian Mountains, 158
celestial prime meridian, 62
celestial sphere, 62, 64, 97, 111
CelesTrak, 347
centaur, 188
  Chariklo, 189
  Chiron, 189
  Kuiper Belt, 188
  Phoebe, 189
  Pholus, 189
Centaurus, 131
central meridian, 30
Central Standard Time zone, see CST
Ceres, 3, 187, 188, 223, 226
  Dawn space probe, 224
  distance from the Sun, 223
  Herschel Space Observatory, 224
  mass, 223
  Occator crater, 224
  orbital period, 223
  Piazzi, Giuseppe, 223
  rotational period, 223
  trojans, 188
Cernan, Eugene, 159
Chariklo, 189, 226
  rings, 189
Charon, 131, 222
  Christy, James, 222
  Harrington, Robert, 222
Chelyabinsk asteroid, 228
Chiron, 189
Christy, James, 222
civil time, 31
civil year, 26
Clarke orbits, 306
Clarke, Arthur C., 265, 318
Clementine, 160
Collins, Michael, 159
comets
  Hale-Bopp, 231
  Halley, 230
  Hyakutake, 231
  Shoemaker, Carolyn, 230
  Shoemaker-Levy 9, 208

conjunction, 178
constellations, 2
  Canis Major, 129, 338
  Centaurus, 131
  Cygnus, 130, 192
  Orion, 115, 336
  Pegasus, 192
  Scutum, 130
  Virgo, 192
coordinate systems
  Cartesian, 277
  ecliptic, 91
  equatorial, 54, 84, 92, 94
  galactic, 54, 84, 96
  horizon, 54, 84, 88, 90
  latitude-longitude, 53
  spherical, 279
  topocentric, 281
Coordinated Universal Time,
  see UTC
Copernicus crater, 157
corona, 129
crescent Moon, 177
CST, 33
culminate, 64
Curiosity Rover, 201
Cygnus, 130, 192

dawn, 140
Dawn space probe, 219, 224, 225
day, 22, 64
  mean solar, 24, 64
  sidereal, 24, 28, 36, 64
  solar, 22
day number, see days into the year
daylight saving time, see DST
days into the year, 44
DE, 348, 349
December solstice, 143
declination, 85, 88, 90, 92
DefaultObsLoc.dat, 123
Deimos, 200
Development Ephemeris, see DE
Dimidium, 192
Dione, 188, 213
DMS format, 15–17
draconic year, 27
Dragon Storm, 211
DST, 33
Duffett-Smith, Peter, 135, 332
dwarf planet, 3, 219
  Ceres, 3, 187, 188, 223
  Dysnomia, 3
  Eris, 3, 188, 219, 225
  Haumea, 3, 224
  Makemake, 3, 224
  Pluto, 189, 219, 220
Dysnomia, 3, 225

# Index

Earth, 189
  escape velocity, 128
  orbital eccentricity, 67
  orbital speed, 126
  radius, 272
  trojans, 188
Earth Centered Inertial, see ECI
Eastern Standard Time zone, see EST
eccentric anomaly, 69, 70, 75–80
eccentricity, 66, 273
ECI, 278
eclipse, 181
  annular, 182
  conjunction, 183
  lunar, 181
  opposition, 183
  penumbra, 181
  Saros cycle, 183
  solar, 181
  total lunar, 181
  total solar, 182
  umbra, 181
ecliptic coordinate system, 91
  latitude, 91
  longitude, 91
ecliptic plane, 82, 83, 132
ellipse, 65
  apoapsis, 68
  apsis, 66
  eccentricity, 66, 273
  focus, 65
  linear eccentricity, 272, 274
  major axis, 65
  minor axis, 65
  occupied focus, 67
  periapsis, 68
  semi-latus rectum, 68, 72, 272, 274
  semi-major axis, 65, 272, 273
  semi-minor axis, 65, 260, 272, 274
elongation, 177, 251
Enceladus, 213
ephemerides, see ephemeris
ephemeris, 332, 340, 342
Epimetheus, 213
epoch, 92
equation of the center, 69, 70, 74, 75, 79, 134
equation of time, 24, 147, 148
equatorial coordinate system, 54, 84, 92, 94, 112
equinox, 26, 140, 145
  autumnal, 26, 141
  fall, 141
  March, 142
  September, 142
  spring, 141
  vernal, 1, 26, 83, 141

Eris, 3, 188, 219, 225
  Brown, Michael, 225
  diameter, 225
  distance from the Sun, 225
  Dysnomia, 225
  mass, 225
  orbital period, 225
  rotational period, 225
EST, 33
Europa, 207, 209
Europa Multiple-Flyby Mission, 209
Evening Star, 194, 196
exoplanet, 192
  Dimidium, 192
  J2126-8140, 193
  Kepler-186f, 192

fall, 142
fall equinox, 141
First Point of Aries, 62, 83, 84, 91
  Hipparchus, 62
First Quarter Moon, 176
FIX, 6
Flamsteed numbers, 340
Flamsteed, John, 214, 340
focus, 65
FRAC, 7
Full Moon, 26, 174, 176, 183
  lunar eclipse, 183
functions
  ABS, 7
  FIX, 6
  FRAC, 7
  INT, 6
  MOD, 7
  ROUND, 8

galactic coordinate system, 54, 84, 96
  latitude, 97
  longitude, 97
Galactic North Pole, 97
galactic year, 27
Galileo, 207, 209
  Neptune, 217
  Saturn's rings, 209
Galileo space probe, 198, 205, 207
Galle Crater, 201
Galle, Johann, 217
Ganymede, 207, 208
Gas Giants, 203
GEO, 305, 306
geocentric, 8
geocentric coordinate system, 278
geostationary orbits, 306
Geosynchronous Earth Orbit, see GEO
gibbous Moon, 177
GMT, 34

great circle, 57, 60
Great Dark Spot, 218
Great Orion Nebula, 340
Great Red Spot, 206
Great White Spot, 212
Greenwich Mean Time, see GMT
Greenwich Sidereal Time, see GST
Gregorian calendar, 39
GST, 36, 112, 113
Guinan, Edward, 218

habitable zone, 192
Hale-Bopp Comet, 231
Halley's Comet, 230
Happy Face Crater, 201
Harrington, Robert, 222
Haumea, 3, 187, 224
  diameter, 224
  distance from the Sun, 224
  mass, 224
  orbital period, 224
  rotational period, 224
  visual magnitude, 224
Helene, 213
heliocentric, 8
Hellas Planitia, 201
HEO, 305, 307
Herschel Space Observatory, 224
Herschel, Sir William, 214
Hewish, Antony, 192
Highly Elliptical Orbit, see HEO
Hipparchus, 214
HMS format, 15, 16, 86
Hohmann transfer orbit, 320
horizon coordinate system, 54, 84, 88, 90, 112
HORIZONS system, 345, 349
hour angle, 86, 87, 89, 112
HST, 207, 216, 218, 294, 307
  catalog number, 300
Hubble Space Telescope, see HST
Huygens, Christiaan, 209
Hyakutake Comet, 231
Hydra, 222

Iapetus, 213
IAU, 3, 35
IJK, see ECI
inner planet, 187, 193
INT, 6
Intelsat I, 265
International Astronomical Union, see IAU
International Space Station, see ISS
Io, 207, 209
Ishtar Terra, 198
ISS, 297, 306
  catalog number, 300
Ithaca Chasma, 213

J2000, 93
J2126-8140, 193
Janus, 213
Jovian planets, 203
JPL
  DE, 349
  HORIZONS system, 345, 349
JUICE, 209
Julian calendar, 38, 40
Julian date, 40
Julian day number, 40–42, 44, 45, 109, 111–113
  modified, 44
Julian year, 27
June solstice, 142
Juno, 209, 226
Jupiter, 189, 204
  atmosphere, 204
  Callisto, 207, 208
  Cassini, 207
  diameter, 204
  distance from the Sun, 204
  Europa, 207, 209
  Europa Multiple-Flyby Mission, 209
  Galileo, 207
  Galileo space probe, 205, 207
  Ganymede, 207
  Great Red Spot, 206
  Io, 207, 209
  JUICE, 209
  Juno, 209
  mass, 204
  Mayr, Simon, 207
  New Horizons, 207
  orbital period, 204
  Pioneer, 206, 207
  rotational period, 204
  surface temperature, 204
  trojans, 188
  Ulysses, 207
  visual magnitude, 204
  Voyager, 207

KBO, 229
Kepler Space Telescope, 192, 194
Kepler's equation, 69, 70, 75, 79, 82, 134, 137
  Newton/Raphson method, 81
Kepler's laws, 22
  first, 71
  second, 68, 258
  third, 257
Kepler-186f, 192
Keplerian elements, 286, 287, 347
  convert to state vector, 294
  TLE data, 301
Kerberos, 222

# Index

Kuiper Belt, 2, 187, 219, 229
  centaurs, 188
  Eris, 3
  Haumea, 187
  Makemake, 187
  New Horizons, 219
  Pluto, 187
Kuiper Belt object, *see* KBO

Large Magellanic Cloud, 130
Larissa, 218
Last Quarter Moon, 176
latitude, 53, 59, 60, 84
latitude-longitude system, 53, 56
*Law of Universal Gravitation*, 69, 253, 262
LCROSS, 158, 160
LCT, 31, 33, 36, 112, 113
Le Verrier, Urbain, 191, 217
  Vulcan, 191
leap year, 26, 38, 39
Leibnitz Mountains, 158
LEO, 305, 306
Lescarbault, Edmond Modeste, 191
  Vulcan, 191
light year, 15
linear eccentricity, 274
LMT, 33
local celestial meridian, 64
local civil time, *see* LCT
Local Mean Time, *see* LMT
Local Sidereal Time, *see* LST
local time, *see* LCT
longitude, 30, 53, 60, 84
longitude of the ascending node, 83
look angle, 283
Low Earth Orbit, *see* LEO
Lowell, Percival, 221
LRO, 153, 160
LST, 36, 49, 86
Luna, 155, 159, 161
Lunar Crater Observation and Sensing Satellite, *see* LCROSS
lunar eclipse, 181
lunar libration, 155
Lunar Prospector, 160
Lunar Reconnaissance Orbiter, *see* LRO
LuxSpace, 161

Magellan space probe, 198
major axis, 65
Makemake, 3, 187, 224
  distance from the Sun, 225
  orbital period, 225
  rotational period, 225
  visual magnitude, 225
March equinox, 142
maria, 155

Mariner, 195, 198, 199
Mars, 189, 199
  2001 Mars Odyssey, 201
  Arandas Crater, 201
  Argyre Planitia, 201
  Arsia Mons, 201
  Ascraeus Mons, 202
  atmosphere, 199
  atmospheric pressure, 199
  Bamberg Crater, 201
  caves, 201
  Curiosity Rover, 201
  Cydonia region, 201
  Deimos, 200
  diameter, 199
  distance from the Sun, 199
  Face on Mars, 201
  Galle Crater, 201
  Happy Face Crater, 201
  Hellas Planitia, 201
  Mariner, 199
  Mars Express, 201
  Mars Orbiter Mission, 201
  mass, 199
  MAVEN, 201
  MRO, 201
  Olympus Mons, 202
  Opportunity Rover, 201
  orbital period, 199
  Pavonis Mons, 201
  Phobos, 200
  polar caps, 199
  rift valley, *see* Valles Marieneris
  rotational period, 199
  Schiaparelli, Giovanni, 199
  seasons, 199
  Spirit Rover, 201
  surface temperature, 199
  Tharsis region, 201
  trojans, 188
  Valles Marieneris, 202
  Vastitas Borealis basin, 201
  Viking, 199
Mars Express, 201
Mars Odyssey, 201
Mars Orbiter Mission, 201
Mars Reconnaissance Orbiter, *see* MRO
MAVEN, 201
Maxwell Montes, 198
Mayr, Simon, 207
mean anomaly, 24, 69, 70, 72–79
mean orbit, 72
mean orbital elements, 287
mean sidereal time, 25, 36
mean solar day, 24, 25, 64, 86
mean solar time, 25, 35, 36
mean Sun, 24, 64

Medium Earth Orbit, *see* MEO
Meeus, Jean, 79, 138, 142, 143, 332
MEO, 305, 307
Mercury, 189, 194
  atmosphere, 195
  BepiColombo mission, 195
  Caloris Basin, 195
  diameter, 195
  distance from the Sun, 195
  Evening Star, 194
  Mariner, 195
  mass, 195
  MESSENGER, 195
  Morning Star, 194
  orbital period, 195
  rotational period, 195
  temperature, 195
  visual magnitude, 197
  water, 195
meridian, 57, 86
  celestial prime, 62
  central, 30
  local celestial, 64
  observer's, 64
MESSENGER, 195
Messier Catalog, 340
Methone, 213
micromoon, 172
Milky Way Galaxy, 97
  center, 97
Mimas, 213, 214
  Herschel, 213, 214
minor axis, 65
MOD, 7
modified Julian day number, 44
modulo, *see* MOD
month, 25, 151
  sidereal, 25
Moon
  age, 173, 175
  age in days, 175
  Aitken Basin, 157, 158
  Alps, 158
  angular diameter, 172
  annual equation correction, 163, 165
  apogee distance, 250
  apogee velocity, 260
  atmosphere, 153
  Cabeus crater, 158
  Caucasian Mountains, 158
  Chandrayaan-1, 158, 161
  Chang'e 3, 161
  Chang'e 5, 161
  Clementine, 160
  conjunction, 178
  Copernicus crater, 157
  craters, 156
  crescent, 177
  distance, 152, 172
  ecliptic longitude, 165
  ecliptic longitude of perigee, 165
  elongation, 177
  equation of the center, 165
  evection correction, 163, 165
  first quarter, 176
  full, 174, 176
  gibbous, 177
  Google lunar prize, 161
  last quarter, 176
  LCROSS, 158, 160
  Leibnitz Mountains, 158
  longitude of the ascending node, 165
  LRO, 153, 160
  Luna, 155, 159, 161
  Lunar Prospector, 160
  LuxSpace flyby, 161
  maria, 155
  mass, 152
  mean anomaly, 165
  mean anomaly correction, 163, 165, 170
  micromoon, 172
  new, 174–176
  opposition, 177
  percent illumination, 173, 178
  perigee distance, 250
  perigee velocity, 260
  phase angle, 178
  phases, 151, 173, 176
  quadrature, 178
  radio transmission delays, 257
  Ranger, 159
  rise, 169
  SELENE, 161
  set, 169
  SMART-1, 160
  supermoon, 172
  Surveyor, 159
  temperature, 153
  terrae, 155
  tides, 153
  total eclipse, 181
  true ecliptic longitude, 166, 170
  Tycho crater, 157
  variation correction, 163, 165
  visual magnitude, 127
  waning, 176
  water, 158, 159
  waxing, 176
  weight on, 254
moonrise, 169
moonset, 169
Morning Star, 194, 196
Mountain Standard Time zone, *see* MST

# Index

MRO, 201
MST, 33

nadir, 64
near Earth asteroids, 226
NEAR Shoemaker, 228
Neptune, 189, 216
  Adams, John C., 217
  atmosphere, 218
  diameter, 218
  distance from the Sun, 218
  Galileo, 217
  Galle, Johann, 217
  Great Dark Spot, 218
  Hubble Space Telescope, 218
  Larissa, 218
  Le Verrier, Urbain, 217
  mass, 218
  orbital period, 218
  rings, 218
  rotational period, 218
  Scooter, 218
  Small Dark Spot, 218
  temperature, 218
  Triton, 219
  trojans, 188
  Voyager, 218
  wind speeds, 218
  Wizard's Eye, 218
*New General Catalog*, see NGC
New Horizons, 207, 219, 220
New Moon, 26, 174–176, 183
  solar eclipse, 183
Newton, Sir Isaac, 126, 265
  *Law of Universal Gravitation*, 69
Newton/Raphson method , 81
NGC, 214, 340
Nix, 222
North Celestial Pole, 62, 64
North Star, 62
Northern Electrostatic Disturbance, 211
nutation, 101, 102, 112

Oberth, Hermann, 265
obliquity of the ecliptic, 92–94, 144, 148
observer's meridian, 64
Occator crater, 224
occultation, 181
occupied focus, 67
Olbers, Wilhelm, 227
Olympus Mons, 202
Oort Cloud, 188, 230
  Hale-Bopp Comet, 231
  Hyakutake Comet, 231
  Oort, Jan, 231
  Öpik, Ernst, 230

Oort, Jan, 231
Öpik, Ernst, 230
Opportunity Rover, 201
opposition, 177
orbital elements, 68, 284, 287
  angle of inclination, 83
  argument of perihelion, 83
  ascending node, 83
  ecliptic plane, 82
  First Point of Aries, 83
  longitude of the ascending node, 83
  orbital plane, 82
orbital period, 72, 73
orbital plane, 83
orbital propagation, 308
orbital velocity, 261
Orion, 115, 336
osculating orbital elements, 287
outer planet, 187, 203

Pacific Standard Time zone, *see* PST
Pallas, 226
  Olbers, Wilhelm, 227
Pallene, 213
Pan, 213
Pandora, 213
parallax, 101, 112
parsec, 15
Pavonis Mons, 201
Payne-Gaposchkin, Cecilia, 126
Pegasus, 192
penumbra, 181
periapsis, 68, 71, 250
  distance from focus, 250
perifocal coordinate system, 293
perigee, 68, 71, 250
  height, 274
  radius, 275
perihelion, 68, 71, 74, 83, 145, 249
phase angle, 178, 251
phases of the Moon, 173, 176
Phobos, 200
Phoebe, 189, 213
Pholus, 189
Piazzi, Giuseppe, 223
Pioneer, 128, 206, 207
planet
  angular diameter, 245, 246
  aphelion distance, 249
  aphelion velocity, 259
  average velocity, 260
  distance from Earth, 245
  Earth, 189
  elongation, 251
  equation of the center, 232
  escape velocity, 262
  geocentric ecliptic latitude, 236

planet (cont.)
 geocentric ecliptic longitude, 236
 heliocentric ecliptic latitude, 232, 235
 heliocentric ecliptic longitude, 232, 235
 inferior, 231, 236
 inferior planet, 236
 Jupiter, 189, 204
 length of year, 257
 locating, 231
 Mars, 189, 199
 mean anomaly, 232
 Mercury, 189, 194
 Neptune, 189, 216
 orbital period, 258
 orbital plane, 82
 orbital velocity, 261
 perihelion distance, 249
 perihelion velocity, 259
 phase angle, 251
 phases, 250
 Planet 9, 192
 radio transmission delays, 255
 radius vector length, 235
 rise, 244
 Saturn, 189, 209
 semi-major axis, 258
 set, 244
 superior, 231, 236
 superior planet, 241
 true anomaly, 235
 Uranus, 189, 214
 Venus, 189, 196
 visual magnitude, 251
 Vulcan, 191
 weight on, 253
Planet 9, 192
 Batygin, Konstantin, 191
 Brown, Michael, 191
Pluto, 131, 187, 189, 219, 220
 Charon, 131, 222
 Christy, James, 222
 diameter, 221
 distance from the Sun, 221
 Harrington, Robert, 222
 Hydra, 222
 Kerberos, 222
 Lowell, Percival, 221
 mass, 221
 New Horizons, 220
 Nix, 222
 orbital period, 221
 rotational period, 221
 Sputnik Planum, 221
 Styx, 222
 surface temperature, 221
 Tartarus Dorsa, 222
 Tombaugh Regio, 221
 Tombaugh, Clyde, 220

polar angle, 119
Polaris, 62, 115, 118
 visual magnitude, 197
Pole Star, *see* Polaris
Polydeuces, 213
*Practical Astronomy with your Calculator or Spreadsheet*, 135
precession, 92, 102, 103, 112
prime meridian, 57, 62
Prometheus, 213
propagation, 308
Proxima Centauri, 131
PST, 33
Ptolemy, 214

quadrature, 178
quasars, 34

R136a1, 130
RAAN, 285
radio transmission delays, 255
radius to the satellite, 274
Ranger, 159
README.TXT, 9, 123, 263
refraction, 101, 112
retrograde orbit, 285
Rhea, 213
Rigel, 336, 340
 Flamsteed number, 340
right ascension, 85, 87, 89, 92, 112
right ascension of the ascending node, *see* RAAN
Roche Limit, 219
ROUND, 8

Sagittarius, 97
Saros cycle, 183
satellite catalog, 297
satellites
 altitude, 273
 apogee, 72
 apogee distance, 313
 apogee height, 274
 apogee radius, 272, 275
 argument of latitude, 288
 Clarke orbits, 306
 converting between state vector and Keplerian elements, 347
 distance from center of the Earth, 278
 distance from Earth, 72, 313
 eccentricity, 273
 footprint, 322
 GEO, 305, 306
 geostationary orbits, 306
 ground track, 323
 HEO, 305, 307
 Hohmann transfer orbit, 320

# Index

Intelsat I, 265
Keplerian elements, 286, 287, 347
LEO, 305, 306
linear eccentricity, 272, 274
look angle, 283
maximum latitude, 303
mean orbital elements, 287
MEO, 305, 307
orbit propagation, 308
orbital decay, 268
orbital elements, 284–285, 287
orbital period, 315
orbital velocity, 318
osculating orbital elements, 287
parking orbit, 303, 307
perigee, 72
perigee distance, 314
perigee height, 272, 274
perigee radius, 272, 275
radius to the satellite, 272, 274
rising and setting times, 311
semi-latus rectum, 272, 274
semi-major axis, 272, 273
semi-minor axis, 272, 274
Sputnik, 265
state vector, 288, 347
station keeping, 268
TLE, 288
true anomaly, 271, 275
true longitude at the epoch, 288
true longitude of perigee, 288
velocity for circular orbit, 319
velocity for elliptical orbit, 319
vis-viva law
   circular orbits, 319
   elliptical orbits, 319
Saturn, 189, 209
   Anthe, 213
   atmosphere, 211
   atmospheric pressure, 211
   Calypso, 213
   Cassini Division, 210
   Cassini-Huygens, 210, 211, 214
   Christiaan Huygens, 209
   density, 211
   diameter, 211
   Dione, 213
   distance from Sun, 211
   Domenico Cassini, 210
   Dragon Storm, 211
   Enceladus, 213
   Epimetheus, 213
   Galileo, 209
   Great White Spot, 212
   Helene, 213
   Iapetus, 213
   Janus, 213
   mass, 211
   Methone, 213
   Mimas, 213
   moons
      Dione, 188
      Phoebe, 189
      Tethys, 188
   Northern Electrostatic Disturbance, 211
   orbital period, 211
   Pallene, 213
   Pan, 213
   Pandora, 213
   Phoebe, 213
   Polydeuces, 213
   Prometheus, 213
   Rhea, 213
   rings, 209
   rotational period, 211
   storms, 211
   Telesto, 213
   temperature, 211
   Tethys, 213
   Titan, 212
   Voyager, 210, 211
Scattered Disc, 188, 229
   Eris, 188
   Halley's Comet, 230
   Sedna, 230
Scattered Disc Object, *see* SDO
Schiaparelli, Giovanni, 199
Schmitt, Harrison, 159
scientific notation, 12–14
Scooter, 218
Scutum, 130
SDO, 229
seasons, 144
Sedna, 230
semi-latus rectum, 68, 72, 274
semi-major axis, 65, 273
semi-minor axis, 65, 274
September equinox, 142
SGP4, 311, 350
Shoemaker, Carolyn, 208, 230
Shoemaker-Levy 9 comet, 208
sidereal day, 24, 25, 28, 36, 64, 86
sidereal month, 25
sidereal time, 24, 36
sidereal year, 27
Simplified General Perturbation, *see* SGP4
Sirius, 131, 338
   visual magnitude, 197
Small Dark Spot, 218
Small Missions for Advanced Research in
   Technology-1, *see* SMART-1
Smart, W. M., 148
SMART-1, 160
solar day, 24, 25

solar eclipse, 129, 181
  maximum time, 184
solar flares, 128
Solar Probe Plus, 129, 198
Solar System
  barycenter, 130
  binary system, 131
solar time, 22
solstice, 140, 142, 145
  December, 143
  June, 142
  summer, 142
  winter, 142
South Celestial Pole, 62
space probes
  2001 Mars Odyssey, 201
  Cassini, 207
  Cassini-Huygens, 210, 211, 214
  Chandrayaan-1, 158, 161
  Chang'e 3, 161
  Chang'e 5, 161
  Clementine, 160
  Curiosity Rover, 201
  Dawn, 219, 224, 225
  Europa Multiple-Flyby Mission, 209
  Galileo, 198, 205, 207
  Hayabusa, 228
  JUICE, 209
  Juno, 209
  LCROSS, 158
  LRO, 153, 160
  Luna, 155, 159, 161
  Luna 9, 159
  Lunar Prospector, 160
  Magellan, 198
  Mariner, 195, 198, 199
  Mars Express, 201
  Mars Orbiter Mission, 201
  MAVEN, 201
  MESSENGER, 195
  MRO, 201
  NEAR Shoemaker, 228
  New Horizons, 207, 219, 220
  Opportunity Rover, 201
  Pioneer, 128, 206, 207
  Ranger, 159
  SELENE, 161
  SMART-1, 160
  Solar Probe Plus, 129, 198
  Spirit Rover, 201
  STEREO, 128, 191
  Surveyor, 159
  Ulysses, 207
  Venera, 198
  Venera-D, 198
  Viking, 199
  Voyager, 207, 211, 216, 218

space shuttle, 306
speed of light, 256
spherical coordinate system, 279
  converting to Cartesian coordinates, 279
spherical coordinates, 119
Spirit Rover, 201
spring, 142
spring equinox, 141
Sputnik, 265
Sputnik Planum, 221
standard epoch, 92, 111
standard gravitational parameter, 259
Standard Time, 31, 33
star
  azimuth at rising, 118
  azimuth at setting, 118
  Bayer letters, 338
  Betelgeuse, 115, 336, 338
  culminate, 64
  Flamsteed numbers, 340
  Greek alphabet numbering, 336, 338, 339
  North Star, 62
  Polaris, 62, 115, 118
  Pole Star, 62, 115
  Rigel, 336, 340
  rising time, 115
  setting time, 115
  Sirius, 338
  transit, 64, 86
star catalog, 84, 92, 332, 336
star chart, 119, 123, 332–334, 336
star map, 332
state vector, 288, 347
  convert to Keplerian elements, 289
station keeping, 268
STEREO, 128
Styx, 222
summer, 142, 145
summer solstice, 142
Sun, 71
  angular diameter, 144, 146
  corona, 129
  density, 127
  diameter, 126
  distance, 126, 144
  ecliptic latitude, 132, 135
  ecliptic longitude, 135
  equation of the center, 135
  escape velocity, 128
  locating, 131
  mass, 126
  mean anomaly, 133, 134
  mean orbit, 132
  orbital elements, 132
  Pioneer, 128
  rise, 138
  rotational speed, 125

# Index

set, 138
solar eclipse, 129
solar flares, 128
Solar Probe Plus, 129
STEREO, 128
sunspots, 128
temperature, 127
tides, 153
total eclipse, 182
true anomaly, 134, 135
visual magnitude, 127
sunrise, 138
sunset, 138
sunspots, 128
 temperature, 128
supermoon, 172
Surveyor, 159
synodic month, 26

Tarantula Nebula, 130
Tartarus Dorsa, 222
Telesto, 213
terrae, 155
terrestrial planets, 193
Terrestrial Time, see TT, 35
Tethys, 188, 213
 Ithaca Chasma, 213
*Textbook on Spherical Astronomy*, 148
tides, 153
time
 apparent sidereal, 36
 civil, 31
 DST, 33
 equation of time, 24, 147
 GMT, 34
 GST, 36, 47, 48, 50
 LCT, 36, 46
 LMT, 33
 local civil, see LCT
 LST, 36, 49, 50
 mean sidereal, 25, 36
 mean solar, 25, 35, 36
 standard, 31
 UT, 36, 40, 46–48
time zone, 29, 33, 36
 adjustment, 36
 boundaries, 31, 33
Titan, 212
TLE, 288, 297
 data format, 297, 348
 Keplerian elements, 301
 mean motion, 301
 mean motion to semi-major
  axis, 301
TNO, 229
Tombaugh Regio, 221
Tombaugh, Clyde, 220

topocentric coordinate system, 281
total lunar eclipse, 181
total solar eclipse, 182
transit, 64, 86
 star, 86
Triton, 219
trojan, 188
 2010 TK7, 188
 Ceres, 188
 Jupiter, 188
 Mars, 188
 Neptune, 188
 Uranus, 188
 Venus, 188
 Vesta, 188
tropical year, 26
true anomaly, 24, 69–72, 74, 77, 78, 83,
 271, 275
true longitude at the epoch, 285, 288
true longitude of perigee, 285, 288
TT, 166
twilight, 140
Two-Line Elements, see TLE
Tycho crater, 157

Ulysses, 207
umbra, 181
Universal Time, see UT, 34
uranographer, 338
*Uranometria Omnium Asterismorum*, 338
uranometry, 338
Uranus, 189, 214
 *Almagest, The*, 214
 atmosphere, 216
 diameter, 216
 distance from Sun, 216
 Flamsteed, John, 214
 Herschel, Sir William, 214
 Hipparchus, 214
 orbital period, 216
 orbital plane, 216
 Ptolemy, 214
 rings, 216
 rotational period, 216
 temperature, 216
 trojans, 188
 Voyager, 216
UT, 34, 36, 40, 112, 113
UTC, 34
UY Scuti, 130

Valles Marineris, 202
Vastitas Borealis basin, 201
vector
 length, 269
 magnitude, 269
 norm, 269

velocity at aphelion, 259
velocity at perihelion, 259
Venera, 198
Venera-D, 198
Venus, 189, 196
  Aphrodite Terra, 198
  atmosphere, 197
  atmospheric pressure, 197
  BepiColombo mission, 198
  diameter, 197
  distance from the Sun, 197
  Galileo space probe, 198
  Ishtar Terra, 198
  Magellan space probe, 198
  Mariner, 198
  mass, 197
  Maxwell Montes, 198
  Mayas, 1
  Morning Star, 196
  orbital period, 197
  rotational period, 197
  Solar Probe Plus, 198
  surface temperature, 197
  trojans, 188
  Venera, 198
  Venera-D, 198
  visual magnitude, 197
vernal equinox, 1, 26, 83, 133, 141
Vesta, 226
  Olbers, Wilhelm, 227
  trojans, 188
Viking, 199
Virgo, 192
vis-viva law, 319
visual magnitude
  comparing magnitudes, 252
von Braun, Wernher, 265
Voyager, 207, 210, 211, 216, 218
VSOP, 348
Vulcan, 191
VY Canis Majoris, 129

WGS84, 268
winter, 142, 145
winter solstice, 142
Wizard's Eye, 218

year
  anomalistic, 27
  Besselian, 27
  civil, 26
  draconic, 27
  galactic, 27
  Julian, 27
  leap, 26, 38, 39
  sidereal, 27
  tropical, 26
yellow dwarf star, 125

zenith, 64, 101
zodiac, 39
Zulu Time, *see* UT, 34